Solving Problems in Technical Communication

SOLVING PROBLEMS IN TECHNICAL COMMUNICATION

Edited by
Johndan Johnson-Eilola
and Stuart A. Selber

THE UNIVERSITY OF CHICAGO PRESS
Chicago and London

Johndan Johnson-Eilola is professor of communication and media at Clarkson University.
Stuart A. Selber is associate professor of English at the Pennsylvania State University.

The University of Chicago Press, Chicago 60637
The University of Chicago Press, Ltd., London

Printed in the United States of America

22 21 20 19 18 17 16 15 14 13   1 2 3 4 5

ISBN-13: 978-0-226-92406-9 (cloth)
ISBN-13: 978-0-226-92407-6 (paper)
ISBN-13: 978-0-226-92408-3 (e-book)
ISBN-10: 0-226-92406-8 (cloth)
ISBN-10: 0-226-92407-6 (paper)
ISBN-10: 0-226-92408-4 (e-book)

Library of Congress Cataloging-in-Publication Data

Solving problems in technical communication /
edited by Johndan Johnson-Eilola and Stuart A. Selber.
pages ; cm
Includes bibliographical references and index.
ISBN-13: 978-0-226-92406-9 (cloth : alkaline paper)
ISBN-13: 978-0-226-92407-6 (paperback : alkaline paper)
ISBN-13: 978-0-226-92408-3 (e-book)
ISBN-10: 0-226-92406-8 (cloth : alkaline paper)
[etc.]
1. Communication of technical information.
2. Technology—Information services. I. Johnson-Eilola, Johndan. II. Selber, Stuart A.
T10.5.S638 2013
601.4—dc23
2012037860

♾ This paper meets the requirements of ANSI/NISO Z39.48-1992 (Permanence of Paper).

Dedicated to our families:

KELLY JOHNSON-EILOLA,

CAROLYN EILOLA,

KATE LATTERELL,

AVERY SELBER, *and*

GRIFFIN SELBER

Contents

JOHNDAN JOHNSON-EILOLA & STUART A. SELBER

Introduction

This book is for students who are learning about the field of technical communication. For both newcomers and people with some experience, it provides a coherent approach to understanding and solving problems and developing strategies that work in different types of communication situations. Because problems in this field are complex, multifaceted, and rooted in local settings, they do not lend themselves to one-size-fits-all solutions. There are, however, *heuristic frameworks* (more on this phrase later) that can help you address communication tasks in a systematic and thoughtful manner. This book offers original heuristics for problem contexts that are central to the field. Taken as a whole, they constitute a broad-based perspective for both education and work that is sensitive to the contingent nature of technical communication.

Although technical communication has a long and rich past—researchers have documented cases of technical communication in ancient Roman culture, for example—the field began to mature as an organized profession in the mid-twentieth century. This ongoing process, modern historians tell us, continues to parallel in close ways the development of complex technical systems and the proliferation of consumer markets for all things technical and scientific. More than ever, the inventions and processes of contemporary society call for a great deal of explanation, instruction, and careful design, in large part because their audiences (users) have expanded to include not only technical specialists but also nonspecialist audiences.

Computers provide an instructive case in point. Early computer systems were built by and for technical professionals, and they were employed in highly specialized settings such as scientific labs and research and development sites. (Think IBM or Westinghouse in the 1950s, or the U.S. Department of Defense in the post-cold-war period.) Communication about and through these early systems occurred in relatively stable work cultures and involved limited groups of workers and activities. Nowadays, of course, computers come in all shapes and sizes, and are used in a wide variety of settings by a wide variety of people. This very real diversity requires more and different approaches to technical communication: not only does computer documentation need to be prepared for a range of abilities and goals, but computer programs themselves need to be adapted to support different types of activities. As computer-program design has matured, technical communicators have increasingly been enlisted to help create systems to support specific types of complex tasks, taking into account not only the functions of a program itself ("Select 'Import . . .' from the File menu.") but also broad social concerns ("What about users who want to save paper? Can we let them print to PDF files?") and workplace issues ("Rather than invite users to read the FAQ file, can we provide a way for them to suggest changes? Or simply make changes themselves?"). Technical communication is no longer simply communication *about* technology; it is also often communication *as* and *in* technology. In other words, technical communication, in certain contexts and cases, has become both a process and a product, something no longer separate from, or secondary to, the fundamental task at hand. (Think wikis and how they can both structure and support communication activities.) As technology has become more ubiquitous, more complicated, and more important to all aspects of work and education, the field of technical communication has responded in increasingly complex ways.

This increased complexity carries over to the process of *learning* how to be a technical communicator. Although many technical communicators still prepare printed user manuals, many more develop wildly divergent types of computer support materials: online help systems, self-paced tutorials, in-person training guides, automated wizards, document templates, tool tips, multimedia demonstrations, and more. How does one person learn to develop not only excellent writing skills but also expertise in task analysis, document design, HTML, CSS, Flash, audio and video production, single-sourcing, usability, and more?

For that matter, computer-user support materials are only a small fraction of what is involved in the ever-expanding world of technical communication. Professionals participate in nearly any area that involves technology, which is just about every field today. Technical communicators work in medical industries designing interfaces for patient health records; in aviation settings testing the usability of cockpit controls; in corporations writing annual reports that explain profits, losses, and trends to shareholders; in social service agencies preparing grant applications and proposals to support community outreach programs; in government agencies writing safety guidelines for field researchers; in publishing houses editing technical specifications for outdoor power and agricultural equipment; and in mining companies creating warnings and risk bulletins for the public. This list is more suggestive than exhaustive: technical communicators can now be found in nearly any and all work locations, handling a vast array of projects and responsibilities.

Can one person become an expert in all of these areas? The short answer is no. Nobody does it all, and certainly not at the same time. Some technical communicators start in one field, such as computing, and work in it for several years, building up their experience first in a specific area (e.g., reference manuals) before moving on to more complicated roles (e.g., participatory design). Others move from one industry to another, applying the skills they have learned to new contexts and challenges. The career paths and future skill sets of technical communicators are anything but predictable: technologies develop and change, industries rise and fall, and current events, especially crises and disasters, help shape cultural expectations for communication and documentation. To use a cliché, the only constant in the field is *change*. You can confirm this fact by talking to people who have practiced technical communication for more than a year or two. You will undoubtedly hear stories about how their jobs evolved over time.

Adaptability is therefore crucial to success in the field. Technical communicators do not merely learn skills; they must also *learn how to learn new skills*, upgrading and augmenting their abilities as they mature in careers, analyzing the matches and mismatches between what they currently know and what a communication situation demands. In other words, technical communicators must learn to become reflective problem solvers. This book will help you with that critical professional enterprise.

TECHNICAL COMMUNICATION AS A PROBLEM-SOLVING ACTIVITY

This book encourages you to think about technical communication as a problem-solving activity. Of the numerous images for the field and those who work in it, "problem solver" is an especially productive characterization to consider, for it acknowledges the extent to which technical communicators contribute to the development and use of technology. It also suggests the challenges of working with language—verbal and visual—and of developing communication artifacts that represent tasks, processes, procedures, and more in a manner that is useful and usable. Because there are always multiple ways to understand and solve problems in this field, technical communicators are constantly interpreting use situations and weighing possible responses. The solution to a problem in technical communication is never the only available solution, but one among several competing alternatives that balances issues on both the development and use sides of the equation. There are always constraints that have a bearing on technical communication.

For our purposes here, problem solvers possess several important characteristics, including the ability to sense a problem, diagnose what forces within a context are causing the problem, and develop and implement a change within the context that addresses the problem. You might have noticed that two of these characteristics depend more on awareness than activity. Technical communication, like many complex forms of communication, relies heavily on the ability to analyze a situation before responding to it. Technical communicators who provide value to organizations do not simply fill in templates or follow rigid procedures. Instead, they constantly move back and forth between analysis and action, checking their assumptions against reality and adjusting (sometimes drastically) when they detect a mismatch. In addition, they understand that even the act of defining a situation can profoundly affect the situation itself.

Consider, for example, the simple list of categories for race on U.S. census forms (designing forms is a common task for technical communicators). You might think the list is pretty straightforward, but it is extremely complicated for many reasons, not the least of which are the realities that respondents pick a single category and that the results are used in creating and supporting social programs of many types. In the late 1990s, the Bureau of the Census proposed adding a new category, "multiracial," to those traditionally listed, in acknowledgment of the growing number of people who identify themselves with more than one of the traditional categories. While some argue that this new category more accurately reflects the current population, others point out that it would also undoubtedly reduce the number of people who would have previously picked a minority category, such as "Black" or "Asian." Both perspectives, of course, contain an element of truth. A heuristic for this situation would help a technical communicator step back and analyze how the situation is being defined and what the (sometimes competing) goals are, in an attempt to balance the factors in the problem space before constructing a solution.

At their base, complex problems like this are

- subjective phenomena open to analysis and interpretation,
- open to change over time and rarely solved permanently, and
- engaged by multiple actors in a social space.

Heuristics, rough frameworks for approaching specific types of situations, help technical communicators solve problems not by providing straightforward answers but by providing tentatively structured procedures for understanding and acting in complex situations. The three bulleted points above suggest a general heuristic, one that might work for other social issues:

1. Acknowledge that racial categories (and how they are used) in the United States change over time.
2. Identify the social actors to involve in the analysis and interpretation of those categories (and, perhaps more importantly, how they are used).
3. Create civil social spaces for stakeholders to engage in productive discussion.

As you might suspect, merely following these steps does not guarantee a definitive solution. Rather, the heuristic provides an initial framework for how to approach the problem space. Over and over during the process of addressing the problem, everyone involved needs to periodically step back to see how the heuristic is working and ask if the heuristic needs to be modified. Ultimately, solving the problem is often an idealized goal: we do not solve technical communication problems once and for all, but instead constantly engage in the *process* of solving them.

In fact, technical communicators continually face both old and new problems. There are few hard and fast rules in the field because projects vary by context and circumstance, sometimes dramatically. In addition, technologies change at an extremely rapid pace, mutating as they are adopted in different ways by different groups of people. As you gain expertise, you will learn to identify tendencies, trends, and common threads from one technical communication situation to another, but you will also learn that some features of every situation are unique. So you will gain the ability not merely to *do* things but to *think about* how you are doing things, to analyze your previous experiences, attempting to come up with common approaches that seem to work from place to place—at least in some cases and to some degree. Put in different terms, you will be learning how to move back and forth from theory to practice, from practice to theory.

In reality, you probably already make this important recursive move but perhaps without knowing it. For example, you probably already understand multiple ways to communicate. In a formal memo to your internship supervisor, you might write, "Please let me know when the final report should be submitted." In an e-mail to the same person, you might say, more casually, "Can you tell me when the final report's due?" In a text message to another student in the class, you might write, "co-op report due this wk?" These shifts in register signal a rhetorical sensitivity to audience and purpose and to the conventions and constraints of various communication media.

Many people develop a similar set of different approaches based on intuition (itself based on experience), almost as if they were natural and obvious ways of communicating. They are not. We learn them, often without thinking about that learning. At certain points, however, you can make yourself a better, more effective technical communicator by moving consciously and explicitly from simple practice to theory. What happens if you discover that, unlike some of your professors at school, your internship supervisor is comfortable texting. Do you text her the same type of sentence you would have written in a formal report? You can see here how you would be moving back and forth between doing and thinking—not merely thinking concretely about what you are going to do but also thinking in complex ways about your past actions, why you took them, how well they worked, how other people responded to them, and more. In such situations, you are revising your practice based on theories about practice, developing heuristics based on past experiences and thinking about those experiences (in both formal and informal ways).

The need to continually revise practice and theory never goes away for most people in the field: technical communicators are problem solvers, and the classes and kinds of problems keep changing. As you get better at addressing more complicated situations and develop productive ways to respond to them, you will move on to ever more complex problems requiring new solutions. As you advance as a professional, you will also develop ways of learning new things, heuristic approaches you can take to different problem contexts in order to help you understand and solve them.

HEURISTICS AND PROBLEM SOLVING: A COMPLEX, RECURSIVE RELATIONSHIP

Effective problem solvers develop and adapt heuristics in a complex, recursive set of procedures that involve not only actions and heuristics but a second level of analysis, including revising heuristics and thinking based on theories, which are themselves open to revision. These processes are illustrated in figure I.1. Because they can be abstract and difficult to grasp, we elaborate with a simple household task: doing the laundry.

When you are doing laundry, you start by sorting clothes into piles based on color: whites, dark colors, light colors, and so forth. That sorting pattern is a *heuristic*. What do you do, however, when you find a white T-shirt with a bright red logo on the front? Cut the logo out and put it on the red pile, then put the remaining white part of the shirt on the white pile? Your heuristic is now your problem. So you resort to "meta-activity" to reconsider the heuristic: you think about the purpose of sorting laundry. One purpose is to keep bright colors (like red) from running. The shirt, however, has been washed before, and the red logo did not run

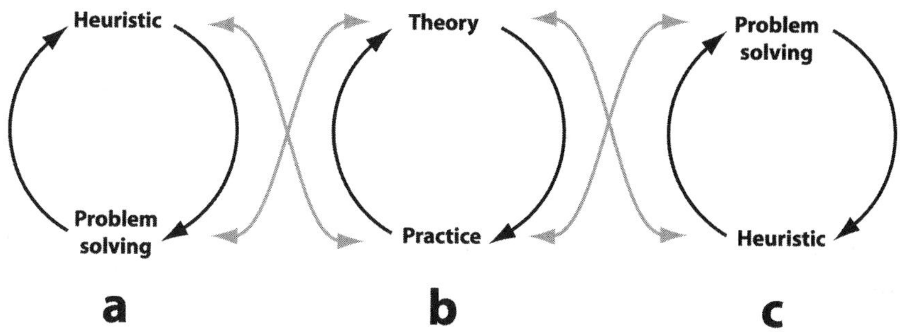

Figure I.1. Recursive procedures for adapting heuristics

on the white background. Therefore, you can probably assume the red is colorfast. You can now modify your heuristic with a backup rule: bright colors on white can safely go into the white pile, assuming a previous washing. This process of modification continues on an as-needed basis.

This book offers you a range of heuristics for different problems that technical communicators tend to grapple with, such as designing information, collaborating with others in an organization, addressing global audiences, researching and writing documents, managing projects, making ethical and legal judgments, and more.

Figure I.1a explores the intersection of concrete problems ("I need to wash my clothes") and heuristics ("I need to sort them into piles"). If you are a novice clothes washer, you may not have learned heuristics for this task. So you might, for example, throw a new red shirt into a pile of white laundry (a course of action that can result in a freshly laundered load of pink clothing). As you work through approaches to solving such problems, you will inevitably consider the fit between a heuristic and a concrete situation, then modify the heuristic to match the particulars of that situation. In other words, as problem situations change, your trusty heuristic may itself become a problem to be addressed: how do I adapt my heuristic to respond to the specific challenge in front of me?

As your intellectual toolbox expands in complexity and size, you will begin forming broader sets of heuristics that are interconnected to each other. In many cases, these sets can be so broad that they begin to feel rather abstract. This is not a problem, because their abstract nature is what makes heuristics powerful. Heuristics do not deal so much with concrete reality. Instead, they are used to connect abstract theories to individual, concrete practices (figure I.1b). As people gain expertise in situations, they typically start trying to see how and why these connections are made. You may have learned how to write a memo, for example, but you are not going to get very far writing the same type of memo for every situation, or writing a memo in situations where a phone call might be more effective. In the laundry example, the white shirt with red logo suggests you might step back to see how your "separate reds from whites" heuristic fits between your theory and your practice: What is the theory that leads to the heuristic? How might you translate that theory differently to address this unique problem?

At some level, you will also start using abstract theories to develop and modify your heuristics to adapt to emerging problems. Again, in a complex, recursive manner, you are now thinking of heuristics themselves as problems to be solved (figure I.1c). You learn, for example, how to use theory to adapt your heuristics to fit new types of problems. You take what you know about the theory of red colors in laundry bleeding into lighter colors, analyze the specifics of the situation and your own experiences, and develop a new heuristic that suggests you can wash reds and whites together in certain circumstances.

Similarly, in the workplace you learn that while e-mail messages are often an avenue to obtain information quickly, you also learn (or should learn) that the ease with which e-mail can be misread in volatile situations suggests that, in some cases, calling for a face-to-face meeting will be slightly less efficient but more likely to be productive in the long run. You learn, that is, to apply problem-solving skills, at a secondary level, to the heuristics themselves. Over time, your problem-solving skills, your heuristics, and your understanding of theory should all continually work together to make you a more effective technical communicator.

Here is another way to think about these relationships: heuristics are relatively static procedures or frameworks to be applied to a problem, rules of thumb, in a sense; problem solving is a messier, more contingent, and higher-order activity. Working effectively means going back and forth between heuristics and problem solving. Heuristics are never completely effective, because they are more abstract than the immediate context. This abstraction is both their strength and their weakness. They are also never completely effective because things change over time and across contexts. Heuristics fit well in some cases but poorly in others. So technical communicators need to resort to problem solving in order to fit the heuristic to the local context. What you learn and apply as a heuristic in one project may very well need to be adapted for another project.

This book provides important resources for engaging in this recursive process of doing and thinking; it provides extensive, concrete examples that address a wide range of standard technical communication practices. Chapter authors describe specific experiences, analyzing the problems posed by those experiences and developing rich heuristics for solving the problems. The full range of problem-solving heuristics constitutes a broad set of approaches you can integrate into your own work. Ultimately, you will adapt and improve these heuristics as you gain your own set of experiences and learn more about the field of technical communication.

LEARNING ABOUT THE FIELD: A FOUR-PHASE HEURISTIC

How does one learn to become a technical communicator? How does a technical communicator continue to advance and learn on the job? In addition to numerous original heuristics for work practices essential to the field, this book provides an overarching framework for imagining a process for lifelong education, including the instructional practices of academic programs in technical communication. The macrolevel heuristic in figure I.2 depicts an approach you can use as a student in a technical communication program or as a professional interested in keeping up with new developments in the field. Although there are many valuable approaches to technical communication education, this heuristic has something of a universal quality that can be applied, in an ongoing fashion, to a shifting, dynamic field.

On a conceptual level, the heuristic emerges from the recursive relationship between practice and theory discussed in the previous section. In slightly different terms, it asks you to work back and forth between the contexts and applications of technical communication, taking both specific and broad views of the work you are doing or will do in the future. Notice that four interconnected phases organize the heuristic: Mapping the Field, Situating the Field, Understanding Field Approaches, and Developing Field Knowledge. Each of these phases, not coincidentally, also represents a different part of the book, and the questions anchored to them are also chapter titles. We will elaborate further on the nature of this arrangement, but for now it is useful to know that the structure of the book matches the structure of the heuristic.

In figure I.2, the four interconnected phases are superimposed on a grid with quadrants that delimit their focus and scope. The quadrant Mapping the Field includes fundamental questions about what technical communicators do, where they work, and how they progress as both students and professionals. These questions explore the work contexts of technical communication in very specific ways, providing concrete snapshots of the field. After working through this part, you should have a clear sense of the responsibilities, job settings, and development paths of technical communicators.

Although the quadrant Situating the Field continues to address the work contexts of technical communication, it foregrounds theories and

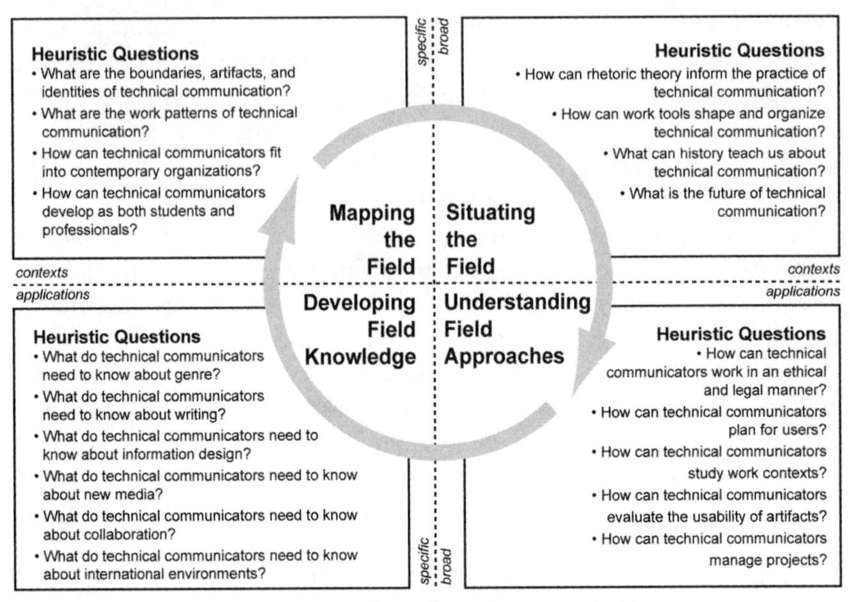

Figure I.2. An educational heuristic for technical communication

histories rather than processes and settings. Discussions associated with the previous part are certainly informed by theory and history, but the questions in this part consider broader perspectives on the contexts of technical communication. After working through this part, you should see the value of theories for practices, including theories that derive from outside the field; understand how the work tools of technical communication can influence practices; recognize ways in which the past can be leveraged to improve practices (the field does not operate in a temporal vacuum); and recognize ways in which the present can be leveraged to predict future possibilities. To reiterate, both phases address technical communication contexts, but oscillate between specific and broad views of the field. Employed in either order—mapping and situating or situating and mapping—the phases enact the recursive processes of doing and thinking that are required for effective work in technical communication.

The next two quadrants retain the specific-broad distinction but change the focus from contexts to applications. In our heuristic, applications are equivalent to the routine practices technical communicators engage in on the job and the routine products technical communicators create for users. Although the other quadrants discussed thus far mention a variety of practices and products, they do so in order to help define and explain the field, to help map and situate it among other fields and endeavors. In more direct terms, the two quadrants in the bottom half of the heuristic help you learn how to do the work of technical communication.

The quadrant Understanding Field Approaches offers a broad examination of the applications of technical communication. It poses and responds to a series of questions that concentrate on methodological procedures that have become routine in technical communication settings. After working through this part, you should be familiar with approaches that people in the field use to conduct research; to plan, evaluate, and assess projects; and to manage projects. You should also be sensitive to the ethical and legal dimensions connected to these and other tasks. There are other field approaches worth learning about, but the approaches reviewed here are commonplace (if complex) and will enable you to initiate and perform key aspects of technical communication projects.

You will need more than methodological procedures, however, in order to create applications that are helpful to users. You will also need to cultivate a series of skills specific to the writing and communication tasks of technical communicators. The quadrant Developing Field Knowledge distinguishes technical communicators from journalists, creative writers, and other types of writers. After working through this section, you should be familiar with how people in the field write and communicate, design artifacts, collaborate, address international audiences, develop and deploy new media, and more. What you learn here will help explain why field knowledge in technical communication has expanded to include many skill areas beyond those traditionally associated with words and written texts.

Although you can begin with any of the four phases in the heuristic, with either broad or specific perspectives, and with either contexts or applications, we encourage novices to start with Mapping the Field, which covers definitions of technical communication and work processes and contexts. This basic information will serve as a useful scaffold for discussions in other parts of the book.

THE STRUCTURE OF CHAPTERS

As our examples in this introduction suggest, work in technical communication, even in settings with mature organizational approaches, is neither formulaic nor entirely predictable. Nor is it prescriptive or generic. Although the field has developed best practices and principles that can help practitioners produce useful and usable documents (print and online), technical communication problems are—at their core—ill-structured, complex, and messy, defying easy or pat solutions. Technical communication work, in other words, is rhetorical work: professionals create as much as they report, inventing solutions to communication problems through various deliberative activities.

Given this reality, students and professionals need to be equipped with a variety of frameworks for approaching ill-structured problems. Although educational materials in the field teach process and often include real-world cases with an emphasis on the contingent, technical communicators need access to a series of systematic frameworks that attend to the numerous complex elements of communication situations. The frameworks in this book—at least one per chapter—show you how to approach ill-structured problems in an organized and methodical fashion.

Although not always labeled as such, each chapter in the book includes the same seven sections.

- Summary
- Introduction
- Literature review
- Heuristic
- Extended example
- Conclusion
- Discussion questions

In broad terms, the *summary* discusses the need to focus on the question for the chapter. The *introduction* offers a scenario that helps clarify the problem context for the chapter question. Except in a few cases where it makes more sense to focus on students and their settings, the scenarios come from the nonacademic workplace and involve practicing technical communicators. Each introduction ends with an overview of the different chapter sections.

The early sections of chapters also include a *literature review*. In these sections, authors elaborate on the problem contexts for their questions by discussing what the field already knows about those contexts, being sure to include the latest thinking and research in the field. Authors organize their discussions by topics or issues rather than by sources.

The heart of each chapter provides a research-based *heuristic*. Drawing on the literature review, authors offer heuristics that address the problem contexts for the chapter questions. That is, authors provide frameworks that enable you to investigate, conceptualize, or engage the problem contexts in some productive fashion. The heuristics take a number of forms—for example, a series of questions to ask and answer; a model that characterizes work or activity; a taxonomy that prioritizes relationships; a flowchart that indicates processes or interactions; a grid that connects aspects of a context. In ways that make sense for the topic at hand, authors formulate suggestive structures you can apply to varying communication situations.

After the heuristics have been explained, authors demonstrate their use by applying them in an *extended example*. The objectives here are to show you how to employ the heuristics and to provide an illustration of how the heuristics can guide technical communication work. Finally, authors find meaningful ways to *conclude* their chapters, revisiting key points revealed by the literature reviews, heuristics, and extended examples.

As an appendix to each chapter, authors provide *discussion questions* to challenge and stimulate your thinking. You can use these questions to facilitate conversations about the use of heuristics for problem solving in technical communication.

A FINAL WORD

The heuristics in this book function as a vital bridge between thinking and doing. They afford coherent methods for undertaking both basic and advanced responsibilities in technical communication. In significant and meaningful ways, they will help you learn about the field and keep up with the rapidly changing demands of the twenty-first-century workplace. However, heuristic approaches, by definition, are dynamic and contingent approaches, requiring ongoing attention and evaluation. We encourage you to view the heuristics in this book with a discerning and critical eye. As you employ them and as you gain more experience as a technical communicator, you will begin to notice ways that you can modify the heuristics, adapting them to new problem contexts and to new purposes. Modifying heuristics for specific purposes is an important skill for reflective problem solvers.

ACKNOWLEDGMENTS

The development of this book benefited from assistance from a range of editors, authors, and students. David Morrow at the University of Chicago Press was a stalwart supporter of the project. From start to end, he offered an important perspective that improved the book in tangible and notable ways. The production staff at the University of Chicago Press deserves special recognition for its efforts. We want to thank authors for their hard work and willingness to revise for the chapter template. Their goodwill enabled coherence across a large volume with many authors. Jim Porter read a draft of our introduction and provided useful revision suggestions. We also want to thank Steve Benninghoff and his students at Eastern Michigan University for testing a draft of the book: John Allinder,

Graham Parker-Finger, Jeffery Price, Phyllissa Ren, Brenda Romig-Fox, and Beth Sabo. Their revision suggestions were helpful to several authors. Finally, we want to thank Clarkson University and Penn State University for continuing to support our collaborative work. It has made a difference to us.

PART 1 MAPPING THE FIELD

As you learn how to function and advance as a technical communicator, you will develop a range of strategies for investigating and mapping the contours of the field, for creating representations of work activities that bound your professional territory in clear and tangible ways. Although this "explorer" metaphor oversimplifies the process, it is still a highly useful image: you will be looking around with an attentive eye, listening to stories from experienced professionals, making quick copies of the maps other people have created, training yourself to spot important and interesting features of the terrain you will be moving through, and more. And because the terrain of the field is constantly shifting and presenting new challenges, the task of mapmaking will be an ongoing professional activity.

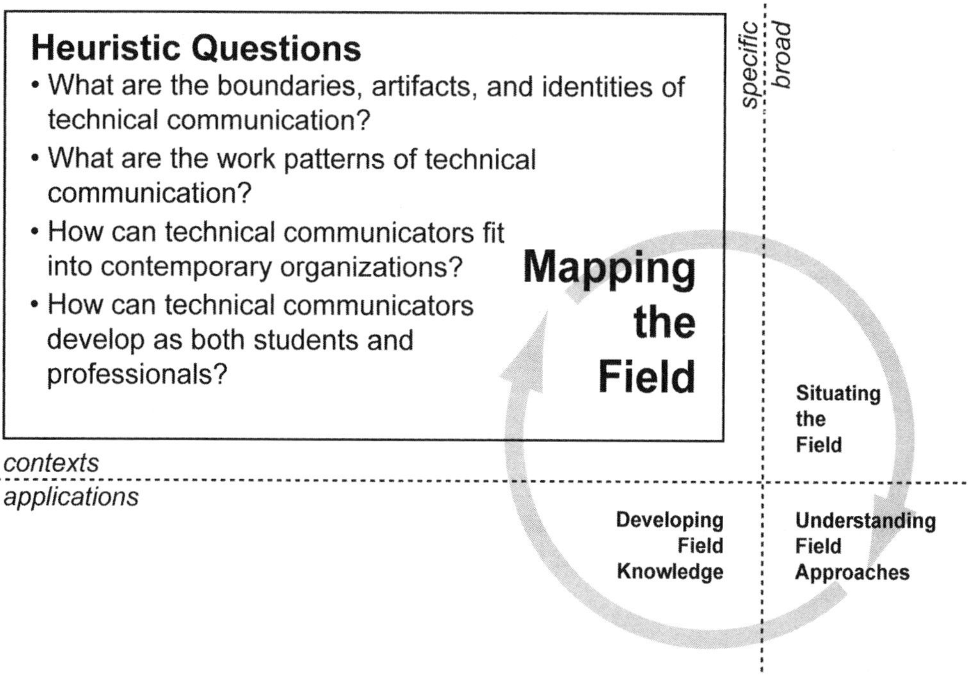

Heuristic Questions

- What are the boundaries, artifacts, and identities of technical communication?
- What are the work patterns of technical communication?
- How can technical communicators fit into contemporary organizations?
- How can technical communicators develop as both students and professionals?

Figure P1.1. Heuristic questions for Mapping the Field

Of course, in most cases, your work as a new technical communicator will not take place in the wild. You will not be the noble, strong-chinned explorer striding in to bring civilization to the natives (that decidedly Western vision always seems to lead to chaos anyway). Instead, you will be working in communities and organizations that are familiar in certain respects and unfamiliar in others. At times, people will talk to you in everyday terms you can understand; at other times, they may seem to speak in another language, referring to strange types of reports and information-gathering strategies, using jargon in dizzying patterns. You will suddenly find yourself in unfamiliar territory and in need of a map (or two). This situation is typical for those new to the field.

Facing new professional challenges can be both energizing and unsettling. Although the chapters in this section (figure P1.1) cannot remove all the uncertainty, they provide heuristics you can use to generate maps to help you find your way around and conceptualize work activities.

In the opening chapter, Richard Selfe and Cynthia Selfe offer an approach to answering one of the first questions you will face: "What Are the Boundaries, Artifacts, and Identities of Technical Communication?" Selfe and Selfe review methods that are commonly employed to help students understand what technical communicators do in the workplace. These methods include examining histories of the field and compiling lists of skills that employers say they want their workers to have. Although these methods are valuable, Selfe and Selfe advocate looking at what technical communicators say to each other about their work, filtering that large and ongoing body of discourse through the information visualization techniques of text clouds to create informative maps of the field. The examples developed by Selfe and Selfe use articles from key journals for practitioners and researchers to create instructive snapshots of professional practice. You can also use these techniques on texts from your own contexts to create different maps of work activities.

In "What Are the Work Patterns of Technical Communication?," William Hart-Davidson also draws on research by and about practicing technical communicators to define three primary tasks for people in the field: designing information across numerous media and genres, advocating for the needs of users, and stewarding information development in organizations. As Hart-Davidson demonstrates, mapping the functions and value of these activities brings technical communication in from the periphery to occupy a central role in the mission of contemporary organizations. You can use the heuristics offered by Hart-Davidson to develop work processes that encourage and support substantive roles and responsibilities.

This active remapping is demonstrated in detailed ways in "How Can

Technical Communicators Fit into Contemporary Organizations?" In this chapter, Jim Henry discusses strategies you can use as you find your way within a new company, community, or organization in your role as a technical communicator. Relying on the experiences of a student team writing an annual report for a community health agency, Henry sketches out heuristics you can draw on as you learn about your first technical communication job and as you move from one organization to another. The heuristics incorporate a variety of qualitative research methods that enable you to map the dynamics and conventions of workplace settings.

The first part of the book concludes with the reminder that mapping the field is an ongoing process that never ends: technical communicators on the job are also always students, learning new ways of working and advancing as situations, goals, and technologies change. Kelli Cargile Cook, Emily Cook, Ben Minson, and Stephanie Wilson address the important question, "How Can Technical Communicators Develop as Both Students and Professionals?" In this chapter, Cook, Minson, and Wilson all look back from their current workplace roles to their time as students in a technical communication class taught by Cargile Cook. As all three demonstrate, effective classroom learning can be useful when you enter the workplace, but you will need to think about the skills and strategies you are transferring, reflecting on how they fit (or fail to fit) with the demands of your current work, adapting and building on them as your own professional abilities evolve and expand. This chapter offers a heuristic that encourages you to constantly update and refine your sense of the field.

1

RICHARD J. SELFE & CYNTHIA L. SELFE

What Are the Boundaries, Artifacts, and Identities of Technical Communication?

SUMMARY

Understanding your field and being able to map the territory of its boundaries, artifacts, and identities is one mark of an informed professional and an important indication of expertise in the workplace. There are numerous ways to define the landscape of technical communication, a field that involves practitioners, researchers, and theorists in a broad range of activities. Some efforts to identify the boundaries of the field rely on historical accounts of how it was born and grew into a recognized area of research and practice. Others describe the research base of technical communication, identifying the topics and issues that provide a focus for investigations and studies. And still other efforts identify the general kinds of skills and understandings needed by technical communicators in the workplace. Each of these approaches has its strengths and limitations, and each produces a very different map of the field.

This chapter focuses on text clouds as a way of mapping technical communication and of describing the boundaries, artifacts, and identities that constitute the field. In the following pages, we create text clouds and use them as heuristics to help us discuss the landscape of technical communication.

INTRODUCTION: MAPPING TECHNICAL COMMUNICATION AS A FIELD

In 2006, Amanda Metz Bemer, a student of technical communication at the University of Washington, Seattle, came face to face with an interesting fact about her chosen field: nobody knew what technical communication was. When she talked to her fellow students and friends outside her major, nobody knew what it was that technical communicators really did, nobody knew what research was done in the field, and nobody could imagine what issues interested technical communicators.

Amanda tried to give her friends and family an understanding of the field by listing the classes she had taken: "technical writing, instruction-manual writing, communication theory, usability testing, document design, rhetorical theory." But, as Amanda noted, she generally got a "blank look and an 'oh'" for her trouble. So Amanda—asking the question "What the heck is technical communication, anyway?"—wondered if there was a better way to talk about her field than by giving a "laundry list of classes."

After doing her own research on technical communication—the boundaries of the field, its artifacts, and its identities—Amanda learned that the matter was more complex than she had thought. Indeed, no single source she read had been able to identify a definition of the field that was both comprehensive and specific enough to do justice to the field and help others comprehend what went on within its boundaries.

As Amanda herself noted in "Technically, It's All Communication: Defining the Field of Technical Communication," a 2006 article she wrote for *Orange*, there had been no shortage of attempts to define the boundaries, artifacts, and identities of technical communication. However, the success of each of these attempts, she realized, had been necessarily limited, perhaps because a good map had to serve so many audiences (students of technical communication, scholars and practitioners in the field, nonspecialists and members of the public interested in what technical communication is and isn't) and perhaps because the field itself covered so much ground. No one map of the territory that the profession occupies had emerged as fully capable of representing so much ground in a concise and understandable way to so many audiences. This is not a flaw of maps as descriptive tools but, rather, a function of their inevitable biases and perspectives. All maps, including the text clouds in this chapter, highlight certain things and not others, depending on the interests and goals of the mapmakers.

The same problem that Amanda identified is shared by many others who study and practice technical communication (Jones 1995), and who argue for the significant benefits of defining the field more clearly. In the following sections, we'll look at the three primary approaches to mapping the field with words, and then outline a fourth approach—text clouds—that may offer a useful way of responding to Amanda's question, "What is technical communication anyway?" which, for the purposes of this chapter, we will restate as "What are the boundaries, artifacts, and identities of technical communication?"

The chapter begins with a literature review that describes what scholars and practitioners have already done to define technical communication with words: looking at the history of the field, defining its objects of research, and identifying the skills and understandings that practitioners need to demonstrate. The chapter then looks at text clouds as a heuristic for mapping the field, one that takes advantage of both words *and* visual information. Finally, the chapter provides an extended example that shows how text clouds might help students like Amanda make sense of technical communication as a field, one that is both complex in scope and dynamic in its practices. Every approach to mapping technical communication, however, has its strengths and weaknesses. As Carolyn Rude (2009, 178) notes, any map of such a large and diverse field is bound to be inherently biased because "some meanings and practices are chosen for emphasis and others are excluded or repressed." This caveat stands true, as well, for text clouds.

LITERATURE REVIEW: MAPPING THE FIELD OF TECHNICAL COMMUNICATION WITH WORDS

Previous attempts to map the identity of technical communication as a field have generally fallen into three categories: maps that focus on *the history* of technical communication, maps that describe *the research base* of technical communication, and maps that identify *the skills and understandings* needed by technical communicators in the workplace.

HISTORICAL MAPS OF TECHNICAL COMMUNICATION

One way of answering the question, "What are the boundaries, artifacts, and identities of technical communication?" involves tracing the roots of technical communication, creating a map—often in the form of an edited collection of works—that focuses on the historical context of scientific and technical writing and the eventual emergence of the field as we now know it. The strength of historical maps is the careful way in which they capture the social, political, economic, and institutional contexts in which technical communication has been practiced, the motivations and conditions of these practices, the preparation of practitioners, and the various forms and genres that have been developed and deployed by technical communicators. With the information that historical maps of the field provide, we can trace why and how particular genres emerged, learn more about the contributions of individual communicators, and better understand the role that technical communication has played in larger social and cultural movements. These investigations of the past accomplish more than simply providing insights into how the field has changed over the years, as Kynell and Moran (1999) note; they also suggest possible vectors along which the discipline might continue to change in the future.

In *Three Keys to the Past: The History of Technical Communication* (1999), for example, Teresa Kynell and Michael Moran trace the roots of the profession to the work of natural philosophers, scientists, and educators in past centuries. In this important historical collection, Charles Bazerman writes about the contributions of Joseph Priestley in describing electricity during the seventeenth century; James Zappen explores the science writing and rhetoric of Francis Bacon in the eighteenth century; R. John Brockmann chronicles Oliver Evans's descriptions of mills and steam engines in the pre- and post–Civil War period of the eighteenth and early nineteenth centuries; and Teresa Kynell tells the story of Sada Harbarger's work on promoting technical communication through the Society for the Promotion of Technical Communication in the 1920s.

Despite the many strengths of historical approaches to mapping the field of technical communication, however, histories do have some limitations. As Jo Allen (1999, 227) has pointed out, historical efforts can appear "haphazard": "Should the work focus on the rise of technical communication as a career; as an academic field of inquiry; or as a centuries-old endeavor . . . ? Should the work examine the subjects, the concept, or the writers of technical communication? And which writers should it examine—those who practiced technical communication or those who have studied it?"

Similarly, when historical accounts focus on key figures, they can encourage what R. John Brockmann (1983, 155) calls a "generals-and-kings" understanding that "history consists of the work of the famous and influential." In addition, when historical studies focus on *key moments* of technological innovation (e.g., the invention of the Astrolabe or electricity, the operation of modern mills and steam engines, the publication of the first books on midwifery written by women), they can occasionally encourage disjointed, episodic understandings of technical communication that may, as Jo Allen (1999, 227–228) points out, fail to provide fully situated understandings of how movements develop and are tied to one another.

Those historical accounts that *do* provide a picture of the long sweep of history, moreover, can suffer from limited detail. Frederick O'Hara's "Brief History of Technical Communication," published in 2001, for instance, covers technical communication from the twelfth century to 2005 in four pages. Although short pieces like this one provide valuable thumbnails of broad historical movements, they provide neither the depth of detail nor accurate representations of the many social, cultural, and economic factors that some people might want. If we were to consider this particular brief piece a representative historical map of technical communication, for example, we would only see the largest of landmarks and these only from a distance: the emergence of mathematical writing among the Aztecs, Egyptians, Chinese, and Babylonians; the development of astronomy in the Middle East; the explosion of scientific, medical, and mechanical arts in the Renaissance; the invention of movable type and the growth of scientific publishing in the fifteenth century; the emergence of scientific journals and patents in the eighteenth century; the introduction of federal research contracts in the nineteenth century; and innovations in military technology and the computer industry during the twentieth century.

Such a map is valuable for the major landmarks and boundaries it can identify within the field of technical communication; at the same time, however, it may provide limited information about the pragmatic concerns that demand the attention of practicing technical communicators.

RESEARCH MAPS OF TECHNICAL COMMUNICATION

A second common approach to mapping the boundaries, artifacts, and identities of technical communication focuses on landmarks identified within scholarly and research studies. These maps focus on investigations of the texts (documentation, online exchanges, reports), textual practices (editing, writing, revising), textual environments (digital spaces, organizations, workplaces), and intellectual approaches (theoretical frames, disciplinary perspectives, research methods) associated with the work of technical communication. The strengths of such research maps is that they direct our attention to the questions that have structured the study of technical communication, the methodologies that investigators have found valuable in exploring these questions, and the information that these investigations have yielded. For students, teachers, and practitioners, research maps trace the field of technical communication as a socially constructed, intellectual endeavor and sometimes offer pragmatic information for practitioners.

Central Works in Technical Communication, edited by Johndan Johnson-Eilola and Stuart Selber (2004), represents an important recent collection that offers a scholarly research map of technical communication. As Johnson-Eilola and Selber acknowledge, their own bounded take on the field is "informed by contemporary social theories" and offers a map focused on the "research and theoretical portions" (xvi) of technical communication's landscape from their position as scholars and faculty members responsible for creating curricula and teaching courses in technical communication that are aimed at preprofessional students. Thus, Johnson-Eilola and Selber note, they exclude "how-to" research projects from this collection in favor of research that is "conceptual in nature" (xvi) and that provides a "way into the scholarly conversation[s]" that constitute the field from an academic perspective. As a result of the boundaries that Johnson-Eilola and Selber set for their project, each of the chapters in the collection is authored by faculty scholars teaching in technical communication programs at colleges and universities around the United States.

The main sections of this germinal and influential collection indicate the topics of concern to the academic research scholars who contributed chapters: philosophies and rhetorical theories of technical communication, issues of ethics and power, examinations of research methods, and pedagogical directions for technical communication programs, to name just a few of these topics.

Although these topics do provide "one map among several" of the field of technical communication, it is a map purposefully influenced by humanistic disciplines (rhetoric, philosophy, ethics) and the social theories that now inform academic studies in composition, history, and English programs. Only six of the collection's thirty-two chapters focus on pragmatic "how-to" concerns of workplace professionals, and all the chapters deal with programs of technical communication based in the United States.

A similar map of technical communication as a field can be found in Tim Peeples's *Professional Writing and Rhetoric: Readings from the Field* (2003), a collection aimed at undergraduate students of technical communication preparing themselves as professional communicators. Peeples's collection, which "aims to be as representative as possible of the issues that define the field" (3), provides a map bounded by three binaries (two terms or topics generally used as polar opposites) that have historically helped structure technical communication: practice versus theory, production versus practice, and school versus work. In describing his collection, the editor notes that these binaries represent misunderstandings of the field and argues for redefinitions of each area that complicate such understandings. Among just a few of the chapter topics represented in this collection are the ethical dimensions of professional writing, the role of professional writers in shaping the social contexts associated with technical communication, and strategies for students who plan to move into the professional ranks of technical communicators.

Because this map of the profession is committed to complicating the three binaries identified above, this list of topics (and the specific articles within each chapter) suggests several key landmarks of technical communication as Peeples perceives the field. First, the collection reveals the belief that workplace practices (writing within organizations and document production) must be placed in conversation with theoretical perspectives that have typically informed academic discussions of technical communication (the social theories that inform participatory design and user-centered communication, rhetorical and ethical theories of communication, and postmodernism) and argues that "theory and practice cannot be separated from one another: good practice requires theoretical knowledge, and good theorizing is not only a practice but also requires a responsiveness to practice" (3). Second, the collection argues, in Peeples's words, that a "focus on the products of writing not only hides the social interaction that is integral to writing," but also distracts from an understanding that writing is a form of "social interaction" or the "means by which we mediate social interaction" (4). Finally, the chapters in this collection are committed to a belief that "rhetorical reasoning" is characteristic of both workplace practitioners and academic scholars of technical communication.

Despite the attempt to establish direct links between academic-based and practitioner-based perspectives on technical communication, however, this collection contains works authored only by academic faculty, rather than by workplace practitioners, and only one work by authors outside the United States (a chapter by Canadian scholars).

Although each of these extensive collections offers a valuable set of contributions to the field of technical communication, one that is especially useful for students of technical communication, as maps of the profession they suffer from being both too large and too small. They are too large, for instance, to provide a map of the profession that can be communicated concisely—in either words or images—to members of the public or nonspecialists. Individuals who hope to make some sense of these maps must read all the chapters within them. And the collections are too small, in that they focus only on works authored by academic scholars and thus necessarily reduce technical communication to a certain kind of theory-and-practice research while providing little how-to research. Even the most extensive collections can contain only a relatively limited and representative number of publications: in the case of *Central Works*, thirty-two pieces were chosen to represent the entire field; in the case of *Professional Writing and Rhetoric*, twenty-five pieces were included. Other studies—which may have gone unnoticed, given the venues in which they were published, the environments within which they were circulated, or the methodologies they deploy—may not be chosen for inclusion in such collections.

SKILLS MAPS OF TECHNICAL COMMUNICATION

A third approach to mapping the field of technical communication attempts to describe the skills and understandings needed by practicing technical communicators in the workplace. The strength of such skills maps is their focus on corporations and public-service organizations as environments for communicative exchange. In response to the dynamic nature of these environments and the changes that shape them, technical communicators are continually required to develop new and different skills—to produce, manipulate, and deploy linguistic and visual elements in different ways, for different purposes, and for different audiences. Thus, skills maps can be understood as direct reflections of larger social, cultural, economic, and ideological movements that influence technical communication as a field. Because skills maps are predicated on a basic responsiveness to contemporary trends, they also offer a timely description of the boundaries, artifacts, and identities of technical communication.

In 2000, for instance, George Hayhoe, as the editor of *Technical Communication*, a journal aimed primarily at practicing technical communicators, sketched a relatively standard set of job requirements, maintaining that all communicators, regardless of their specific jobs, needed foundational skills in "writing, editing, visual communication, multimedia, document design, audience and task analysis, usability testing of products and documents, and interpersonal communication" (151); a mastery of "one or more subject domains in the sciences, medicine, engineering, or another technical field"; and knowledge of "how to use the software tools required for a specific task" (152).

Other experts, however, have argued that the transition from a manufacturing society to an information culture in the later twentieth century has necessitated a change in the description of technical communicators' jobs. As Johndan Johnson-Eilola noted in 1996, technical communicators are no longer engaged in simply translating technical information for nonspecialist audiences or supporting the product development and manufacturing sectors of corporations; rather, he continues, they are doing what Robert Reich calls "symbolic-analytic work," engaging in the "manipulation and abstraction of information" (Johnson-Eilola 1996, 253). In such environments, Johnson-Eilola continues, technical communicators need the "ability to identify, circulate, abstract, and broker information" (255). In a similar vein, Corey Wick (2000) describes the work of technical communicators as "knowledge management," noting that practitioners have to "grasp the immeasurable complexities of knowledge, language, and communication" (524) and "facilitate cross-functional collaboration" (525), as well as serving as "expert communicators" (526).

Such works, while instructive on a general level, offer relatively abstract maps of technical communication as a field of practice; they do little, for instance, to identify the specific locations of technical communication work within a range of profit and nonprofit workplaces, or to describe the specific documents, texts, objects, or discourses that occupy the attention of technical communicators. Such maps are also *future focused* in that they try to anticipate the skill sets and understandings emerging within a range of workplace contexts, given larger social, cultural, and economic trends. Because of this perspective, they may be most valuable to academic teachers of technical communication, who need to anticipate such trends so they can shape curricula that will help students prepare themselves to meet the needs of emerging work environments. These maps, however, may be of less pragmatic help to practicing technical communicators who *already* inhabit positions within the field, whose work is shaped by immediate demands of a specific industry, or whose efforts are shaped by the uneven nature of change in the large, varied, and far-reaching field of technical communication.

Finally, skills maps of technical communications are usually formulated in words—as book chapters, journal articles, or magazine features. Audiences must often read these genres from beginning to end in order to apprehend the specific particulars of authors' linear, propositional arguments. Visual maps of the field, in contrast, while not necessarily suited for presenting ideas in linear propositional logic, can present information economically and in ways that readers can apprehend through relatively quick visual examinations, much the same as they do when looking at photographs or graphs.

A HEURISTIC: MAPPING TECHNICAL COMMUNICATION WITH TEXT CLOUDS

So what other types of maps can technical communicators employ to provide a sense of the large, diverse, and dynamic professional field? In recent years, one approach technical communicators have come to rely on when they want to make sense of large amounts of information—especially when dealing with complex ideas and numerous documents and data that change over time—is the use of tag clouds and the related variant of text clouds (Nielson 2007; Rivadeneira et al. 2007), visual representations of words, typically a set of "tags" that describe different pieces of information contained in extensive websites, databases, or blogs. The visual attributes of these words—size, weight, and color, for example—are used to "represent features, such as the frequency of the associated terms" (Rivadeneira et al. 2007, 995). In describing the value of text, information designer Joe Lamantia (2007) notes that text clouds "tweak the eye-brain conduit directly," functioning like "the common executive summary on steroids and acid simultaneously" to help human readers process meaning quickly and economically. Lamantia continues: "Text clouds are meant to facilitate rapid understanding and comprehension of a body of words, links, phrases, etc." Daniel Steinbock (2008), the inventor of TagCrowd (a free text-cloud-generating program) notes that "when we look at a text cloud, we see not only an informative, beautiful image that communicates much in a single glance, we see a whole new perspective on text." Among other functions of text clouds, Steinbock notes that they provide "topic summaries" of a text, a means of "data mining" a corpus, and a tool for reflection.

Although text clouds and tag clouds have been used for making sense of large data sets, they have not been used as a heuristic for understanding and mapping the field of technical communication. In the section that follows, we demonstrate how to use a text cloud as a heuristic for reflecting on the field of technical communication. This heuristic includes five steps.

CREATING TEXT CLOUDS

The basic process involved in creating a text cloud is represented in figure 1.1. It involves selecting a text or texts that can be used as a data set, and then employing a computer application to create a text-cloud representation of the language in the document(s). While making a simple text cloud is not particularly difficult, however, following a systematic process for focusing, refining, and interpreting a cloud can help create a visualization that is more likely to meet the rhetorical needs of an audience.

Step 1: Identify Focusing Question(s) for the Text Cloud and Its Rhetorical Context

The best place to start when creating a text cloud is to consider the rhetorical purpose, audience, and content for the cloud and the context in which it will be used. To accomplish this step of the process, ask yourself the following questions and make notes on your answers:

- What is the question I want to explore? On what subject do I want to reflect?
- What is the purpose of constructing this text cloud?
- Who is the audience who will read/look at the text cloud?
- What documents will provide the content for the cloud?

The more work you can do at this stage of your process, the more specific you can get about the question that focuses your text cloud and the rhetorical context in which the text cloud will function, the easier it will ultimately be to construct a cloud that provides an effective visual representation.

Step 2: Identify and Refine a Document/Data Set Appropriate to the Rhetorical Context

The next step in the process involves choosing a set of documents that will help you answer your focusing question and accomplish the rhetorical

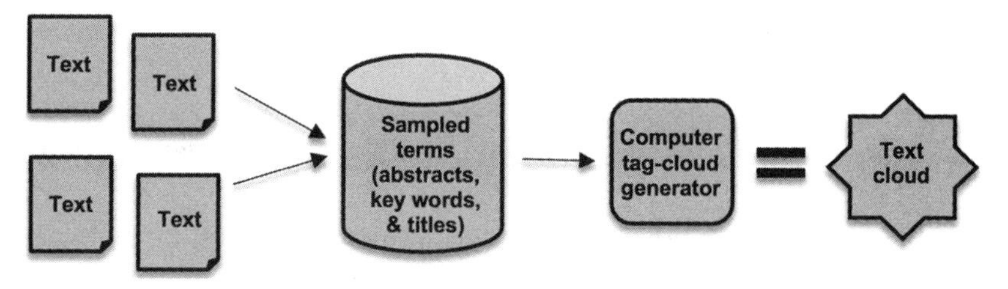

Figure 1.1. Process of creating a text cloud

purpose of your text cloud. To accomplish this step, ask yourself the following questions:

- What kind of documents and what specific documents will provide the best/most appropriate information in terms of my focusing question?
- How many documents do I need in order to accomplish my rhetorical purpose? Are the documents readily available in digital form?
- What parts of the documents contain the content most appropriate to my focusing question?

Some questions and some rhetorical contexts will require larger sets of data than others. For instance, creating a text cloud focused narrowly on one question (e.g., What are the priorities expressed in the language of Corporation X's annual report?) may require only one document as a data set (e.g., the annual report itself), especially if the text cloud is being created for a narrowly focused audience (e.g., the corporation's management) and for a particular rhetorical context (e.g., reflecting on the corporation's priorities as expressed in a draft of the annual report in order to polish the language for the document that gets sent out to shareholders). Other clouds that focus on questions more broadly conceived (e.g., What are the boundaries, artifacts, and identities of technical communication?) and are meant for broader audiences (e.g., students, scholars, practitioners) and contexts (e.g., reflecting on the boundaries, artifacts, and identities of technical communication) may require many more documents (e.g., multiple articles from journals about technical communication).

Step 3: Identify Rules for Structuring Terms and Generate a Text Cloud

Once you have collected the digital documents you are going to use, you can submit them to a computerized cloud generator like TagCrowd, the free online text-cloud generator that we used to create the text clouds in this chapter, or Wordle, another free online text-cloud generator. These programs create clouds that give greater prominence (represented by the size and/or color of the word, or the number next to the word) to words that appear more frequently in the original source text. Figure 1.2 is a small word cloud (limited to twenty terms) that uses this paragraph as a source text.

Thinking of text clouds as *wholly* determined by computers, however, can mask a number of important issues involved in generating a text cloud and much of the work that must be done to make text clouds useful to a particular audience. To make good use of computerized text-cloud generators, you need to make certain decisions about the rules that structure the terms within the cloud. Often these rules are determined by the text-cloud generator that you use. TagCrowd, for instance, allows you to decide on the size of the cloud you generate (e.g., the maximum number of terms it can include), whether the frequency counts of words should be displayed, whether to group similar words (e.g., *focus*, *focused*, *focusing*), and whether some words should be ignored (for the text cloud in figure 1.2, we directed the program to ignore the articles *a*, *an*, and *the*, as well as the coordinating conjunctions *and*, *but*, *or*, *nor*, *for*, and *yet*, so that the word cloud would focus on nouns, pronouns, verbs, adverbs, and adjectives, which are more informative in this context).

Depending on the rhetorical context within which a text cloud will be used, creating a *rhetorically useful* cloud may involve additional steps that require manual manipulation of the data either before it is submitted to a text-cloud generator or after the text cloud is generated. To determine these rules, which should always be considered in terms of rhetorical context (e.g., purpose, audience, information, situation), ask yourself the following questions and derive rules from the answers:

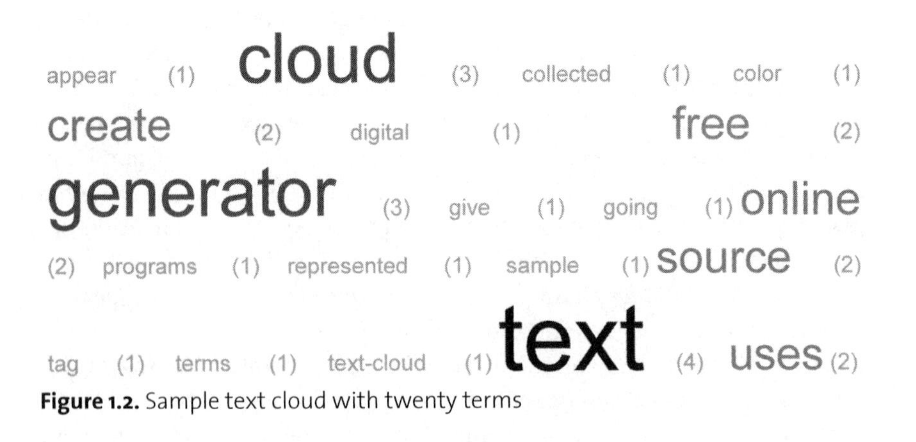

Figure 1.2. Sample text cloud with twenty terms

- Given the rhetorical context in which the text cloud will be used, is it important to include or exclude proper nouns (e.g., Texas, Exxon) or people's names (e.g., Jane Smith, Thomas Wolfe)?
- Given the rhetorical context in which the text cloud will be used, is it important to hyphenate nouns and the words that directly modify these nouns (e.g., analysis-*factor*, tags-*computerized*).
- Given the rhetorical context in which the text cloud will be used, is it important to manipulate word order (identifying subordinate and superordinate, or parent, terms) or to use hyphens to preserve semantic relationships (e.g., text-computer-tagging-of, analysis-corpus, analysis-factor, analysis-textual)?

Step 4: Adjust the Granularity in Text Clouds in Light of Their Rhetorical Context

After you have focused your text cloud and identified its rhetorical context, assembled the digital source documents and submitted them to a computerized text-cloud generator, and determined the rules that will help structure the most useful cloud for a particular rhetorical context, the next step is to determine the optimal size and level of granularity for the text cloud. Often these strategies will need to be applied in combination to create text clouds that are appropriate for a particular rhetorical purpose.

The size of text clouds, for instance, depends to a large extent on the rhetorical purpose for which they are designed and the rhetorical context within which they will be used. In some situations, it may seem impossible to create a text cloud *small enough* to make sense to readers seeking a quick overview and, at the same time, *detailed enough* in terms of granularity to represent all the important information that they need in order to make decisions. In such cases, you can design and compare several text clouds of different sizes (small text clouds with a limited number of terms that are easier to read and larger text clouds that include a larger number of terms but that are harder to read) and compare them in terms of their utility.

Similarly, techniques for structuring the level and granular focus of text clouds—by using brackets to cluster related terms—can also improve their rhetorical value. Text clouds structured almost exclusively by alphabetic order, for instance, provide a reliable way to locate individual words, but they give readers little help in identifying related words (e.g., *woman* and *female*) unless they occur next to each other in alphabetic order or they happen to share a similar root (e.g., *user* and *usability*).

To focus on the size and granularity of text clouds, we can suggest several additional questions to ask.

- Given the rhetorical context in which the text cloud will be used, what is the appropriate level of granularity or detail? Is the text cloud more informative if it is larger and more encompassing or smaller and more focused?
- Given the rhetorical context in which the text cloud will be used, should related terms be grouped within brackets that highlight certain relationships (e.g., is it useful to group terms like *hospital-costs*, *drug-costs*, *salaries-doctors* within a single set of brackets labeled *Health Care Costs*)?

Step 5: Interpret and Compare Text Clouds

Generating text clouds is relatively easy, but *interpreting* them is a more difficult task. To make sense of the text clouds that you create requires active reading and interpretation. The following questions can help in this task. Some of these questions may necessitate creating additional text clouds for purposes of comparison.

- What insights are suggested by the relative frequencies of terms? What are the boundary terms at the low end of frequency, and why might they be mentioned so infrequently? What are the boundary terms at the high end of the frequency scale, and why might they be mentioned so frequently?
- What insights are suggested by the larger clusters of terms? What issues/topics/terms connect these larger clusters? What insights are suggested by smaller clusters of terms? Why might these clusters of terms be significant?
- What patterns and trends emerge from the text cloud? What repetitions and singularities?
- What terms emerge as key for particular users? Do users' interpretations of the text clouds differ, or are they more or less congruent? Why?
- What terms seem to be missing and why?
- When you change the size and the rules that govern a text cloud, what new patterns, trends, repetitions, and frequencies emerge? Why?
- What happens when a time-based series of text clouds is created from comparable data (e.g., text clouds of annual reports over a period of ten years or text clouds of annual reports from every five

years)? What patterns, trends, repetitions, and frequencies change? Which remain stable? Why?

AN EXTENDED EXAMPLE: CREATING TEXT CLOUDS

In this section, we create some examples of text clouds that could help Amanda get a fix on the field of technical communication and the boundaries, artifacts, and identities that characterize the profession. It is important to acknowledge at the outset, however, that it is impossible to represent, in any summarized form, the *entire* field of technical communication—to represent all of the technical communicators who are practitioners, all of the different kinds of businesses and organizations that employ technical communicators, all of the genres associated with the academic study of technical communication, all of the different ways of researching technical communication and its effectiveness. Nor is it possible to examine every practice involved in technical communication, every piece of technical communication produced in this country and others, the history of the field, or all of the research investigations that help make up technical communication. It is also true that any map of the field will necessarily be limited to the field as it exists at a specific time and place—it will provide a snapshot of technical communication, rather than a movie. In the final analysis, *all* maps have their biases and shortcomings—but that recognition shouldn't keep us from trying different representations and experimenting with different kinds of text clouds to see how and why they serve our rhetorical purposes.

Step 1: Identify Focusing Question(s) for the Text Cloud and Its Rhetorical Context

To begin answering Amanda's question "What is technical communication anyway?" we rephrased her query more specifically as "What are the boundaries, artifacts, and identities of technical communication?" The next step of our process involves focusing more narrowly on the rhetorical context of the task. For our sample text-cloud exercise, we have identified the following context:

Purpose: To create maps of technical communication that provide informed overviews of the field's work and general understandings of its contours, and that are easily comprehended in terms of visual display and content.

Audience: Students of technical communication, like Amanda Metz Bemer.

Step 2: Identify and Refine a Document/Data Set Appropriate to the Rhetorical Context

Creating any text cloud requires sampling key terms from a document or set of documents. To answer Amanda's question, we have chosen to focus our sample text clouds on the research conducted in technical communication as published in two journals in that field: *IEEE Transactions on Professional Communication* and *Technical Communication Quarterly*. *IEEE Transactions on Professional Communication* focuses on research projects undertaken primarily by workplace practitioners and in workplace contexts. *Technical Communication Quarterly*, in contrast, not only focuses on research about technical communication practices in the workplace, but also features research undertaken by both academic scholars of technical communication and workplace practitioners. Because we want to focus on technical communication *practices* rather than *curricular and classroom issues*, we can eliminate those articles from the two journals that have a curricular or teaching focus.

Both of these journals are accessible in digital format from 1996 to 2006, a ten-year window on the research within the profession and on the practice of technical communication. During this ten-year period, we can identify 168 articles from *IEEE Transactions on Professional Communication* and 116 articles from *Technical Communication Quarterly* as a preliminary data set. However, if we use the entire text of all 284 published articles, the text clouds we generated would be far too large. Fortunately, both journals provide a short list of controlled indexing terms or subject terms for each article as well as a brief abstract and a title. These important elements—the indexing terms, subject terms, abstracts, and the titles for all 284 articles—provide a smaller and more manageable data set for text clouds designed to explore the boundaries, artifacts, and identities of technical communication, and to address Amanda's original question about the field of technical communication.

Step 3: Identify Rules for Structuring Terms and Generate a Text Cloud

The next step in creating a useful text cloud involves using an online cloud-generating program to process the data set, or source text(s), and turn it into a text cloud. Even though we have reduced the size of our data set, it still needs to be structured in ways that make it useful before we employ a cloud-generating program. In particular, we need to "clean" the source text to eliminate words that don't help us accomplish Amanda's task of exploring the boundaries, artifacts, and identities of technical communication.

For this purpose, we used the following rules to focus and structure our combined source text:

- Keep nouns and noun phrases (e.g., text, tagging, analysis), but eliminate most other parts of speech (e.g., verbs, pronouns, conjunctions) that are likely to provide less information about the boundaries, artifacts, and identities that characterize technical communication.
- Exclude proper names of people (e.g., the names of authors, the names included in textual citations).
- Include words that directly modify nouns (e.g., *factor* analysis, *computer* tagging) to provide additional context for understanding.
- Manipulate word order (identifying subordinate and superordinate, or parent, terms) and use hyphens to preserve semantic relationships.

Once we apply these rules to our source text, we can then submit it to TagCrowd to generate a text cloud that will address Amanda's original question about the field of technical communication.

Step 4: Adjust the Granularity in Text Clouds in Light of Their Rhetorical Context

The text cloud that TagCrowd generates from our focused source text is still too large and unwieldy, containing approximately 2,240 terms. This enormous text cloud is almost impossible to make sense of and will not help us address Amanda's question by providing an *economical* visual overview of the field of technical communication, in reflecting on the field, or in constructing a collective social sense of the boundaries, artifacts, and identities of the profession. Figure 1.3 shows a very small piece of this larger cloud, focused on terms associated with the word *language*.

This portion of the text cloud, containing twenty-eight different terms, may still be too detailed to provide the kind of economical overview of the patterns and trends of *language* use in technical communication that

language (3) language-biomedicine (1) language-boring (1) language-challenges (1) language-disorders (1) language-documentation (2) language-impact (1) language-impaired-people (1) language-impaired-users (1) language-intensity (1) language-markup-extensible (2) language-markup-hypermedia (1) language-needs (1) language-proficiency-English (1) language-restricted (1) language-skills-teaching-course-English (1) language-skills (1) language-support-tool (1) language-theory (1) language-translation (5) language-universal (1) language-usage-contemporary (1) language-use (2) languages-local (1) languages-natural (4) languages-object-oriented (1) languages-page-description (3) languages-visual (1)

Figure 1.3. One small portion of the large and detailed text cloud focused on the boundaries, artifacts, and identities of technical communication

language (3) language-documentation (2) language-markup-extensible (2) language-translation (5) language-use (2) languages-natural (4) languages-page-description (3)

Figure 1.4. The same portion of the text cloud as in figure 1.3, focused on language but excluding words mentioned only once

might be useful for Amanda's task. By adjusting the granularity (level of detail) of this portion of the text cloud, we can eliminate those words occurring less than two times to generate a smaller and much more manageable set of seven terms with their variants (figure 1.4). At this level of granularity, with fewer terms competing for attention, two key focal points in the field of technical communication emerge more clearly: first, the recent emphasis on and need for *language translation* in technical communication resulting from patterns of globalization, and, second, the emphasis on *natural-language* processing, which characterizes work in artificial intelligence and references the digital work that describes so much of technical communicators' labor in the twenty-first century, both trends that we will discuss later in this chapter. This small portion of text cloud also has its shortcomings, however; twenty-one terms have been dropped completely because they occur only once.

This kind of problem becomes even more evident if we think about the entire text cloud in the context of Amanda's rhetorical task and how it would help her characterize the boundaries, artifacts, and identities of technical communication. Another example will help us explore this point. If we eliminated those terms appearing only once from the entire text cloud, we would lose key words like *cancer*, *Alzheimer's disease*, and *diabetes*. Further, because two of these terms (e.g., *cancer*, *diabetes*) were used in the data set without being next to a word like *disease*, they were not linked to any parent term (a superordinate term that serves to collect several items under one umbrella) that would help Amanda spot their connection. To solve this problem, we can group those terms associated with a common parent term in a set of brackets within a text cloud.

[disability (1) accessibility (4) Americans-with-disabilities-act (1) curb-cuts (1)]

Figure 1.5. Grouping related words within a text cloud, using brackets

Consider, for example, the terms in figure 1.5: *curb-cuts*, *accessibility*, *Americans-with-Disabilities-Act*, and *disability*. In an alphabetized and unclustered text cloud, these terms would be separated, and, as a result, readers like Amanda might not recognize them as forming a semantically related cluster. Grouping related terms using brackets allows us to identify the parent term of *accessibility* and gives a clearer picture of the topics that the field of technical communication addresses in connection with that term.

In this small portion of the text cloud, what becomes visually clear is a concern about accessibility which arose, in part, from the work of disabled activists and, in part, from the legislation of the Americans with Disabilities Act that was passed on July 26, 1990 (Public Law 101–336 [42 U.S.C. sec. 12101 et seq.])—both cultural phenomena were reflected in the topics taken up by the field of technical communication from 1996 to 2006 (Wilson 2000).

Another example of a focused text cloud can be seen in figure 1.6. Here we have created a relatively small text cloud that focuses on the boundaries of technical communication, interpreting *boundaries*, in this particular case, as having to do not only with geography and geographical borders, but also with larger topics related to geography, like *globalization*. This approach results in a text cloud that would offer students like Amanda both focus and detail, identifying all of the geographical locations—and the terms associated with geography—mentioned in the abstracts, titles, and indexed terms of *IEEE Transactions on Professional Communication* and *Technical Communication Quarterly* from 1996 to 2006.

In part, this focused text cloud reflects how the field of technical communication, in the period between 1996 and 2006, had become increasingly concerned with communication practices in countries outside the United States, including China, Japan, and Korea. These Asian nations experienced rapid growth and made related technological advancements during the decade on which our text cloud focuses, and, thus, had come to influence the practice of technical communication in the United States. Fatemeh Zahedi, William Van Pelt, and Jaeki Song (2001, 83) trace one perspective of this trend as it relates to web design for technical communicators: "Since the web can now link diverse regions and communities across the globe that were relatively isolated by time and space, the growth of culture(al)(ism)/cross-cultural (48)

foreign-engineers (1) foreign-scientists (1) [global(ized)(ization) (11)

[global-communication-business (1) global-marketplace (2) global-online-access (1) global-pharmaceutical-industry (1) global-reach (1) global-strategies-business (1) global-strategies-corporate (1), global-team (1) globalization (1) globalization-of-rhetoric (1)] [local(ized)-(6) local-services (1) localization (1) localization-document (1) localization-practices (1) localization-user (1) localized research (1)] transnational (1)] international (19)

[language (42) translation (14) Chinese (2) [Chinese (1) Chinese-native-speakers (1)] English (13) [native-English-speaking-countries (1)] Finnish (1) Japan(ese) (5) [Japanese (2) Japanese-native speakers (1) Japanese-readers (1) Japanese-speaking-countries (1)] Spanish-native-speakers (1)]

[North-America(n) (3) United States (16) American (6) [American (2) American-international-health-alliance (1) American-Medical-Association-Journal (1) American-West (1)] Alaska (1) Louisiana (1) North-Carolina (2) Roanoke-Island (1) Virginia (1) Mexico (1) Canada (3)] Australia (2) [Britain-colonies (1) Britain (2) colonists-English (1)] [Far-East Asian-countries (1) China (1) Japan (1) Korea (2) Malaysia (1)] [Europe(an) (6) [Europe (2) Europe-eastern (1) European-commission (1) European-union (1) European-Union-member-states (1)] Finland (4) France (3) German (1) Scotland (1)] [India (1) Indians (1)] New-Zealand (1) Russia (1) [South America (1) Equador-Quito (1)]

Figure 1.6. Text cloud focusing on the geography of technical communication

global communications has increased and intensified the need for learning to communicate successfully with a multitude of diverse, localized cultures. No single model of cultural understanding is sufficient for communicating effectively with all web audiences."

The text cloud in figure 1.6 also employs several kinds of brackets that provide additional insight into the field of technical communication for students like Amanda. The larger brackets are used to group terms that are topically related, a rhetorical decision on the part of the text-cloud creators. Note, for instance, that the terms associated with *globalization* and terms associated with *localization* are grouped together to indicate that they both relate to discussions of changing markets and trade, transportation systems, and communication patterns on a global scale. The smaller brackets are then used to group terms that provide some detail and focus within this same discussion. For example, the term *global*, its variants, and associated hyphenated terms—used a total of eleven times in the data set—are contained within a set of smaller brackets. Smaller brackets are also used to group the term *local* and its variants, used a total of six times.

Both of these important parent terms (*global* and *local*) in the text cloud shown in figure 1.6—as well as the mention of nations (Ecuador,

Finland, France, China, Korea, Mexico, Malaysia, India, Scotland, Russia, New Zealand, among others) and languages (Chinese, Japanese, Spanish, Finnish)—clearly reference the increasingly international scope of technical communication as a profession during the period 1996–2006. Importantly for students like Amanda, the visual prominence of the key terms *global* and *local* also reflects the tension between globalization and localization that was influencing technical communication during this period when companies who used extended computer and telecommunication networks, an expanded system of international trade, and extended global transportation routes were struggling with issues of how to leverage the power and reach of globalization while addressing the specific linguistic and cultural needs of local audiences.

Kirk St. Amant (2005, 73) references this important boundary tension between globalizing and localizing forces, noting that "website designers . . . find themselves in the position of creating online materials that will be used by a broad international audience. In an ideal situation, the designer works with a localizer who revises materials to meet the expectations of a particular cultural audience. The speed and cost with which localizers can revise materials, however, is often related to the items provided by the original website designer."

We can undertake a similar exercise by creating a text cloud focused on the identities of technical communication as a field, looking at the terms used to name or refer to the profession. The cloud in figure 1.7, for instance, uses two levels of brackets to group terms.

Focused in this way, the text cloud in figure 1.7 offers some additional perspective with regard to Amanda's original question "What is technical communication anyway?" First, the text cloud reflects a snapshot of a field. During the period identified by the sampled texts, the years 1996 to 2006, technical communication, as scholar Saul Carliner (2010) documents, was focusing not only on *writing* (variations of which appear 109 times in this text cloud with such terms as *technical writing*, *editing*, *technical editing*) as a primary meaning-making activity of technical communicators but also on a broader conception of *communication* (variations of which appear 560 times with terms such as *technical communication* and *communication of technical information*, *professional communication*, *organizational communication*, *business communication*), by way of acknowledging the role that visual design, images, and other modalities of expressions played in meaning making. From this perspective, the two major clusters of words that anchor this text cloud (*writing* and *communication*) reflect the rapid growth of technical communication as a field after World War II and through the information-age decades of the 1980s, 1990s, and the beginning of the [communication (339) communication-technical (29) communication-technical-information (45) communicator(s)-technical (26) communication-organizational (7) communication-professional (92) communication-business (22)] [design(ers) (58) [visual (22) image (7) illustration (6) graphic(s) (7)] [media(-) (um) (21) media-choice (4)] multimedia (10) photo(s) (graphy) (8) video (8) hyper(text) (media) (9)] information (83) presentation-technical (41) [knowledge (28) [knowledge-management (6)]] publishing (21) read(er) (ing) (17) [professional (19) [professional (4) professional-aspects (10)]] rhetoric(s)(al)(ician) (69) technology (24) [tele-(work) (communication) (conferencing) (presence) (15) [teleconferencing (5)]] work(ing)(ers) (36) [writer(s)(ing) (81) [writers-technical (10) writing-technical (18)] edit(ing) (ial) (ors) (19)[editing-technical (4)]]

Figure 1.7. Text cloud focusing on identity terms for technical communication

twenty-first century. During this time, the artifacts that technical communicators designed and produced and the work in which they engaged changed dramatically. Where the primary effort in the field had focused early in this period on printed and written manuals to accompany mass-produced manufactured goods and on texts to explain scientific and engineering innovations to nonspecialist readers (Pringle and Williams 2005), technical communication in the postindustrial age became a much more broadly practiced and defined profession, whose members were involved in communication that employed a variety of modes and genres and that occurred in a range of contexts. Technical communicators were engaged in the dynamic design, management, and manipulation of information (Slattery 2005); the kinds of experimentation, collaboration, abstraction and system thinking required of "symbolic-analytic workers" (Johnson-Eilola 1996); the interpretation of end users' needs (Pringle and Williams 2005, 364); and scholarship, teaching, and research within the profession of technical communication.

If Amanda looked at figure 1.7, for instance, she would see a text cloud that provides a glimpse of this changed landscape, in which the profession's identity had come to include both *writing* and a broader conception of *communication*. Part of this landscape is evident in the text cloud's focus on key terms like *information* (mentioned 83 times in the source texts), *knowledge* (mentioned 28 times), and emergent terms like *knowledge-management* (mentioned 6 times). By 2005, for instance, Pringle and Williams described the profession of technical *communication* without using the word *writing*: "As technical communicators begin to articulate and understand our own professional identity . . . we will be recognized as ones who approach technology from a users' perspective and who possess expertise in 'communicating.' If there's one thing that the stunning speed of technological innovation has made clear, it's that communication is no longer just an adjunct to business, it *is* business" (369).

From this perspective, the text cloud in figure 1.7 might help readers like Amanda perceive the centrality that terms like *design* (variants and related terms used 65 times), *visual* (variants and related terms used 42 times), and *media* work (variants and related terms used 60 times) had come to play in the field of technical communication by the end of the twentieth century.

In the field of technical communication, this historical trend was accelerated by the first mass-produced personal computers in the late 1970s and 1980s, which made possible the subsequent development of software applications that allowed for what-you-see-is-what-you-get (WYSIWYG) page design and layout, graphic design, photo manipulation, and digital video and audio editing (Pringle and Williams 2005). If Amanda looked at figure 1.7, for instance, she might see multiple references to the results of this trend not only in the key term of *visual*, for example, but also in related terms that reflect visual communication (*image*, mentioned 7 times; *illustration*, mentioned 6 times; *graphics*, mentioned 7 times). Similarly, in connection with the key term *media(um)*, Amanda might get a sense of technical communication's interest in *media choice*, mentioned 4 times; *multimedia*, mentioned 10 times; *photography* and *video*, mentioned 8 times each; and *hypertext*, mentioned 9 times.

A final related perspective offered by the text cloud in figure 1.7 is that technical communication as a profession was increasingly practiced throughout the period within digital environments and often within distributed networks. One specific aspect of these practices is evident in the cluster of terms around *telework*, *telecommunication*, *telepresence*, and *teleconferencing*. The same technologies and computer networks that supported multimodal and multimedia communication practices at the end of the twentieth and the beginning of the twenty-first century also supported the growth of telecommuting, telework, and teleconferencing, making these activities increasingly prominent landmarks within the field of technical communication during the period represented in the text cloud. As Scott and Timmerman (1999) observed, information workers who telecommute rely heavily on a variety of communication technologies not only to do their work, but also to maintain their relationships with coworkers and supervisors within the companies that employ them and to exchange information with contacts outside those companies.

Although telework represents only a small part of the online and digital information work done within the field of technical communication (as evidenced by the relatively few terms and the relatively low frequency of these terms in the text cloud in figure 1.7), the text cloud provides Amanda with a visual snapshot of these practices in distributed and online work environments between 1996 and 2006.

BY WAY OF CONCLUDING, BUT NOT FINISHING

Technical communication scholars and practitioners like Amanda Metz Bemer, as contemporary "knowledge workers," "must make sense of huge amounts of unstructured textual data" (Havre, Hetzler, and Newell 2002, 9). One valuable heuristic for making sense of these large data sets involves "exploring multiple visual presentations, or visualizations of the data," each version of which may well "lead to important insights and/or a better global understanding of the collection" as a whole (1077). Text clouds, as Mogens Nielson notes, allow users to "quickly and intuitively get an overview of the most used tags in a tag space." This kind of representation functions "like a satellite image of an area" (Nielson 2007, 7), to provide a particular perspective on the field and to offer what Keng Siau and Tan Xin (2005, 275) call a visual "frame of reference."

So let's conclude by creating a final 10,000-foot view of the field of technical communication for students like Amanda. Toward this end, we can generate a text cloud that includes only those terms used ten times or more in the sample; clustering them under the focal categories of "boundaries," "artifacts," and "identities"; and using brackets to group terms into clusters around related ideas. Importantly, this text-cloud map does not give us some of the detail we've identified in the more focused text clouds in figures 1.3 to 1.7. In addition, this text-cloud map is highly interpretive in that it depends both on authorial judgment to determine how terms are grouped and to suggest how they might be read, and on readers' understanding that the contexts within which terms were originally used in the sample may vary widely from the ways in which they are interpreted in a text-cloud map. At the same time, this cloud can offer Amanda one example of how students might compose their own visual-verbal map of technical communication as a field, by composing an overview that, while interpretive and limited in its own ways, is also informative.

In figure 1.8, readers will see that we have grouped those terms we associate with **boundaries** (supplemented with the terms *borders* and *locations* to suggest dimensions of geography and focus, as well as location) into four sets of brackets by way of signaling what we consider to be related clusters of words from our sample texts. We have bracketed the first set of terms to indicate a field that seems, from our reading of the sample

BOUNDARIES (BORDERS/LOCATIONS) [business (19) corporate (13) industry (10) organization(s) (47)] [science(ists)/scientific (49) engineers(ing) (26) environment (13) medicine(-)(al-) (23) health (12) socio(al)(ology) (52) psychology(ical) (16) cognition(tive) (12) usability (20)] [cultural(ism)/cross-cultural (48) English (13) global(ized)(ization) (11) international (19) language (42) linguistics/lingual (11) translation (14) United-States (16)] [automation (14) computer(s)/ computing (27) data (45) database (18)] (e-) electronic (26) e-mail (email, electronic mail) (33) groupware (18) human-factors (37) information (83) information-technology (13) internet (42) interaction(s)(ive) (15) interface(s) (23) interfaces-user (19) media-(um) (21) message(s) (10) multimedia (10) online (16) software (32) system(s)- (62) technology (24) visual- (22) tele- (17) virtual-(22) web-work(ing) (ers) (36)]

ARTIFACTS (APPROACHES/ACTIVITIES) [communication (335) discourse(s)/discursive (27)] [document(s) (32) documentation- (43) help (10) manual(s)(-) (20) text (25) report (19)] [method(s)(ology) (15) analysis(es)/analytical (44) history(ical) (24) metaphor(ic) (10) model(ing) (20) problem (13) process(es) (ing)/procedure- (12) product(s) (14) resource(s) (13) task(s) (10) theory(etical) (ies) (40)] [research (69) study(ies) (19) query- (11)] [author(ship) (11) publishing (21) presentation (46) edit(ing)(ial)(ors) (19) error (11)] [collaborate(ing)(tion)(ive) (19) group(s) (35) team (33)]

IDENTITIES [communication-business (22) communication-of-technical-information (36) communication-professional (92) communication-technical (29) communicator(s)-technical (26) design(ers) (58) rhetoric(s)(al)(ician) (69) writers-technical (10) writing-technical (18)] [read(er) (ing) (17) writer(s) (ing) (81) learner(s) (ing) (14) innovation (13) knowledge (28)] [employee (s)(er)(ment) (18) human (43) management(-) (25) professional- (19) user(s) (66)]

Figure 1.8. A text-cloud map of technical communication, based on both data and interpretation

texts, to have its feet planted firmly within the borders of the private sector (*business*, *corporate*, and *industry*) and related organizational contexts (*organizations*); and the second set of terms to suggest that the field's primary interests remain located in the arenas of *science* and *engineering*, with strong interests, as well, in *environment*, *medical*, and *health* areas during the period in which the texts for this chapter were sampled. The text cloud also indicates technical communication's interest in *sociology*, *psychology*, *cognition*, and *usability* as locations of work. With the third set of brackets, we point to what we see as technical communication's border-crossing activities: the production of texts in the *United States* and in *international* settings, a focus on communications that are *cross-cultural*, and an interest in projects that involve *language* and *translation*—all work that is *globalized* (in terms of communication systems and standards) yet marked by localized user requirements arising from the needs of readers in various cultural and social settings. In the final set of brackets, we have included terms that we believe describe the location of technical communication in the boundary-crossing spaces of the *internet*, *electronic* networks, *infor-*

mation networks, and *telework*. Terms within this set of brackets, we believe—*multimedia*, *interface*, *virtual*, and *web work*, among others—point to the work that now captures so much of technical communicators' attention within the boundaries of digital spaces.

The terms we've associated with **artifacts** that characterize the field of technical communication (to which we have added *approaches* and *activities* to capture a sense of process) include six bracketed clusters that relate to work processes, approaches, and products. We have included terms in the first set of brackets, for example, because we believe they signal the field's focus on a core set of activities surrounding *communication* and *discourse*, and we have included terms that describe the most common kinds of artifacts associated with the work of technical communication in the second set of brackets: *documents* and *documentation*, *texts* and *reports*, *manuals* and *help* texts. In the third set of brackets, we have included terms that could suggest the methodological approaches used by technical communicators to create these documents, including *analytical* and *theoretical* understandings; *historical* and literary approaches (*metaphor*); *modeling* and *problem* investigations; and attention to *procedures*, *tasks*, and *products*. We believe these approaches are deployed by technical communication specialists, as the fourth set of brackets indicates, as they (and sometimes their audiences) engage in *research* and *studies* of communication or attempt to answer or pose *queries* involving communication. We have used the fifth and sixth sets of brackets in the artifacts category to signal the continuing importance of authoring (*authorship*), *publishing*, and presenting (*presentation*) within the field of technical communication, as well as the ongoing focus on *editing* and *collaboration* among *groups* and *teams* to produce finished communication products.

We have grouped terms in the **identities** category for the field of technical communication into three sets of brackets. The first we use to signal both the range and the focus of the field, pointing out identity terms that are traditional (*technical communicators*, *technical writers*, *rhetoricians*) and those that are emergent descriptors (*designers*) for specialists in the field and the work they do (*business communication*, *technical communication*, *professional communication*, and *technical writing*). Because this text cloud cannot show the context in which these words originally occurred, the second set of brackets may be more ambiguous. Here we have grouped identifying terms that could be associated with technical communication specialists, the work they do, or the audiences they attempt to reach (*reader-reading*, *writer-writing*, *learner-learning*). Similarly, we have included terms in the second set of brackets to characterize the contexts in which we see technical communicators working, highlighting *innovation* and *knowledge* work as suggested by scholars like Johndan Johnson-Eilola in his germinal "Relocating the Value of Work: Technical Communication in a Post-Industrial World" (1996), which documents the fundamental changes that computer-based communications introduced to the U.S. labor market in the twentieth century, shifting the alignment of technical communication from a service tied to the manufacturing of products to the kind of broad-based knowledge work characterizing a postindustrial world. The third set of brackets have been used to identify the different kinds of people shaping the communicative exchanges within and around technical environments: *employees*, members of *management* teams, and *users*, among them.

The final text cloud we have composed in figure 1.8, then, is a map based both on data from our sample and on our own interpretive understanding of that data. Based on our reading of the sample texts, we have clustered the frequently appearing terms to provide what we consider to be a reasonable visual and verbal map of a field, one that describes technical communication as a profession focused on the production and study of texts of all kinds (print, digital, multimedia) and related communication practices. We have composed a map of a profession exploring the recent changes that globalized and localized trends like digitization and information/knowledge work have introduced into our lives. We have also created a map that marks the involvement of technical communication in a wide range of arenas ranging from science and engineering to sociology and human factors.

We end our chapter, then, by turning to Julie Fisher (1998, 186), who acknowledges realistically that "[t]he profession of technical communication is not easily defined in part because the profession encompasses a wide range of skills and crosses many professional boundaries. Even among researchers of technical communication there are disparate views of who technical communicators are and what they do. Beck suggests 'Perhaps one reason for this lack of definition comes from the inherent diversity within the field, a diversity that expands as the membership increases.'" Despite the difficulties Fisher describes, however, the act of composing, focusing, and interpreting text clouds like the ones we've featured in this chapter could help students like Amanda Metz Bemer identify for themselves a set of informed—albeit not definitive—answers to her question "What is technical communication anyway?" and help her explain to friends and family the importance, the complexity, and the range of the field she is studying.

DISCUSSION QUESTIONS

1. Think about the textual work (e.g., websites, poetry, grocery lists, e-mail, text messages) that you do in the course of your life at home, in the community, and at work. Choose one of these data sets. How could you create a text cloud that would help make sense of the data you produce?
2. How could you use text clouds to show how your interests and activities have changed over time—for instance, comparing your interests in grade school to your interests today? What would these comparative text clouds show about you and the ways you have changed?
3. What is your own best learning style? Do you learn best through images? Printed words? Aural sources? How might this approach to learning suit you as you enter a career as a technical communicator?
4. Try creating a text cloud from a lecture that one of your teachers gives. Take notes on the lecture or use an audio recorder and then transcribe the recording. Structure your notes or the transcription and submit it to an online text-cloud generator. What key terms seem to anchor the text? What words are used most often? What words could be eliminated without much loss of information? How could you structure clusters of words to help you make more sense from the lecture?
5. Try creating a text cloud from your own résumé. What kind of a visual picture of your skills and interests emerges? What key terms seem to anchor the text? What words are used most often? What words could be eliminated without much loss of information? How might you change the entries on your résumé to give a more professional visual image of yourself?
6. Try creating a text cloud that captures the key threads of a chapter or an article that you are assigned to read for class. What kind of a visual picture emerges? What key terms seem to anchor the text? What words are used most often? What words could be eliminated without much loss of information? How could you structure clusters of words to help other readers make more sense of the chapter or article?
7. Using the heuristic we have provided, have each member of the class create a cloud from their own résumé. Then, in teams of two or three, examine these text clouds and see what they reveal about the professional preparation of the class as a whole. What are the strong points of class members' preparation? The weak points? What areas of technical communication as a field seem well represented? Less well represented? When teams report to the whole group, see if you can add to or challenge the findings of other groups.
8. If you are a practicing technical communicator, find a digital docu-

ment in your organization that you can use to create a text cloud of the company's concerns/foci (e.g., an annual report, a strategic plan, a mission/goals statement). If you are a student, pick a company you'd like to work for that has such a document online where you can capture it in digital form. Submit the document to an online text-cloud generator. Do some interpretation of this text cloud to identify the key patterns or trends you find. How accurate is the text cloud you produce? How could you structure or focus it to be more informative? What impression does this text cloud give of the company? Do you think the company would be pleased with this impression? Why or why not?

9. Choose any text/document and its accompanying text cloud to examine as a class. Identify and discuss the cloud's weaknesses and gaps. What is missing? Why? How could these weaknesses be corrected? How might you create a revised, and more productive, version of the text cloud?
10. What are the differences between an index and a text cloud drawn from the same source text? Discuss for what rhetorical circumstances and for what audiences each might be preferable. Which would you prefer as a reader and why?

WORKS CITED

Allen, Jo. 1999. Refining a social consciousness: Late 20th century influences, effects, and ongoing struggles in technical communication. In *Three Keys to the Past: The History of Technical Communication*, ed. Teresa C. Kynell and Michael G. Moran, 227–246. New York: Praeger.

Bazerman, Charles. 1999. How natural philosophers can cooperate: The literary technology of coordinated investigation in Joseph Priestley's *History and Present State of Electricity*. In *Three Keys to the Past: The History of Technical Communication*, ed. Teresa C. Kynell and Michael G. Moran, 21–48. New York: Praeger.

Bemer, Amanda Metz. 2006. Technically, it's all communication: Defining the field of technical communication. *Orange: A Student Journal of Technical Communication* 6, no. 2. http://orange.eserver.org/issues/6-2/bemer.html.

Brockmann, R. John. 1983. Bibliography of articles on the history of technical communication. *Journal of Technical Writing and Communication* 13, no. 2: 155–165.

———. 1999. Oliver Evans and his antebellum wrestling with rhetorical arrangement. In *Three Keys to the Past: The History of Technical Communication*, ed. Teresa C. Kynell and Michael G. Moran, 63–91. New York: Praeger.

Carliner, Saul. 2010. Computers and technical communication in the 21st century. In *Digital Literacy for Technical Communication: 21st Century Theory and Practice*, ed. Rachel Spilka, 21–50. New York: Routledge.

Fisher, Julie. 1998. Defining the role of a technical communicator in the development of information systems. *IEEE Transactions on Professional Communication* 41, no 3: 186–199.

Havre, Susan, Elizabeth Hetzler, and Lucy Newell. 2002. ThemeRiver: Visualizing thematic changes in large document collections. *IEEE Transactions on Visualization and Computer Graphics* 8, no. 1: 9–20.

Hayhoe, George. 2000. What do technical communicators need to know? *Technical Communication* 47, no. 2: 151–153.

Johnson-Eilola, Johndan. 1996. Relocating the value of work: Technical communication in a post-industrial world. *Technical Communication Quarterly* 5, no. 3: 245–270.

Johnson-Eilola, Johndan, and Stuart A. Selber, eds. 2004. *Central Works in Technical Communication*. New York: Oxford University Press.

Jones, Dan. 1995. A question of identity. *Technical Communication* 42, no. 4: 567–569. http://find.galegroup.com.proxy.lib.ohio-state.edu/gtx/start.do?prodId=AONE.

Kynell, Teresa C. 1999. Sada A. Harbarger's contributions to technical communication in the 1920s. In *Three Keys to the Past: The History of Technical Communication*, ed. Teresa C. Kynell and Michael G. Moran, 94–102. New York: Praeger.

Kynell, Teresa C., and Michael G. Moran, eds. 1999. *Three Keys to the Past: The History of Technical Communication*. New York: Praeger.

Lamantia, Joe. 2007. Text clouds: A new form of tag cloud? *Joe Lamantia.com blog*. http://www.joelamantia.com/blog/archives/tag_clouds/text_clouds_a_new_form_of_tag_cloud.html.

Nielson, Mogens. 2007. Functionality in a second generation tag cloud. Master's thesis, Gjovik University College, Gjovik, Norway.

O'Hara, Frederick M. 2001. A brief history of technical communication. In *STC's 48th Annual Conference Proceedings*, 500–504. http://www.stc.org/confproceed/2001/PDFs/STC48-000052.pdf.

Peeples, Tim, ed. 2003. *Professional Writing and Rhetoric: Readings from the Field*. New York: Longman.

Pringle, Kathy, and Sean Williams. 2005. The future is the past: Has technical communication arrived as a profession? *Technical Communication* 52, no. 3: 361–370.

Rivadeneira, A. W., Daniel M. Gruen, Michael J. Muller, and David R. Millen. 2007. Getting our head in the clouds: Toward evaluation studies of tag clouds. In *Proceedings of the SIGCHI Conference on Human Factors in Computing Systems*, ed. Mary Beth Rossen and David Gilmore, 995–996. San Jose, CA: ACM. doi.acm.org/10.1145/1240624.1240775.

Rude, Carolyn. 2009. Mapping the research questions in technical communication. *Journal of Business and Technical Communication* 23:174–215.

St. Amant, Kirk. 2005. A prototype theory approach to international website analysis and design. *Technical Communication Quarterly* 14, no. 1: 73–91.

Scott, Craig, and C. Erik Timmerman. 1999. Communication technology use and multiple workplace identifications among organizational teleworkers with varied degrees of virtuality. *IEEE Transactions on Professional Communication* 42, no. 4: 240–260.

Siau, Keng, and Tan Xin. 2005. Technical communication in information systems development: The use of cognitive mapping. *IEEE Transactions on Professional Communication* 48, no. 3: 269–284.

Slattery, Shaun. 2005. Technical writing as textual coordination: An argument for the value of writers' skill with information technology. *Technical Communication* 52, no 3: 353–360.

Steinbock, Daniel. 2008. TagCrowd. http://www.tagcrowd.com/.

Wick, Corey. 2000. Knowledge management and leadership opportunities for technical communicators. *Technical Communication* 47, no. 4: 515–529.

Wilson, James C. 2000. Making disability visible: How disabilities studies might

transform the medical and science writing classroom. *Technical Communication Quarterly* 9, no. 2: 149–161.
Zahedi, Fatemeh "Mariam," William V. Van Pelt, and Jaeki Song. 2001. A conceptual framework for web design. *IEEE Transactions on Professional Communication* 44, no. 2: 83–103.
Zappen, James P. 1999. Francis Bacon and the historiography of scientific rhetoric. In *Three Keys to the Past: The History of Technical Communication*, ed. Teresa C. Kynell and Michael G. Moran, 49–63. New York: Praeger.

WILLIAM HART-DAVIDSON

2 What Are the Work Patterns of Technical Communication?

SUMMARY

Technical communication is more than just writing. Technical communicators make videos, diagrams, websites, and many other types of information resources. They often create the material for all of these formats at once so users can access them online, on demand, and simultaneously. Technical communicators advocate for users and work to ensure that information resources meet users' needs. And as more and more workers create information, it falls to the technical communicator to oversee writing and editing practices, helping their coworkers communicate more effectively and ensuring that their organization, as a whole, does so as well.

INTRODUCTION: CHANGING PATTERNS OF WORK IN TECHNICAL COMMUNICATION

During the rise of technical communication as a career category in the U.S. industrial economy, the work of technical communicators was difficult to evaluate in explicit terms because it was seen as ancillary to the production of manufactured goods and, as Faber (2002) and others note, disconnected from the service-oriented economy of the professions. But the work of technical communicators today is more readily visible and vital to the core mission and the bottom line of organizations of all types.

To create a vivid picture of the work of technical communicators today, consider the example of Elena.[1] Elena is a technical writer, or at least that's how she thinks of herself. But in truth she doesn't do much of what she, or anyone else for that matter, would call "writing" these days. She has had several job titles in the last few years: information developer (a title meant to equate her role on product development teams in the technology company she worked for with software developers), digital content analyst, and lately senior information designer. That last one is also her current title with the documentation contracting services firm she works for most often: Great Lakes Information Solutions (GLIS). GLIS finds Elena contract work with clients who need help with a variety of writing, editing, information design, content management, and electronic publishing problems. In exchange, they take a cut of the client fee.

This week Elena needs to bill hours on four projects for three different clients. Three of those projects are for clients of GLIS; the fourth is a project for an information services company for which she used to work as an in-house employee, but since a layoff, she now works for it on a contract basis, doing projects a lot like the ones she used to do when she was a digital content analyst there. She gets paid more than she used to for this work, but she also has to pay for her own health insurance and fund her own retirement account. GLIS makes this a bit easier than it might be, though, by providing its stable of contract workers access to group policies and financial planning services.

The goal of this chapter is to provide an overview of the work practices of folks like Elena: contemporary technical communicators. Three major work patterns are highlighted that are characteristic of technical communication today: information design, user advocacy, and content and community management. The chapter draws from the research literature in the field to reveal trends that contribute to the growing responsibility of technical communicators in knowledge-intensive organizations. A set of heuristic exercises drawn from the literature will help you practice the three work patterns. The patterns are made concrete in a story about one particular technical communicator, Elena. Her story helps illustrate how the patterns animate the day-to-day work of a technical communicator.

LITERATURE REVIEW

The kinds of work Elena does from day to day reflect three overall patterns of work that emerge from research in technical and organizational communication and that speak to the activities, responsibilities, and skills of technical communicators at the beginning of the twenty-first century. This chapter presents research on the core issues of each work pattern, highlights key patterns and skills, and presents strategies that you can implement. Here are brief descriptions of the three work patterns.

- *Pattern 1: Technical communicators work as information designers.* Technical communicators must create information that no longer stays neatly within the boundaries of a single genre or even a single medium, but is published in multiple formats for multiple audiences, using multiple display formats and technologies.
- *Pattern 2: Technical communicators work as user advocates.* Technical communicators work to ensure the usability of products in all phases of the user-centered design process. Often the technical com-

municator acts as the voice of the user, advocating design features that ensure users' needs are met and that the user experience gives the product a competitive advantage relative to others in the market.

- *Pattern 3: Technical communicators work as stewards of writing activity in organizations.* Technical communicators work in organizations where many other workers, and often customers and users, are writing and creating content. Technical communicators' expertise helps ensure that organizations support content development as a vital component to the organization's success, much as scientists ensure that pharmaceutical companies do good science.

As we might expect, there is considerable overlap among these three patterns of work when they are considered in light of the careers, projects, and even day-to-day routines of technical communicators. So from the outset, it might be best to think of these areas as regions in a Venn diagram. For folks like Elena, a typical day includes a blend of all three.

PATTERN 1: TECHNICAL COMMUNICATORS WORK AS INFORMATION DESIGNERS

A 2007 special issue of *Technical Communication Quarterly* was dedicated to research on the ways technical communicators are dealing with the circumstances of something the editor of that issue called "distributed work." At the organizational level, the term "distributed work" refers to the way individuals and teams often work at different times, from different locations, and across a variety of technological platforms and systems. But Spinuzzi (2007), in the introduction to the special issue, suggests a more precise definition of the term for technical communicators. He notes that distributed work describes both the means and ends of technical communication in that the work is "coordinative" in nature, so that it may enable the required "transformations of information and texts" (266). If this sounds as if technical communicators don't merely make texts from scratch, but instead manipulate many existing texts, images, and fragments of information in order to make new ones, then you are starting to get the picture.

Johnson-Eilola was among the first to sketch such a picture of the writing work that technical communicators do in scenes of distributed work. Writing around the same time that most people were becoming acquainted with the World Wide Web for the first time, Johnson-Eilola (1996) borrowed a category from Reich (1999) to characterize the nature of technical communication as "symbolic-analytic work." Symbolic analysts, to put it simply, solve problems with information, texts, and images. They find, arrange, synthesize, and transform existing texts to meet the needs of diverse users and to address new challenges. As Spinuzzi suggests, the primary act is one of *coordination*, though Johnson-Eilola warns that this should not be understood to be devoid of significant inventional (i.e., knowledge-making) responsibility. Why? Because coordination is merely the means by which technical communicators work. *Transformation* is the end goal. Making something new and adding value are the hallmarks of distributed work in technical communication. A key skill is understanding how information can move and how it should transform when it moves from one system to another, from one user group to another, and even from one culture to another.

One of the more vivid and detailed portrayals of the moment-to-moment and day-to-day work of technical communicators today is offered by Slattery (2007), who studied writers working for a technical documentation services company that contracts with client organizations in a variety of industry segments such as software/IT and pharmaceuticals. Slattery used screen-capture software to record five senior technical writers' work and then conducted a series of follow-up interviews in which the writers watched replays of segments of work. He concludes, among other things, that the documentation they were asked to produce "was not so much written as assembled—a pastiche of contributions from multiple individuals (and sometimes the technologies themselves) over the duration of a project" (315). Close up, this work can seem intimidating and strange when the image one has of "writing" consists of a single writer working with a single, coherent text.

The writers in Slattery's study were using multiple texts as sources, creating multiple texts at one time, and actively managing the links among many, many documents: "In his one-hour session of drafting a new technical document, Doug used 11 different documents across six software programs and one printout. Dirk used 20 documents across 15 programs in an hour and a half while revising a draft. And as Parker updated help files based on e-mail prompts, she used 37 documents across six programs in just under an hour" (318). Slattery gives the skill required to do this kind of work a name: textual coordination. Textual coordination, for Slattery, also involves the expertise required to identify, select, stage, and recombine bits of existing texts in order to form new ones (318).

An equally rich and fascinating portrait of the transformational aim of, in this case, a designer of digital media emerges in a case study by Graham and Whalen (2008). Graham, the researcher, shadowed and interviewed Whalen, the website designer, as he worked on a number of different client projects. One highlighted project was for a manufacturer of bar-code scanners for whom Whalen created an e-greeting card using Flash, which recipients can play like a game. Whalen's design of that project, according to the published account, was a series of one transformation after another as he took familiar elements from the two dominant communication genres (greeting cards and online games) and from the company's corporate identity and marketing materials (bar codes, scanner images) and combined these into an interactive experience intended to carry a positive corporate message, boost the company's reputation, and thank customers. Whalen had multiple audiences, purposes, and contexts to consider, and he had to deal with multiple modalities (visual/textual), media (Flash animation, static websites), and genres (greeting cards, e-cards, games).

Graham and Whalen's illustration of the ways audiences, contexts, and purposes interact with modes, media, and genres helps show just how complex the coordinative work of new media design can be, given its transformational aims. And what seems to be true for both Whalen and for the senior technical writers that Slattery studied is that, despite the complexity of distributed work, there can be order amid the apparent chaos. The heuristics section in this chapter offers some strategies you can use as you learn to make texts that transform.

PATTERN 2: TECHNICAL COMMUNICATORS WORK AS USER ADVOCATES

"Know your audience." It is perhaps the oldest rule in the book when it comes to being a good writer in just about any genre. In technical communication, it is a mantra. When you are creating information meant to guide someone else through a task, for instance, there is just no substitute for being in the shoes of the user. Or, at least, as close to that as one can manage with careful inquiry, observation, and interaction with real users.

But sometimes for technical communicators, the rule might more accurately be phrased this way: "No, your audience." That is, not only must technical communicators lead the research effort to learn about users, they must also represent the knowledge gained about users—their goals, their needs, their preferences—in the design process as well. The results of audience analysis become, in these moments, the evidence required to make good decisions in a user-centered design (UCD) process. But if UCD is something nearly all development teams subscribe to in principle, in practice this way of working requires the users' interests to come first, ahead of those of developers, at least equal to those of clients paying the bill, and so forth. Hence, "No, your audience" is a phrase that represents what technical communicators must do to keep user interests at the forefront of decision making.

One of the more interesting changes that affect the way technical communicators work is that the same technologies that had been used almost exclusively by experts in work settings—computers, hand-held devices, and so forth—are now used by just about everybody in all kinds of contexts. Such a shift dramatically changes the conditions under which a piece of software and the information meant to guide action are usable. Mirel (2004) suggests that because of this kind of shift, even in work settings where problems are complex, *usable* is just not good enough. Systems must be *useful*. Usability is a minimal requirement: necessary, but insufficient. Usefulness is a higher bar, and the one that users who have choices will demand. Although Mirel's research on workers involved in complex problem solving was instrumental in emphasizing the usefulness of systems, the vast number of users engaged in complex activity that is not necessarily related to work is likely to make usefulness itself a minimal requirement and to lead to a yet higher bar: *compelling*.

Sun (2004) produced a detailed account of users whose interactions with technology help illustrate what lies beyond the goal of designing usable products, with her comparative study of Chinese and American mobile-phone text-messaging users. In the first phase of her study, Sun asked groups of young, urban students and early-career professionals to complete questionnaires and detailed use diaries, including a log of all the short message service (SMS) messages sent and received over the course of a given period. She then analyzed both the content and the patterns of messaging to determine what these users were texting about and where and when they were texting. The results were eye opening.

Among the more surprising findings, Sun's study revealed that both the Chinese and American users were willing to put up with what, in both cultures, amounted to a cumbersome interface for text entry in order to take advantage of what she calls the "social affordances" of SMS: unobtrusive, even covert, messaging, the ability to maintain social contact with friends and family despite a hectic schedule, and so forth. In short, usefulness trumped usability (473–474).

So what do Sun's results mean for designers and technical communicators? One implication is that design teams should not try to solve all the usability problems with a given system *before* it ships. Sun's work shows that not only is this impossible, but also that it is not always necessary for a product to succeed. Rather, technical communicators can take the lead in listening to users postadoption and learning from their feedback (478).

Another implication of Sun's work is that technical communicators should be researchers in addition to being writers. One of the most interesting places to see how a technical communicator working primarily as a researcher might make a significant impact is in risk communication. Risk communication, generally, deals with the ways information in a variety of forms is deployed to mitigate negative consequences ranging from immediate physical danger to financial risk.

Beverly Sauer's (1998) research on risk communication in the mining industry is one example of work that, in the end, prescribes new roles for technical communicators. Sauer wrote about the gap that emerges between safety regulations meant to govern miners' behavior and post-accident reports and depositions about miners' actions. One conclusion is that miners get hurt or killed because they fail to follow the rules as written.

But Sauer's research showed that from the miners' point of view, written rules may be inadequate and inaccessible during the moments when workers need to make a critical decision. There is literally no way, in some of the most high-stake situations, for a text to adequately convey what miners need to know to avert disaster. So how does it happen that tragedies are often avoided? Sauer used documentary evidence that included in-depth postincident interviews and depositions with miners to reveal that some of them use sensory information—what the miners call "pit sense"—to avoid injury or death during catastrophic events such as collapses.

Sauer's point is that the official model of how safety information originates and circulates is wrong. Some safety information—perhaps the most critical kind of information needed in the dynamic environment of a pending collapse, for instance—cannot be known in full ahead of time. The information originates with miners in the pit. Safety instructions follow after that—sometimes in time to save lives in an emerging disaster, and others only for "next time" (159).

So what are the implications for the role of the technical communicator in this type of situation? If we assume that the technical communicator is the person whose job is to write safety regulations—a fairly traditional role—then the implications are profound indeed. What is missing, according to Sauer's revised model, is the ability to listen in a timely way and facilitate the spread of knowledge that originates, quite literally, in the body of an individual miner, so that others can use it. A whole new mandate with a familiar ring to it: know your audience.

PATTERN 3: TECHNICAL COMMUNICATORS WORK AS STEWARDS OF WRITING ACTIVITY IN ORGANIZATIONS

Single-source publishing, distributed production, and user-generated content: these trends accompany a broader shift in North America toward an information economy. This shift is potentially positive for technical communicators because, to adopt the language of economics, it moves technical communicators' work higher up the value chain. But what exactly does this mean? Simply put, it means that the most valuable thing that many organizations produce, today, is information. And to the degree that the value of information is directly influenced by how it can be understood and used by others, then the core expertise of technical communicators is creating and maintaining that value.

This shift to information as a valuable commodity is easiest to see and understand in companies that sell information, or access to information, as their primary product. Google is one obvious example of this sort of company; it makes money primarily by providing a variety of services to web users and charging other companies to place ads where users will see them. Clearly, the value proposition for users of Google's search service lies in how it helps them find information.

But the shift that moves technical communicators up the value chain is not limited to companies that sell information directly. Even in the manufacture of traditional goods, it is often the information produced about and along with a product that is the most valuable commodity. It was not always so. A manufacturer of washing machines, for example, used to make most of its money from selling the machine. The information that went along with it—the operator's manual, the repair manual, the design specifications—those were necessary, but largely non-revenue-generating, by-products. Today manufacturers operate in a very different kind of environment that is global in scale and densely networked. And we must think, here, of both transportation and information networks.

What this means is that for the maker of medical devices, for example, there is no longer as much value in manufacturing the plastic housing or even the silicon chipset that goes into the devices, at least not in North America. These operations are lower in the value chain than the design of the devices, a process that involves not only engineering but also clinical testing to ensure that they do what they are intended to do. The company that is likely to sit highest in the value chain is the one that can produce the information needed by others to market the device as effective and to manufacture the components offshore. They likely also own the brand that will go on the finished product.

One implication of this trend is that it is no longer possible for technical communicators to be the only ones doing the "writing"—and here we would need to consider a broad definition of *writing* that can encompass the types of information and genres that an organization might produce. Now, everybody in the information economy writes or contributes to an organization that must write, and write well, to stay competitive and fulfill its goals.

While initially there was considerable concern, if not outright fear, in the technical communication community about the shifts mentioned above and what they were doing to endanger the prospects for writing specialists, today there is less worry. Michael Albers wrote an article published in *Technical Communication Quarterly* in 2000 that was a cause of trepidation because it foretold significant changes in the responsibilities of technical editors. To be fair, Albers's own view was not pessimistic. In retrospect, we can say that he saw the importance of an editor's expertise growing as the possibilities for distributing content creation to more people grew. Writing specialists—those trained in making texts coherent and usable by their intended audiences—would move, we might say, a little higher up the value chain as more subject-matter experts (SMEs) contributed content to repositories from which finished texts could be dynamically assembled. But Albers knew that this message would be unsettling to many who still held a fairly traditional view of editing that involved technical communicators maintaining ownership over a single, coherent document (203).

Ten years later, we can say that Albers was right. Today, companies, even small ones, maintain much of their information as small, reusable chunks in content management systems (CMSs) that allow pieces of information to be reorganized and assembled into new texts on demand. For many organizations, a CMS is the way they manage their website. And for companies that publish information in print or online that they sell or provide as a service, a CMS is likely a critical part of production and quality control. Two key reasons organizations choose to adopt a CMS are to involve more people in producing content and to reuse more of that content whenever possible (Hart-Davidson 2010). Both of these trends, moreover, tend to cause shifts in the role that technical communicators play in organizations.

The more expertise technical communicators have about how to organize writing patterns, and about concepts such as how genres work, the less likely they are to spend their time actually writing. The more experienced and senior they become, the more likely they are to adopt roles as managers: managers of content, yes, but also managers of people. In a more recent article published in *Technical Communication*, Albers (2003) updated his vision of the career track of the technical communicator in light of developments in the world of CMS and single sourcing. He noted that the image of the technical writer controlling all the details of a document from start to finish—from initial outlining to drafting to revising and editing—belonged to a "craftsman" model of document production that was difficult and, in any case, unproductive to sustain in single-sourcing environments (337). As a result, tasks we once associated with technical communicators exclusively, such as "writing" and "designing and manipulating graphic displays of information," belong now only to junior-level technical communicators. Senior-level technical communicators are decision makers, who spend their time defining high-level requirements for a range of information products and training others to produce the content for those products.

It is not that writing has gone away. Nor is it the case that technical communicators are not involved with writing. It is, rather, that the whole organization must write, and write well. And the charge of the senior technical communicator is to look after the quality of writing at an organizational level and to see that it is constantly improving.

We are only just beginning to see research on the ways technical communicators can help whole organizations transform their writing and communication practices for the better (see, e.g., Hart-Davidson et al. 2008). But we do have a rich body of work from researchers who examine communication practices at an organizational level with an eye toward improving the way these organizations function by improving the way they write.[2] Two well-known researchers in this area are JoAnne Yates and Wanda Orlikowski. When they collaborate, they produce studies that are particularly valuable for the way they reveal how organizations work, or fail to work well, as a function of how well they communicate.

In one such study, for example, Kellog, Orlikowski, and Yates (2006) examined a category of activity they called "boundary crossing," a key factor in the success of businesses that must regularly coordinate their own activity (and interests) with those of other businesses in order to succeed. They learned that successful boundary crossing involves coordination at two levels. The first level is strategic and involves sharing knowledge and establishing common ground. The second level is more directly communicative and involves the use of specific methods for "boundary spanning," such as shared "routines, languages, stories, repositories, and models" (24). By carefully observing the project work of a corporate web development firm known as Adweb in their study, Kellog et al. discovered an interesting and useful pattern: "Adweb members *displayed* work across boundaries (i.e., they made their work visible and accessible to other communities), they *represented* work across boundaries (i.e., they expressed their work in a form that was legible to other communities), and they *assembled* their separate contributions across boundaries into an emerging collage of diverse elements (i.e., they reused, revised, and aligned their work over time so as to keep it dynamically connected across multiple communities)" (28).

Displaying, representing, and assembling are, for Kellog and her colleagues, categories of communication practices. Over time, members of the organization learn to do them in conjunction with one another, as a matter of routine, creating an overall structure for coordinating (and boundary crossing) that the researchers call a "trading zone." Borrowing the concept from another researcher, Kellog et al. find the concept of a trading zone to be useful because it emphasizes two key aspects of boundary crossing: exchange and transformation.

By now, you should be hearing something familiar: the essential moves or basic particles of the information economy—the atoms that make up organizational work, if you will—are acts of coordination and transformation. You might recall from the discussion of the first work pattern, earlier in this chapter, that these are also the fundamental moves that make up the work of technical communicators, according to some researchers. Technical communicators work in the trading zone. Or, as Albers might point out, junior technical communicators do. Senior technical communicators are involved in creating the conditions for the trading zone to function and, over time, in improving those conditions so that it functions optimally.

HEURISTICS FOR PRACTICING INFORMATION DESIGN, USER ADVOCACY, AND CONTENT AND COMMUNITY MANAGEMENT

What should you do to prepare yourself for a career that blends information design, user advocacy, and content and community management? You can start by learning to make texts that transform, getting to know your users, and getting users actively involved in shaping information that suits their needs. If you do these things, you may find yourself becoming more valuable to your workplace as you help others execute a coherent communication strategy for the whole company or organization. The heuristics that follow reduce rather complicated strategies to a series of steps, so that you can visualize and try them out. But these heuristics are not formulas for doing the work of technical communication; that work is simply not formulaic. The work patterns described in this chapter have many variations. You can think of the heuristics that follow as similar to those deceptively short, tricky practice tunes called études that musicians use to strengthen their skills in particular areas of their art. You may not be able to get through the whole exercise at first. Work through what you do know how to do, however, and as you learn more, return to these to extend your repertoire.

FOR INFORMATION DESIGNERS: LEARN TO MAKE TEXTS THAT TRANSFORM

As I have written elsewhere, technical communicators today must shape texts that can be readily transformed, and in the case of interactive genres, can themselves transform to meet the needs of their users (Hart-Davidson 2004). The pattern might be complex, but it can, in fact, develop and become a routine and even a shared set of actions for technical communicators. A simplified version of that routine is represented below.

Coordinative Work

1. Select source texts from (or that will be combined to create) the source repository of information required by the audiences you are creating resources for. You will make this project easier if you choose a text genre that already has a semantic structure you are familiar with. (You may switch the order of steps 1 and 2 if needed.)

2. Get to know your audiences. Talk with them and pay special attention to what they do, what specific tasks they need your information to perform. Also pay attention to who they are and how this might influence what information they need and how they prefer to have that information delivered.

3. Analyze source texts and inventory the contents, preferably using some kind of standard labeling system. This labeling system may later become a set of descriptive tags. As you go, use techniques such as Rockley, Kostur, and Manning's (2003) reuse map that will help you define shared versus audience-specific objects in the source. A reuse map is a grid that documents all the places that a single piece of information shows up within or, perhaps, across publications. Definitions for key terms and mission statements are two examples of commonly reused information types. Reusing information not only saves the effort of rewriting; it also ensures consistency. Also note where information is missing and will need to be created.

Transformative Work

4. Model the information, creating an outline that makes the relationships among the information chunks explicit. In this step, your chunks should have semantic names rather than structural ones—labels that emphasize what the information means rather than the form it might take.

Choose labels like "product description" rather than "paragraph." The goal of this step is to represent the information free of style and presentation details, because this will allow you to apply more than one set of presentation rules later.

5. Begin creating the primary transformations: sketch "views" of the information you have assembled and designated that are appropriate for each audience. What will the view look like for mobile-device users? For those with administrative privileges and tasks? Your building blocks for each view are the elements you included in the content outline you created in step 4. You are now defining how they will appear when your users see them.

6. Design the secondary transformations: define how each of the primary views you sketched in step 5 will transform to meet the needs of users as they interact with it. When a user clicks an object's title in a list of products to get more information, what will they see next? You may need to create additional style rules, describe how screen elements will change, and even specify what information will need to be stored and/or retrieved from a database to make the transformations happen smoothly.

7. Go back to your audiences, as often as you can, to test out the decisions you are making. Whenever you can, have them attempt to do the tasks for which you are creating information resources. Watch them carefully. Can they do what they need to do? Seek their feedback. Do they feel confident that they are achieving their goals? Does your information help them?

Coda. Repeat steps 1–6, as needed and in response to feedback from step 7.

FOR USER ADVOCATES: GET TO KNOW USERS OR, BETTER YET, GET THEM INVOLVED

I had occasion to offer some advice to technical communicators a few years ago about a phenomenon called Web 2.0 in the pages of a trade publication of the Society for Technical Communication (Hart-Davidson 2007). In that article, I asked technical communicators to think of Web 2.0 not so much as a change in the way the web works, but rather as a change in the way users behave. Users produce rather than merely consume information in a Web 2.0 model, and they also sort, categorize, label, rank, and rate it. When they do those things, they add value to existing information. And they do it all whether we ask them to or not. So what is a technical communicator to do? Aren't all those things the very same things we used to do? Are we out of work in a Web 2.0 world?

Not by a long shot. It just means, as we saw with Sun and Sauer, that our relationship to users needs to change. They say things; we listen. They write; we help them do that.

Here, then, are four things technical communicators should practice, to become more effective user advocates. Think of the list as something to practice and document examples in your professional portfolio. For each one, consider: What experiences do I have that fit the descriptions here? What can I include in my portfolio—sample documents, evaluations from supervisors, prototypes—that will demonstrate what I've learned?

1. Listen to users (and watch what they do). While this seems like a straightforward recommendation, it is not always seen as the first priority when organizations design, create, or revise information. It should be the first priority. And there should be explicit procedures in place for listening and watching. If there are not, it may fall to the technical communicator to work with managers and designers to develop listening procedures. Many companies currently use social media such as Facebook or Twitter as ways to listen to customer feedback, for instance. Traditional methods of gathering feedback such as surveys, focus groups, and interviews may also be useful.

2. Participate in the development of new tools and services that respond to emerging user needs. Technical communicators are often expected to participate in the full cycle of product development, providing insight and working alongside others on the team. For example, it is not uncommon for a technical communicator to be asked to bring evidence-based recommendations for product features to the team. That evidence should come from the kind of direct interaction with users described in step 1.

3. Curate (i.e., gather, organize, label, etc.) the ever-growing content collections produced by users. If you are successful at building ways for your users to give you good feedback, you may begin to provide the means for users to create their own support information. And if you are successful at that, you will soon have another problem: how to keep a growing repository of information organized, so that it can remain useful and usable. While your users may well provide answers to one another about technical questions on a user forum, chances are they will not monitor forum posts looking for the best solutions or even for the most common questions or problems. But if you can do those things, you will add value to a user-created resource.

4. Create designs that allow users to pursue their goals with minimal technical demands. The most powerful evidence of a successful design comes when users can simply do what they need to do. When that happens, it can look like nothing out of the ordinary has taken place. And so it can be a challenge to document! Think of the way the interface designers

of the iPhone must have felt when they first saw a three-year-old successfully navigating the touch-screen interface. That is powerful evidence of the intuitiveness of the design. We can become so focused on problems when we are attending to users' needs that we sometimes forget to watch for those moments of seamless activity that constitute success.

FOR CONTENT AND COMMUNITY MANAGERS: IMPROVE YOUR COWORKERS' ABILITIES TO WRITE TOGETHER

Building on Kellog, Orlikowski, and Yates (2006), you might consider the three categories of boundary-crossing practices as areas to examine and improve in the organizations where you work. To the degree that you can do this, you will likely become more valuable to the organization. You can start by attuning yourself to the ways that an organization writes. Start by observing writing activities and making those that are routine more visible, just as Kellog and her colleagues did.

Each of the practice categories listed below is something a technical writer might do to improve organizational writing activities. It is common for organizations to do a great deal of writing without actively training or evaluating workers' writing practices.

For each category listed below, the bulleted items represent a genre that a technical communicator might create or an action she might take to help an organization improve their writing practices. These are only examples. Many others are possible, and not all of these apply in all settings.

To work with this heuristic, start with a workplace context, a company or organization with which you are familiar. Survey the bulleted items. Add additional ones that you think might work in the setting you have identified. Delete some that you think are not relevant. For each bulleted item, try to collect an example—a physical artifact like a document or a picture—of what the practice results in or how it is carried out. You may have to make one.

Finally, plan where, when, and how you could observe writing practices in action in the workplace setting you are imagining. Watch several instances. Can you identify where the opportunities are to implement these work display, representation, and optimization practices to make an improvement?

Work-display practices make work done by one or more members of a team visible and accessible to others.

- Send e-mail updates to project team.
- Create project plans.
- Post status information on a project wiki.

- Update project logs.
- Flag finished milestones on a calendar.

Work-representation practices express the work of a group in a way that others may use to adapt their own practices.

- Convert textual process descriptions into diagrams or charts.
- Write and circulate notes from a team meeting.
- Create or update role descriptions for a project.

Work-optimization practices enable work products to be reused, revised, and aligned over time and dynamically connected across multiple communities.

- Reuse project-plan milestones in a progress-report presentation.
- Develop a common format for all presentations in order to allow different team members to contribute sections independently.
- Create templates for common genres and store them in a shared repository.
- Facilitate review by establishing criteria, review work flow, and deadlines.

ELENA'S STORY: INFORMATION DESIGN, USER ADVOCACY, AND CONTENT AND COMMUNITY MANAGEMENT

In this section, I put all three work patterns together in a way that helps those preparing for a career learn what it might be like to work as a technical communicator. I'll return to our fictional character Elena, to see how she shifts among the three work patterns discussed above, and how she shifts roles, tasks, projects, and teams in the course of her everyday routine.

Elena is spending the morning doing "information design." This is a kind of work that she finds nearly impossible to explain to her friends and family when they ask her about projects she is working on. Her computer screen is full of windows, and there are many more minimized in her task bar. It looks chaotic, but she has a system. She is flagging terms in a large repository of text and image files that make up the source for a number of technical manuals. Two of the manuals can be viewed in a format that nonspecialists would recognize as a manual, though neither is printed. A third and much larger "manual" is really, at this point, just a collection of marked-up text files—nearly 10,000 of them in all—that do not yet have an easily visible structure that brings them together into a coherent whole. The other two manuals consist of a subset of these same files, but because there are style sheets already created, she can use a browsing tool to render them rather than reading them in her integrated development environment (IDE). The ability to do that makes it a little easier to find terms that need to be added to the controlled vocabulary for the repository—a kind of glossary, she would tell you if you looked puzzled—because she can see more of the context in which a reader might encounter the term. To flag terms that either need a new definition or may need a revision, Elena moves back and forth between reading the rendered version of one of the manuals, creating links in the corresponding text files in the IDE, and adding the terms to a work list she has created as a separate file. Later she will break that file into separate lists of terms to send to her SMEs, who will write and/or revise the definitions.

All three manuals contain technical information for medical devices. The client designs and markets these devices in the United States and the European Union, but uses a network of manufacturers in Asia who produce hardware, components, and firmware for the devices. The technical manuals that Elena is working on are used by these manufacturers and, occasionally, by other companies who create add-on devices, replacement parts, or other consumable supplies for users of the devices. The manuals are published in three languages: English, German, and Korean. In the next release, a fourth will be added: Mandarin. Roughly two-thirds of the source text is originally produced in German, as the client is headquartered there, and the rest is produced in English. Elena works with English text only—much of it translated from the original German—but she also does some work to prepare the English versions for translation to Korean and Mandarin, mostly smoothing out syntax in English into a straightforward subject-verb-object pattern so that the translators have short, clear phrases to begin with. This service is a value-added option that this particular client is trying out for the first time, so her "agent" at GLIS has asked her to give it special emphasis. They hope to sell this service to the same client for other projects they are bidding.

Elena is mentally rehearsing what she will say in the design-team meeting that starts in a few minutes. She will take the meeting on the phone. The rest of her team is on the other side of the continent, three time zones away. She and one other person on the team, the user-interface designer, have been texting back and forth all morning in preparation for what they both anticipate will be a contentious discussion, if not a honest-to-goodness fight. Her role in the meeting will be to speak on behalf of the users of the product under development, to ensure their needs are met.

At issue in the design meeting will be her recommendation for changing the design of a software product used by medical technicians; the change will prevent a common transcription error that occurs because the modal interface design forces users to switch between screens. To find the information they need, users need to be in search mode, but to enter a new value, they have to switch away from search results to enter data in diagnostic mode. Elena and the UI designer, Chad, will be arguing for a dashboard view that eliminates the modes and allows users to find and enter information without changing views. This, they will say, will avoid a lot of errors. Elena will point out that it will also save them the cost of developing help and training information meant to guide users around this problem and to fix errors in patient records that result from it. This should play well with the project manager, who wants to keep development costs low.

The issue under discussion is not a new idea, or a new problem. If the past is any indication, the product manager will resist the suggested change because it involves significant changes to the existing codebase and, therefore, significant cost. The software developers will likely agree—at least one of the people who will be on the call is the same person who, the last time the issue came up, insisted that this problem was with "stupid users" rather than with the design of the product. Chad pointed out that these errors occurred because modal interfaces force people to adapt themselves to the machine because of the machine's constraints. These constraints no longer exist as practical matters of processing or memory now, of course, but they remain in the codebase of legacy products like this one.

Elena takes a deep breath, sips her coffee, and puts her phone on speaker. If she had time, she'd recite to herself the user-advocacy work patterns she knows so well: "listen, participate, curate, create." This call will be a tour through all four. She's ready for battle.

Elena puts her phone on mute. Waiting a beat to be sure that the red indicator light is glowing, she says, out loud, "Since forever!" This is in response to Dan's snide question to her: "Since when are you an interaction designer anyway?" She smiles. Dan's comment is an acknowledgment, however begrudgingly, that she has persuaded the team lead to go with the UI changes she and Chad have been pushing for.

Later, she will calmly explain that attending to user interaction (UX) is precisely what she was brought in to do. In fact, they paid her to go and watch users working with their product, talk to those users, and bring that information back to inform the next release of the product. UX designer may not be her title on this particular job, but UX research is certainly one of her responsibilities.

She can now turn her attention to her next goal of persuading the project manager—the person who controls the purse strings for this product—that what she has learned from observing users for a limited time in just two work sites is so important that they should be doing this much more often. She heard many, many good ideas for improving the product in the few hours she spent with users. And she has only been able to suggest the most basic of these to the development team so far. What really needs to happen, she thinks, is that those users—those nurse practitioners who, after all, know their work contexts the best—need to be on the next conference call. Sometimes there is no substitute for the voices of users.

The afternoon finds Elena working on the long-term project for her former employer, the company that sells streams of filtered content to libraries and other companies. The company doesn't create the content it sells access to, nor does it do much to repackage it in traditional publication formats. The value-added service they sell to their clients is most visible, in fact, in what is left out of the streams of data rather than what is contained in them. In a world awash with information, they filter out what is unwanted, leaving only the most valuable, trustworthy, and useful information.

Elena is going through a series of client comments—some of them solicited during a recent round of phone interviews and others that have come in unsolicited via e-mail—and trying to synthesize from these a list of priorities for improving their data-filtering offerings for a particular industry sector: pharmaceutical and medical-device manufacturing. This is information-design work, first in coordinative mode, preparing to make transformations that bring value to users. It is Elena's responsibility to monitor the use patterns of the community as they change over time, acting as a "community manager." Her report will make the service the company provides more valuable to this community by ensuring that they continue to see information they need to see and less of what they do not wish to see.

The recommendations that Elena will make will ultimately find their way into code: XML formats that determine which clients see what types of information from the thousands of content providers the client company has relationships with. But Elena will not deliver the recommendations in code. She will write a report and, if necessary, create some diagrams that illustrate how the new filters should work to fine-tune the way the content must flow. She would tell you that it's a lot like plumbing, or maybe more like heating and air-conditioning and ventilation. Air flows rather than liquid flows seem a more apt metaphor for the material she is trying to direct. Elena knows that the more customized a flow is for a given customer, the happier they are likely to be. But, at the same time, the more a given flow (and here we are talking about a filter or combination of filters) can be reused, the more profitable it is to the client. And so she is working to find patterns in customers' comments, requests, and complaints that will allow for the optimum balance between customized and generalized. Those patterns will be the basis of her recommendations in the report.

It is Sunday afternoon, and Elena is working in a local coffee shop, a pile of print documents spread out on the table. From a few feet away, anyone who knows Elena might guess it is just another work project. But this one is a bit different. The papers are a mix of genres: scientific journal articles, EPA standards, and water-quality test reports. Elena is sorting them, highlighting the occasional passage, and looking for evidence.

This project concerns the neighborhood where she grew up and where her mother and stepfather still live. For years now, there have been concerns about contamination in the aquifer that supplies the part of the city in which the neighborhood sits. Recently a citizen action group has formed in an attempt to pressure local and federal government agencies to perform additional testing. Elena learned about the group when a member brought a flyer to her mother's door. The flyer announced a meeting and included a call for volunteers with specific kinds of skills. They wanted folks with science degrees and laboratory experience, lawyers, and health-care professionals. And last on the list: technical writers. She went to the meeting.

Once there, she saw that there was indeed a way for her to help, but it wasn't exactly what the organizers had imagined a technical writer might do. They wanted her to write a brochure to explain to community residents—mostly nonscientists and many without a college degree—what the potential negative effects of unregulated contamination of their community's drinking water might be. She listened and nodded. But as the meeting went on, she heard something that made her think of a different way she might be helpful.

Partially in response to the pressure brought by the group, the regulatory commission had agreed to hold a series of public meetings to discuss the results of previous water-quality tests of the aquifer and whether further tests were warranted. Public comment was solicited for these meetings, and individuals could respond in writing as well as raise questions and concerns during the meeting itself. The group had already decided that it would be most effective if a report could be written explaining the group's position on the need for further testing. But group reports were not allowed at this stage, as they fell outside the strict guidelines established for "public comment." Individuals could submit, but groups could not.

Elena formed a plan right away. She could establish a website where community members writing reports could coordinate what they were doing, share the evidence they were finding, and report on the focus of their individual arguments. Because community members represented different perspectives on the issue—legal, health, economics, and environmental—there would be a range of individual reports. But Elena knew they would be more effective if they were all driving toward the same overall conclusion. At the heart of the website would be two features. One would be a discussion board on which to post questions, share ideas and links to reference material, and coordinate further meetings, reviews, and so forth. The other feature would be a shared repository of scientific source material that provided the best evidence, previous reports by the regulatory agency that gave members a framework for presenting their claims, and templates for the reports themselves to create some common structures and ensure similar overall impact of the members' reports.

She pitched that idea as the meeting came to an end, and the group sat silently for a moment. They had never had a "writing steward" before. They did not quite know what to make of someone like Elena, who could step in and help them execute a communication plan. Then a chorus of laughs broke out. And in another second they were tossing out ideas right and left about what kinds of resources each had to share with the group, digging into briefcases and producing folders stuffed with the documents that now filled the small table at the coffee shop. That was last Saturday. A week or so later, Elena is creating an index of the resources she had collected so far. This index would sit in the "scientific resources" folder of the shared repository and provide the members with a title, author, date, and abstract of each article and report contained there. This metadata format would then be used to create a new entry in the index whenever a new resource file was added.

After she finishes the index, her plan is to start working on the brochure. She looks forward to being able to write again.

CONCLUSION

There are two ideas in this chapter that may come as something of a surprise to those who are new to technical communication, or to those who have not kept up with recent changes in the field. The first is the idea that opens the story of Elena in the introduction: technical communicators do much more than just write; so much more, in fact, that it sometimes seems as if the writing they do is a minor part of their job responsibilities. The second idea is that technical communicators' contributions have moved closer to the bottom line: they routinely contribute directly to the most valuable aspects of a company's business or an organization's mission. Elena's work in the examples above, for instance, bears directly on whether a design will save the company money (e.g., by reducing errors and increasing efficiency). Both of these trends are supported by research such as that reported in the U.S. Bureau of Labor Statistics (BLS) Occupational Outlook Handbook (2010) and on O-Net, the BLS resource site that lists details about work activities, qualifications, and so forth for various jobs in the U.S. economy. In fact, the BLS projects better-than-average opportunity for job growth in technical communication through 2018 in a diverse range of industry sectors, including software, IT consulting, and health care. Technical communication is trending well because technical communicators are becoming more valuable for their companies and organizations.

DISCUSSION QUESTIONS

1. As noted in the introduction, the three work patterns of technical communicators highlighted here frequently intersect. This is most visible when you consider the details of specific projects. Select a project you have worked on or that you have read about, and trace the ways the work involved exhibits one or more of the patterns. It may help to sketch a Venn diagram with three overlapping circles. Place key events from the project in each region.
2. A key coordinative skill that experienced writers in Slattery's study developed was the ability to pre-stage their work environment for a productive session of work. What do you do before you write to make sure that you have what you need to be successful? Does this differ when you are revising? Composing a multimodal text or website? What tools (electronic and physical) do you keep handy? What rituals or routines do you rely upon to keep yourself on track?
3. Make a content inventory for an organizational website with which you are familiar. List the content types present on the site and, as best

you can, who in the organization (by role, if not by name) is responsible for creating, editing, and approving them. For each type, note how frequently it is produced by the organization. For frequently produced content types, how consistent is the overall structure from one instance of the type to another?

4. Design an alternative view of an informational document for display on a mobile phone or other device with a small screen. Try to keep all the information the same—that is, reuse the text and images from the current format—but make the new view as useful and usable for the new display as you can. For an added challenge, tailor the new display for a specific audience. Some ideas for informational documents to use in this exercise: course-catalog information for students, movie reviews, nutrition information for fast-food menu items.
5. Talk with several users of a product that you also use. Ask open-ended questions designed to gather feedback about how they use the product, such as "tell me about a time when you used the product in a way that made you satisfied" and "describe a situation in which you experienced a problem using the product that you were eventually able to solve." What kinds of information from these accounts would be helpful for members of a design or development team to know? What similarities did you hear among the users' responses? What surprised you?
6. When you work with others, what are some of the challenges related to helping them write well together? What have you done to make a team that you've been a part of work more effectively together?
7. Document a day in your life as a writer, using a visual format like a flowchart that allows you to show the flow or sequence of actions. What parts of this flow or sequence do you see as constituting a successful pattern? What aspects would you like to repeat each time you write? What aspects would you recommend to others and why? What aspects constitute patterns you'd like to change or avoid next time?
8. Which of the types of work Elena does is most appealing to you and why? Which do you feel most prepared for? Least prepared for? Looking at your planned program of study for your major or concentration, where will you develop the skills and knowledge to hone your strengths and build your experience?

NOTES

1. Elena is not a real person. She is a composite character whose job titles and work patterns are borrowed from several real people.

2. I can't review all of that valuable work here, but I recommend that readers see the full-length works by Cross (2003), Smart (2006), Winsor (2003), Spinuzzi (2008), Grabill

(2007), and Simmons (2007), to name just a few. Also see the outstanding collection edited by Zachry and Thralls (2007).

WORKS CITED

Albers, Michael. 2000. The technical editor and document databases: What the future may hold. *Technical Communication Quarterly* 9:191–206.

———. 2003. Single sourcing and the technical communication career path. *Technical Communication* 50:335–343.

Cross, Geoff. 2003. *Forming the Collective Mind: A Contextual Exploration of Large-Scale Collaborative Writing in Industry*. Cresskill, NJ: Hampton Press.

Faber, Brenton. 2002. Professional identities: What is professional about professional communication? *Journal of Business and Technical Communication* 16:306–337.

Grabill, Jeffery T. 2007. *Writing Community Change: Designing Technologies for Citizen Action*. Cresskill, NJ: Hampton Press.

Graham, S. S., and B. Whalen. 2008. Mode, medium, and genre: A case study of decisions in new-media design. *Journal of Business and Technical Communication* 22:66–91.

Hart-Davidson, William. 2004. Shaping texts that transform: Toward a rhetoric of objects, views, and relationships. In *Technical Communication and the World Wide Web*, ed. Carol Lipson and Michael Day, 27–42. New York: Routledge.

———. 2007. Web 2.0: What technical communicators should know. *Intercom* 54 (September/October): 8–12.

———. 2010. Content management: Beyond single sourcing. In *Digital Literacy for Technical Communinciation: 21st Century Theory and Practice*, ed. Rachel Spilka, 128–144. New York: Routledge.

Hart-Davidson, William, Grace Bernhard, Michael McLeod, Martine Rife, and Jeffery T. Grabill. 2008. Coming to content management: Inventing infrastructure for organizational knowledge work. *Technical Communication Quarterly* 17:10–34.

Johnson-Eilola, Johndan. 1996. Relocating the value of work: Technical communication in a post-industrial age. *Technical Communication Quarterly* 5:245–270.

Kellog, Katherine C., Wanda J. Orlikowski, and JoAnne Yates. 2006. Life in the trading zone: Structuring coordination across boundaries in postbureaucratic organizations. *Organization Science* 17:22–44.

Mirel, Barbara. 2004. *Interaction Design for Complex Problem Solving: Developing Useful and Usable Software*. San Francisco: Morgan Kaufmann.

Reich, Robert B. 1999. *The Work of Nations*. New York: Vintage.

Rockley, Ann, Pamela Kostur, and Steve Manning. *Managing Enterprise Content: A Unified Content Strategy*. Berkeley, CA: New Riders, 2003.

Sauer, Beverly. 1998. Embodied knowledge: The textual representation of embodied sensory information in a dynamic and uncertain material environment. *Written Communication* 18:131–169.

Simmons, W. Michele. 2007. *Participation and Power: Civic Discourse in Environmental Policy Decisions*. Albany, NY: SUNY Press.

Slattery, Shaun. 2007. Undistributing work through writing: How technical writers manage texts in complex information environments. *Technical Communication Quarterly* 16:311–325.

Smart, Graham. 2006. *Writing the Economy: Activity, Genre, and Technology in the World of Banking*. London: Equinox.

Spinuzzi, Clay. 2007. Technical communication in the age of distributed work [introduction to special issue]. *Technical Communication Quarterly* 16:265–277.

———. 2008. *Network: Theorizing Knowledge Work in Telecommunications*. Cambridge: Cambridge University Press.

Sun, Huatong. 2004. Expanding the scope of localization: A cultural usability perspective on mobile text messaging use in American and Chinese contexts. PhD diss., Rensselaer Polytechnic Institute.

United States Bureau of Labor Statistics. 2010a. Summary report for 27-3042.00: Technical writers. O-Net. http://online.onetcenter.org/link/summary/27-3042.00. Accessed July 20, 2010.

———. 2010b. Technical writers. In *Occupational Outlook Handbook, 2010–2011 Edition*. http://www.bls.gov/oco/ocos319.htm. Accessed July 20, 2010.

Winsor, Dorothy. 2003. *Writing Power: Communication in an Engineering Center*. Albany, NY: SUNY Press.

Zachry, Mark, and Charlotte Thralls, eds. 2007. *Communicative Practices in Workplaces and the Professions: Cultural Perspectives on the Regulation of Discourse and Organizations*. Amityville, NY: Baywood.

JIM HENRY

3 How Can Technical Communicators Fit into Contemporary Organizations?

SUMMARY

Every contemporary organization possesses an organizational culture that distinguishes it from others, and technical communicators who seek to fit into any organization must develop skills as cultural analysts. This chapter discusses a number of studies of technical communication that have drawn on such cultural analysis, then presents a framework for conducting participant observation, compiling fieldnotes, interviewing culture members, collecting and analyzing artifacts, and writing. Many of these methods can dovetail directly with one's work as a technical communicator on any given project, and the chapter presents an extended example as illustration: A team of technical writers charged with composing an annual report for a nonprofit organization interviewed organizational members, collected artifacts, and compiled extensive fieldnotes to help them fulfill the assignment. In the course of their cultural analysis, they perceived a need in the local culture that they knew they could fill—and they did so. Thus the chapter provides a method not only for fitting into the contemporary organization but also for exercising greater agency as a part of it.

INTRODUCTION: PONDERING A WORKPLACE CULTURE

Kate and her two peers, Jin and Cassie, have just met their point of contact for the nonprofit organization with which they'll be working during the next eight weeks to help compose that organization's annual report.[1] They have connected with Sylvia, the director of development of Women's Family Planning Centers (WFPC), through our Service Learning Center on campus to fulfill the major collaborative writing assignment for our technical writing course. Sylvia has arranged this first meeting at a popular off-campus café to explain a bit more about the organization and its mission. All three technical writers already know of the organization (though none has visited it), and none has ever written an annual report. They are eager to learn more and get started on the writing project. At this meeting, Sylvia

	First Meeting: 2/25/07 Sylvia is wearing her WFPC shirt -enthusiastic -teacher persona -very direct about political views -answers questions directly and does not stray off topic too much (less than many professors I know)

Figure 3.1. A typeset version of Kate's handwritten fieldnotes from the first meeting with Sylvia

is welcoming, warm, and full of energy, projecting a persona that the three women suspect might reflect the broader ethos (a rhetorical term referring to the trustworthiness or credibility of a speaker or writer) of WFPC. Figure 3.1 shows an excerpt from Kate's notes at that meeting, typeset here from the original handwritten notes.

Afterwards, Kate, Jin, and Cassie are excited to have scheduled their on-site arrival later that week, yet they are also full of questions: What if Sylvia's persona *doesn't* reflect the ethos of WFPC? What if the organization is rigidly formal, antiseptic, and cold? What if WFPC expects the three technical writers already to know the ins and outs of annual reports? Or alternatively, what if others at the organization don't reflect Sylvia's professionalism and instead are lax, disorganized, or indifferent to the organization's mission? What if they expect the technical writing team to do all the work with no support, or with inadequate information to compose the report? What if there are feuds, infighting, and territorial tendencies? What if the three of them arrive at work later that week to find a workplace that feels alien—and alienating? How can they fit in?

In a way, this last question condenses all those preceding it, and it's the kind of question we have all asked ourselves when pondering a new job (or the first job!) as a technical communicator. Although we often relegate such concerns to the (seemingly secondary) category of social skills, the fact is that we are constantly figuring out how to fit into our workplace in order to accomplish the organization's goals. We daily process hundreds of little clues as to *how things are done here*, often at the nearly subliminal level, to accomplish successful writing projects. This chapter will help you develop expertise in attending to these little clues—and many others—by teaching you how to become an *analyst of your workplace culture*. The chapter is organized as follows: the recent trend among certain scholars intent on bringing cultural studies to bear on technical communication is characterized as "top-down" (originating in theoretical constructs) or "bottom-up" (based on concrete, direct experience) through a brief literature review, the better to situate this chapter's approach to cultural analysis among the latter. The literature review next presents a definition of "organizational culture" provided in the literature of organizational psychology, then surveys pertinent scholarship that elaborates on fieldwork approaches and the disciplines that deploy them, to conclude with scholarship in technical communication focused on the analysis of workplace cultures.

This background sets the stage for a heuristic you can follow to analyze a new (or even familiar) workplace culture, the better to discern "how things are done here" in close detail and to document this work. Such documentation can help you both to fit in and to reshape the organization, however modestly, by providing insights on cultural practices that feed directly into your technical communication practices. An extended example from Kate's team writing project at WFPC illustrates this cultural analysis at work. The chapter concludes with an illustration of how such analysis enables both a better fit *and* taking proactive steps to shape that culture.

HOW RESEARCHERS HAVE ANALYZED WORKPLACE CULTURES AND TECHNICAL COMMUNICATION IN THEM

If the choice of the verb *fit* in my title might suggest a one-way process through which a technical communicator becomes socialized into a (static and preexisting) workplace culture, the reality is anything but. Workplace cultures in the twenty-first century are dynamic and in flux. Their rules, norms, and values vary in strength and direction, and technical communicators who best fit into a contemporary organization are those who become adept at analyzing its culture—the better to shape that culture even as they are shaped by it. Of course, the work of technical communicators is always shaped by cultural forces *beyond* the organization, and recent works in technical communication have tapped the field of cultural studies to signal this important interrelatedness (Scott, Longo, and Wills 2006; Zachry and Thralls 2007). Cultural studies approaches often use specific prechosen theories to illuminate communicative practices, and such top-down theorizing can shed new light on everyday work-life issues such as authority and agency (Herndl and Licona 2007).

The invaluable counterpart to such top-down theorizing about organizational cultures is the bottom-up analyzing offered in this chapter, in which you, the technical communicator, become a cultural analyst. The basis for your analysis will be the everyday practices that you can observe, document, compare, and ponder in order to understand the many aspects of this local culture as they shape writing. Such analysis is common in the fields of sociology, anthropology, folklore, and other disciplines where inquiry is grounded in fieldwork. The methodologies for conducting such analysis often rely heavily on ethnography (the systematic study of a local culture through participant observation) and autoethnography (systematic study of one's own local culture) to produce research. For example, when Stephen Doheny-Farina published "Writing in an Emerging Organization" in 1986, he drew on on-site observations, fieldnotes, interviews, and analysis of cultural artifacts to analyze ways in which writing practices and corporate policies in the organization coevolved. With this orientation to cultural analysis in mind, I discuss below a few key works on the topics of organizational culture, fieldwork, and technical communication in workplace cultures.

ORGANIZATIONAL CULTURE

In 1990, valuable approaches and definitions to help us think about culture as it takes form in contemporary organizations were introduced by the organizational psychologist and management studies expert Edgar Schein, who published "Organizational Culture" in the journal *American Psychologist*. Among the approaches he lists in this article is ethnography, and among the definitions is this one, of *culture*: "*Culture* can now be defined as (a) a pattern of basic assumptions, (b) invented, discovered, or developed by a given group, (c) as it learns to cope with its problems of external adaptation and internal integration, (d) that has worked well enough to be considered valid and, therefore (e) is to be taught to new members as the (f) correct way to perceive, think and feel in relation to those problems" (111). His definition of *culture* is by no means the only one available (Kroeber and Kluckhohn [1963] published an entire *book* on definitions of this contested term), yet it proves valuable for technical communicators stepping into the role of cultural analyst, because of the way he has broken it down into parts: What basic assumptions seem to drive this organizational culture? How were these assumptions developed, and how are they sustained? What problems of external adaptation does this organization face (i.e., what is its market niche, and who are its competitors?), and what mechanisms work internally to maintain cultural cohesion? How are these assumptions taught to new members? As you collect data related to such questions, you are on your way to cultural analysis that will enable you to fit into your organization.

With the spread of the concept of organizational culture across many disciplines in the past two decades, the numbers of publications relying on this concept now total in the hundreds, and a comprehensive review of the literature would be impossible here. Yet as a sign of the potential of the approach offered in this chapter, consider the 2007 publication by

Boyle and Parry in *Culture and Organization*, entitled "Telling the Whole Story: The Case for Organizational Autoethnography." They say, "First, this approach has the ability to connect the everyday mundane aspects of organizational life with the broader political and strategic organizational agendas and practices" (185). By understanding these broader political and strategic organizational agendas and practices as they interrelate with your day-to-day work, you will not only fit more successfully into your workplace culture, you will also shape your professional identity as an active influence in that culture—and in those that will probably follow.

ETHNOGRAPHY, AUTOETHNOGRAPHY, AND FIELDWORK

As mentioned above, fieldwork is vital to cultural analysis in ethnography, defined by Fetterman (1998, 1) as "the art and science of describing a group or culture." Anthropologists and folklorists have provided valuable sources on fieldwork skills that you can apply in your analysis, including the vital practice of capturing good fieldnotes (Emerson, Fretz, and Shaw 1995; Hammersley and Atkinson 1995, 175–186). Beyond fieldnotes, anthropologists have addressed the whole process of ethnographic analysis from start to finish, to address participant observation, interviewing, artifact analysis, and the ethics that should guide the whole undertaking (Hammersley and Atkinson 1995; Fetterman 1998). In most scholarly fieldwork studies, the researcher is not part of the culture under study, yet the subgenre discussed by Boyle and Parry (2007), "autoethnography," refers to this kind of cultural analysis. The prefix *auto* indicates that the analyst is part of the same culture that he or she is analyzing, and if you are studying an organization where you have worked for some time, you fall into this category. To learn about the roots of this subgenre, you can read David Hayano's 1979 article in *Human Organization*, in which he observes the promise of autoethnography for "its potential advisory capabilities in programs of change or development" (103). Early critiques of autoethnography claimed that the autoethnographer could never be "objective," yet recent thinking on qualitative research—the kind you will be conducting—points out that no one can ever be entirely "objective" when it comes to interpreting lived realities. We all filter our interpretations based on ways in which our "subjective" viewpoints have been shaped by factors such as gender, age, ethnicity, and many other influences. At the same time, we can also temper our subjectivity by "triangulating" with other sources (more on this below). The goal is to achieve something like "positioned subjectivity" that takes into consideration our own subjective positioning as part of the analysis.

Researchers such as Doheny-Farina used ethnographic techniques to report on writing projects in specific workplace settings in *Writing in Nonacademic Settings* (Odell and Goswami 1985). In the years following, ethnographic techniques for studying workplace writing became a mainstay that produced knowledge on such topics as teaching and learning in workplace discourse communities (Matalene 1985) and the connections between workplace literacies and the teaching of English (Garay and Bernhardt 1998). Single-authored books using ethnographic methods have probed issues such as the ways in which collaborative authoring takes place in workplace cultures (Cross 2001) and in which workplace culture members improvise (Spinuzzi 2003). Recent articles using ethnographic methods have addressed topics such as the ways that language practices informed by local cultural knowledge shape cultural memberships (Racine 1999) and the ways that writing practices shape change within an organizational culture (Anderson 2004). Autoethnographic accounts of technical writing have been much rarer, given that most practitioners' publications are professional rather than scholarly, yet the dynamics of workplace ghostwriting have been traced through autoethnographic analysis (Henry and "George" 1995); three other publications represent similar autoethnographic collaboration among a university researcher and seventeen technical communicators, conducted over seven years (Henry 2000, 2001, 2006).

A HEURISTIC FOR BECOMING A CULTURAL ANALYST

With the research outlined above in mind, let's equip you to become a cultural analyst by following a careful heuristic as illustrated in figure 3.2.

As the figure shows, your procedure will be grounded in a common technique for fieldwork known as participant observation. This technique requires observing the activities of culture members while participating in them yourself; you will understand them thoroughly because you have not only watched but also learned experientially. Ideally, you will be able to participate in a range of activities required of technical communicators while observing this culture, all the while making notes on the five *W*'s of the culture—who does what, when, and where, and how do they do it?[2] Because this part of your heuristic really grounds all the other parts of it, I have represented it as the initial and governing phase of the flowchart in figure 3.2, interconnected with the other parts of your fieldwork. The bidirectional arrows to each of the other elements of the heuristic indicate this interconnectedness. The notes that you will take are known as fieldnotes, and the pages that follow show you how to compile these fieldnotes

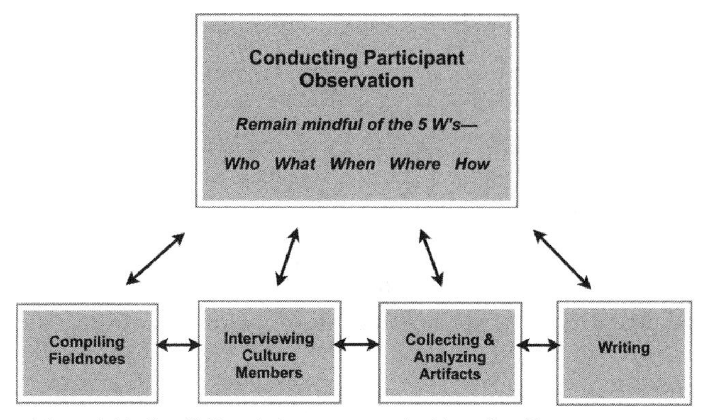

Figure 3.2. A heuristic for fitting into an organizational culture

carefully and systematically to help you capture in writing all that you observe. These fieldnotes give you a basis for interviewing cultural members, and you will read some helpful techniques and hints for doing so. Finally, you will learn to collect samples of as many different kinds of writing as you can find, and then set about interpreting this writing based on your growing knowledge grounded in your fieldwork. As figure 3.2 indicates by the arrows among boxes on the flowchart, these four processes of compiling fieldnotes, interviewing culture members, collecting and analyzing artifacts, and writing are often conducted in close succession or even simultaneously, as one process makes you realize that one of the other processes could add to your knowledge.

This written end product could be as formal as a research report for your class, an article you might submit to a professional journal,[3] or a report you prepare for the CEO (most likely in collaboration with your colleagues). Alternatively, your end product might be as informal as notes to yourself that attempt some synthesis of what you have learned or memos to file that document an issue in its complexity and that can help you articulate your own position on that issue to your colleagues. One technical communicator I once worked with likened his fact-finding with these methods to becoming a kind of "sleuth," and he was quite proud when this sleuthing uncovered a serious flaw in his company's forthcoming style manual: the plans for the manual had been based primarily on input from upper management rather than the frontline technical communicators who would use it, and when he consulted these technical communicators on the proposed outline, they were able to offer suggestions to improve it dramatically. (See Henry 1995.)

Whether you are brand-new to a workplace and thus similar to the ethnographer arriving among indigenous peoples in classic Western ethnographies, or you are already ensconced in a workplace culture yet interested in analyzing it in the tradition of autoethnograpy, you'll be following some standard procedures. Let's consider each of them in some detail.

CONDUCTING PARTICIPANT OBSERVATION

Dating at least as far back as early (Western) anthropologists' trips to non-Western cultures to document their customs in the ethnographic tradition, participant observation entails joining culture members in their daily life and attempting to interpret cultural practices based on such participation. The fieldworker eats, sleeps, works, plays—even runs from the police, if necessary (see Geertz 1973)—alongside culture members. In your case, you won't have to eat and sleep with the culture members—and hopefully you will not have to evade law enforcement—but you will seek to experience the culture exactly as the others do. Unlike them, you'll be monitoring these activities and writing about them, attempting to capture as closely as possible the details of document production and review as this culture does it in its own special way. To keep yourself observing even as you are participating, keep in mind the familiar five *W*'s:

- Who are the people in this culture and what are their job titles and duties? Who reports to whom?
- What is the organization's primary mission, and what are its secondary missions? What is its market niche, and what kinds of writing (often called a deliverable) enable it to survive?
- When do processes ebb and flow? How are deadlines established and enforced?
- Where are people located physically in the organization (office locations and configurations can tell you a *lot* about cultural values), and where are their collaborators, intramural and extramural?
- How do technical communicators compose here? In teams? Individually? How do review processes take place? How does writing expertise take form in this culture?

As you answer such questions, pay particular attention to language use. Inevitably, you will notice specific uses and applications of terms and phrases that might seem familiar to you yet that take on new dimension and significance when understood in their specific organizational culture. To help you capture all of these details as precisely and astutely as possible, let's look at the primary tool you'll be using: compiling fieldnotes.

COMPILING FIELDNOTES

Fieldnotes are entries made in systematic and regular fashion in a log or journal. Historically, ethnographers would fill up journal after journal (often in the form of spiral notebooks), in longhand, with their observations about cultural practices, striving to capture as many details as possible. Very often, traditional ethnographies would include an arrival scene that depicted the local culture, and it's a good idea to do the same in yours: you never get a second chance to encounter a new culture for the first time, and documenting first impressions can provide you with a valuable file that reminds you of the very first day in your "fitting" process. Very often, moreover, that arrival scene will include a tour of the workplace, an event that corresponds remarkably to a field technique that ethnographer David Fetterman (1998, 41) calls the "grand tour."

As you make your fieldnotes, be sure to date each page and leave a wide margin on the left-hand side (figure 3.3). Try to capture as many details as possible, to be able to recreate the smell and feel of this local culture. At times, your notes on the right-hand side might be little more than hastily scribbled jottings as you try to capture details, and you can come back later and fill in the details.

As time passes, you will want to revisit all the notes you have taken and systematically comment on them in the left-hand margin. Adding these comments can help you remember to take certain steps, as in the case of procedural notes, and they can enable you to start connecting the dots among observed events, comments by colleagues, and written artifacts. Eventually, you will cross-reference your notes to other pages of fieldnotes or to other sources of information, to begin identifying recurrent topics or patterns or to leverage these fieldnotes for greater understanding of your position of technical communicator in this workplace culture, as you will see in the extended example below.

You may want to compile your fieldnotes directly in a word-processing program, particularly since the vast majority of your work as a technical communicator will be completed on a computer. In that case, use your spiral notebook only for those moments when you are walking about or on site somewhere and don't have access to a computer, then transcribe these notes as soon as possible to put them with your other notes. (This procedure is the one Kate followed for her fieldnotes.) Or if you feel more comfortable compiling notes in a smartphone or small tablet device or speaking your notes into a smartphone, use that technique. Regardless of your technology during the collection of notes, be sure to transcribe them as soon as possible after you have captured them. Follow the same

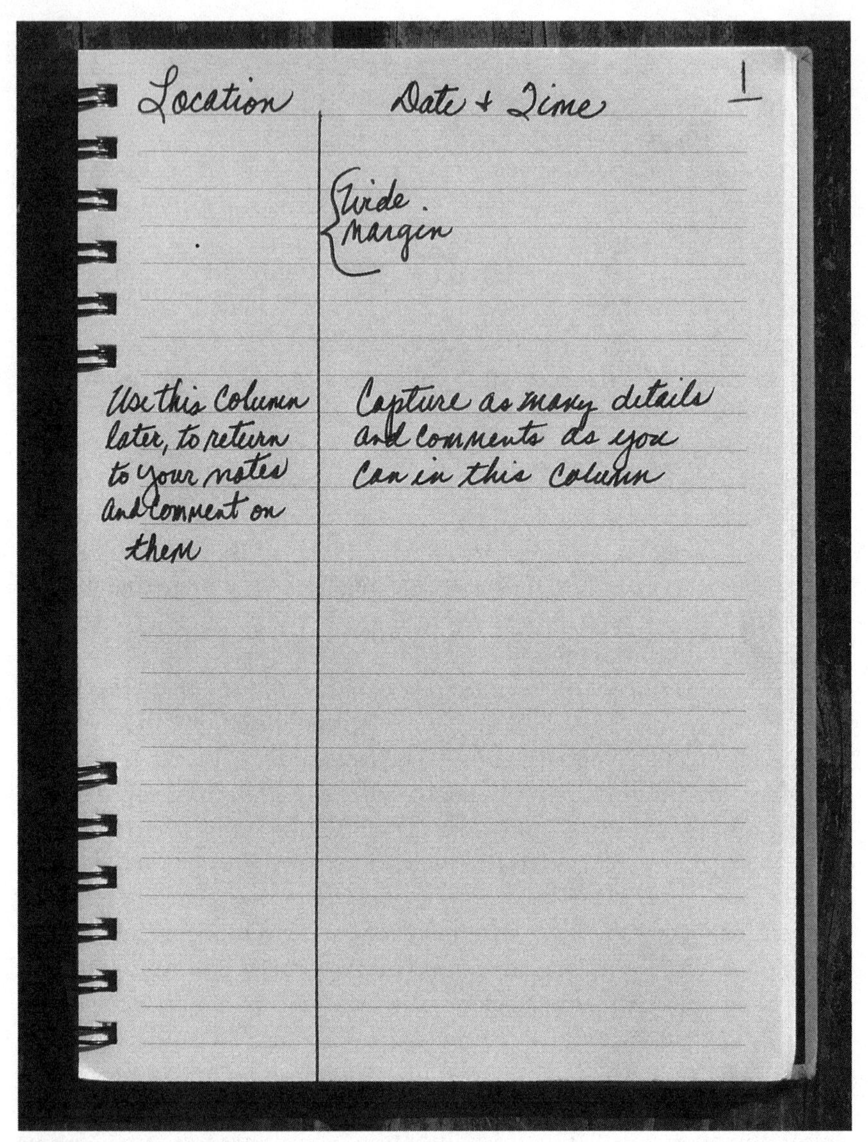

Figure 3.3. How to format a page for fieldnotes

procedure of sequentially dating the entries and do some creative formatting (by inserting a table with two columns, for example) that will allow you to comment on these notes later on. You can of course take advantage of your word processor's insert-comment function to enhance this metacommentary on your notes, and as your commenting reveals patterns, trends, or interesting connections, you may find yourself color-coding them. If your research evolves into very elaborate analysis, you may also take advantage of commercial and open-source software along with other technological resources to enhance this analysis.

INTERVIEWING CULTURE MEMBERS

Interviews are another key ingredient in analyzing a workplace culture, and an activity that dovetails easily with your work assignments. After all, technical communicators regularly ask questions of their supervisors and other coworkers to garner information necessary to complete a task. Think of those ad hoc questions as *spot interviews*, to be captured in your fieldnotes. In fact, you might not refer to them as interviews with culture members but simply as meetings during which you will take notes or audiorecord to help you be as precise as possible in your learning. As your analysis progresses, you might want to schedule more formal interviews with specific organizational players. You will want to wait until your fieldnotes have yielded some specific patterns or trends, or have otherwise prompted specific questions about the local culture, and you will want to schedule these interviews in advance. If you are pursuing a full-blown formal analysis—possibly for a class report or for a scholarly journal—then you may want to ask your interviewee's permission to record the interview digitally or with a simple analog tape recorder. After the interview, you will index the recording during playback by topic, noting the time on the counter as each topic shifts, and you will probably choose to transcribe certain passages verbatim.

Why conduct these extra interviews? Well, fieldworkers often speak of *triangulation* as a valuable research principle. As the term suggests, triangulation consists of comparing (at least) three sources on a given topic to see convergences and divergences. The objective is *not* to determine some definitive "truth" (and therefore who is or who is not telling it) but rather to acknowledge a basic tenet in qualitative research: all social facts are complex, and varying points of view on them enable you to identify certain beliefs or viewpoints held by culture members and see where these accounts point to a trend or to a significant, commonly held assumption or value. When you are able to identify these kinds of things, you are in the process of learning the basic elements of culture as defined by Schein. In the process, you have been able to get to know other culture members better, all the while showing them your serious, professional, engaged workplace identity.

If you have never conducted a formal interview, be sure to do some preparatory work. Consult some sources on interviewing techniques and common pitfalls, such as Edward Ives's *Tape-Recorded Interview* (1995) or David Fetterman's *Ethnography: Step by Step* (1998, particularly pages 37–52). Guides such as these go into depth on the key issues below.

- *Be absolutely ethical.* Never record anyone without his or her permission, and never divulge information given in an interview that the interviewee deems sensitive. If you wish to cite an interview, request permission and, ideally, let your interviewee see the citation in your writing to confirm it. Sometimes a comment in one context requires qualifying when presented in another context.
- *Do your homework.* Find out as much as you can about your interviewee by studying his or her organizational position and cultural role(s). Don't waste your interviewee's time by asking questions that you could have found the answers to otherwise.
- *Prepare questions.* Prepare the interview meticulously by writing down specific questions beforehand. Do *not* think that you can go into the interview and simply wing it. Frame your questions as open-ended, that is, as questions that cannot be answered with a simple yes or no, or if they can, be ready with a follow-up question that cannot.
- *Use probes and elicitors.* Described by Edward Ives (1995, 53) as "a device for eliciting more and better information in an interview, or to put it another way, of helping the interviewee tell a story more completely," a probe can be a pointed question, often introduced by "How . . ." or a follow-up to a *general* response that prompts the interviewee to ground that response in the specifics of everyday work. An elicitor can be any artifact from the workplace culture (see below) that will elicit extended commentary by your interviewee. Previous work such as reports or technical documentation projects (which you would have with you, in hard copy or online) are great elicitors. If you have a very specific question about composing processes, you might even ask your coworker to speak aloud his or her thoughts while in the throes of composing, while you both watch the screen.
- *Strive for mutuality.* Some fieldworkers characterize the interviewer's role as that of a "collaborative listener" (Sunstein and Chiseri-Strater 1997, 345–415), a concept that might help you elaborate your role as you interview writers. The more the process feels collaborative, the more your interviewer is likely to speak candidly.
- *Conclude professionally.* Thank your interviewee for his or her time (or for *their* time if you have chosen to interview more than one

colleague at a time, to tap the synergy that can emerge in such situations), and promise to get back to them if your notes lead you to share their views with others. Then keep your promise.

Though these steps may seem time consuming, by following them you have achieved two key goals: (1) you have garnered information in systematic fashion to the end of understanding your workplace culture in greater depth and breadth, and (2) you have demonstrated a professional and smart workplace identity to other colleagues, thus establishing yourself as someone with whom they will readily collaborate in the future.

COLLECTING AND ANALYZING ARTIFACTS

You might think that an artifact is an object taken from an exotic culture or some priceless piece in a museum, and if so, your understanding reflects the etymology of this word: something that has been made (*factum*) by drawing on skill, craft, and knowledge (*ars*). You can see how this etymology opens the doors for considering all kinds of cultural objects as artifacts. Edgar Schein (1990, 111) offers a very broad interpretation: "When one enters an organization one observes and feels its *artifacts*. This category includes everything from the physical layout, the dress code, the manner in which people address each other, the smell and feel of the place, its emotional intensity, and other phenomena, to the more permanent archival manifestations such as company records, products, statements of philosophy, and annual reports."

When Schein refers to the dress code as an artifact, he's reminding us to probe cultural practices at two levels: on the one hand, the organization you are studying may have an established, printed policy on dress, and on the other hand, you will see employees adhering to this dress code (or flouting it) in ways that tell you something about local cultural practices. You can collect lots of the other formal documents that Schein lists above and compare them with observed practices, too. Treat every document you encounter as data, and make copies (in paper or digital form) that you can study later as part of your cultural analysis. In addition to such formal writing, collect informal writing, too: sticky notes, organizational charts, whiteboard diagrams (use your cell phone to photograph them), *anything*. All artifacts can help you interpret cultural practices and technical communication within them.

One key element in interpreting artifacts is to ask yourself not only what an artifact *says*, as in the case of so many written documents, for example, but also how it *functions*, within and beyond the workplace culture. Think of the difference, for example, between what a report says and how it might get used by different members of an organizational culture for different purposes and according to a variety of "agendas." As soon as you start digging below the face value of cultural artifacts, you uncover more cultural values and assumptions (and the ways in which other cultural members are interpreting, deploying, and subtly reshaping them), the better to determine your own fit.

WRITING

Writing is at once a central tool in conducting your fieldwork and in many cases the culmination of it. Your extensive notes, even if they go no further than your fieldwork journals, will serve as important documentation for you to refer to when you want to validate a hunch about cultural values or refresh your memory on specific practices. You may use these notes to compose *memos to file*, formal memos stored on your hard drive (backed up in paper form and in each case stored in a secure location), which elaborate your understandings of such values and practices. If you are not required by your organization or your supervisor to keep a log of your work activities, or even if you are, these memos can constitute a valuable corpus that goes beyond the mere recording of completed activities to probe many implications of these activities and even ideas for action grounded in such probing. Doctors, lawyers, and CEOs regularly document their work through memos to file, so why shouldn't you? Such memos can remain with you during your entire professional life, constituting a variation on journal entries you might have been asked to complete in writing classes and serving as an invaluable source for elaborating upon your résumé orally throughout your career. Think of these memos as a *working paper* to which you can add as you go and that will serve as an invaluable documentation of your own enculturation at your place of work.

Your research on your workplace culture will likely take other forms, too. You may be asked by a coworker or supervisor to contribute to a report going up the hierarchy on work practices in your unit, or you may be charged with writing a letter of transmittal for a longer document, your goal being to succinctly represent this document's background to the addressee(s). Or you may be tasked with a documentation project that actually *requires* a number of the fieldwork practices above, even if those requirements are not explicitly stated. Finally, you may author or coauthor a research report submitted to a scholarly journal that establishes you as a researcher for a wider audience. Whatever the case, the methodology described above will serve you well, as the extended example below illustrates.

FINDING A FIT AT WOMEN'S FAMILY PLANNING CENTERS

As you could tell from the opening scenario, Kate and her team members had stepped into the role of cultural analyst even before they entered the culture of WFPC—by taking fieldnotes during their first meeting with Sylvia. When they arrived at WFPC the following week to begin their work, they were given a tour of the offices. Such a tour is common when being inducted into a new job as a technical communicator and in many ways resembles the "grand tour" sometimes described by ethnographers. Following our guidelines for taking fieldnotes during this grand tour, Kate scribbled as the team walked. Later, she transcribed these notes into her word-processing program and followed the practice of commenting on the notes in the left-hand column, to produce the set shown in figure 3.4.

As you can see in this example, Kate was capturing copious notes on the physical environment, attempting to capture the smell and feel of the office workspace as it might shape the environment where coworkers collaborate. At the same time, she was conducting those spot interviews mentioned above and capturing verbatim snippets of comments. The reflective comments in the left-hand column not only anticipate a fit between her (and her team) and WFPC, but also already carry implications for the annual report on which Kate will be collaborating: How does an issue of client privacy get handled in such a report? Can the team use the pamphlets, videos, and binders in the lunchroom to get a better sense of the ethos that WFPC will want to project in this report? In a "flat organization," who else is likely to have input on this report? Some of these questions were answered when Kate's team secured another key artifact from the WFPC culture: the organizational chart, shown in figure 3.5.

As you can see, the org chart in its document design wouldn't *necessarily* indicate a "flat organization," as WFPC was described by Sylvia in their second meeting, yet Kate's scribbled note between the CEO and the development director helps confirm such a cultural practice: "direct communication; no formal communication practices." Scribbling such notes on an org chart is indispensable for decoding a written artifact to get at practices that it fails to convey. In the case of their team's work, this artifact also suggested quick turnaround times between drafts and responses to them. Kate's notes also include some valuable information on the workplace values and assumptions: "highest standards for women's clinical services, stringent protocols." Such values and assumptions gave the team a good written touchstone to return to when drafting parts of the annual report, and the note following it ("high-level security computer system houses patient info and statistics") indicates that this local culture strove

	Second Meeting: 3/5/07
	Narrow carpeted hallway in hodgepodge office building, matching blue painted doorframes. Other businesses in building include real estate agencies, doctors, insurance companies, and a travel agency. Building built in 1962.
How important is surveillance?	Surveillance cameras, security doors with keypads (can change without having to change locks or get keys back)
Coworkers seem to reflect Sylvia in dress Use of space: focus on services rather than status Not sure of room #s, may need to get these later	Employees in business casual 4 rooms on the third floor 309 **Business** office waiting area, reception, copy room, 3 office Very small waiting area with just 2 chairs and one coffee table 307 **Finance** office, central area and offices, 3 main people 30? **Lunchroom**, 2 more offices, material storage--lots of pamphlets, videos, binders stored on floor-to-ceiling shelves 30? **Clinic** – reception, waiting room, exam rooms, surgery, recovery
Follow up on check-in procedures	condoms, lube, etc. displayed just as you enter reception (behind door). small, simple and clean exam rooms; comfortable recovery area waiting room has a very small window to reception desk area, this makes me think that check-in is all that happens in the public sector of the waiting room, and high levels of privacy are maintained as all conversations and details are discussed behind closed doors; matching rattan chairs with matching dark blue cushions; freshly painted two-toned walls done in purple to match WFPC logo colors; large windows bring a good deal of light into the area
They have thought about client privacy	More than half of the waiting room is only visible after you turn around a corner after coming in through the door; this means that the door opened to the hallway does not expose those in the waiting room to anyone walking by in the hallway
How will the org being "flat" factor into our work?	Quotes: Sylvia's comments: "We are a very flat organization" "Board of Directors is in charge of overall direction and financial health of organization; they are responsible for paying back any money due and this new strictness has resulting in more careful oversight than in the past"
Do others think of themselves as "jack of all trades?"	Financial person's comment: "Jack of all trades"

Figure 3.4. A typeset version of Kate's fieldnotes from the first visit to WFPC

to make good on its image as a safe and secure place for women by using technology in a way that protected women's private information.

As this team progressed in their work on the annual report, they were able to render their initial observations more nuanced. For example, in her working paper, composed to capture their learning about this culture, Kate returned to the observation about the "flat organization": "While WFPC enjoys emphasizing the 'flatness of the Organization,' there also

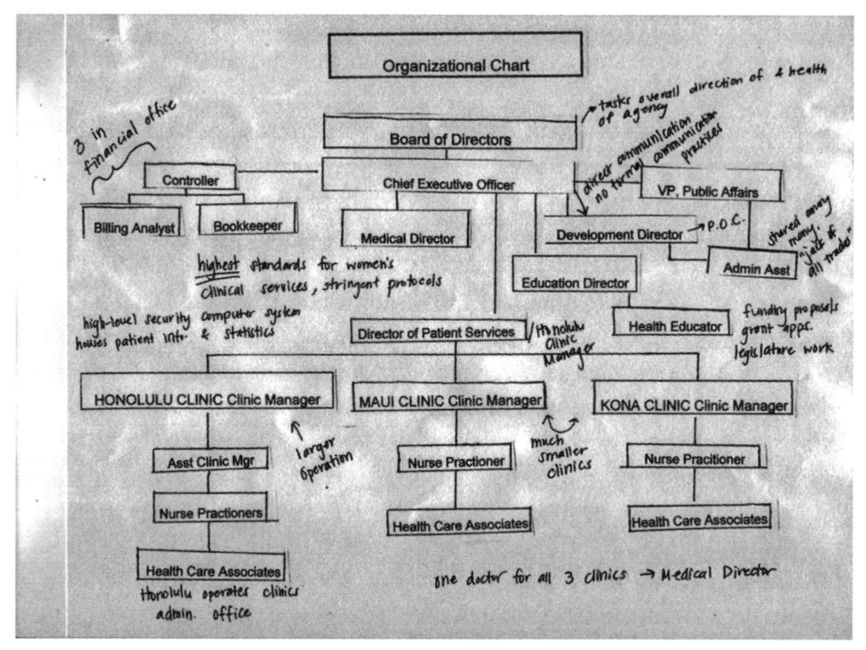

Figure 3.5. WFPC organizational chart

exists a hierarchy and consequential procedures . . . utilized in certain situations. For instance, any published materials coming out of the organization have to be approved by Bob, the CEO, before being sent to the printers." As a very important document, the annual report would certainly require Bob's approval, and the technical communication team of Kate, Jin, and Cassie now incorporated him into their mental picture of implied readers for this report.

They also drew upon an invaluable artifact from WFPC's files: the 2004–2005 annual report. The cover of the report included a horizontal band of images including a mother and child, two senior women in front of a U.S. flag, a nurse, a pro-choice demonstration, and a baby; these images were bordered on the top by a running marquee "STRENGTH IN DIVERSITY" and on the bottom by "UNITED BY PURPOSE." The team's analysis of this document design alone told them that the contents of the report would most likely have multiple appeals and multiple audiences, and that any draft wording that they would supply would inevitably be revised. They had done a lot of research outside of the culture—specifically on annual reports as a genre, compiling a folder of reports available online from other nonprofits—and together with their ongoing work *inside* the culture, they were able to draft a page-by-page outline, then fill it in with suggested copy for each section. Ultimately, their drafts did get revised quite a bit, but as Kate put it in an e-mail reflection on March 7, 2008, "although the final product had many changes from our draft (generally for financial reasons or because of repetition in other mailed materials), it reinforced for them that we were able to 'get' them and thus, as we were [outsiders] in a way, this must mean they have a pretty streamlined and consistent message. That was clearly something that they gleaned from the whole process. It was a sort of two-way reinforcement." When I sought Sylvia's affirmation of Kate's claim, her e-mail of March 10, 2008 concurred: "One of the most satisfying aspects of the project was communicating our corporate culture to the writing team and then the way in which the writing team did indeed 'get it.' It is important to us that the overall message of our organization is clear and communicated effectively across the different layers of our organization." In Sylvia's affirmation, we see a subtle and valuable elaboration on the concept of organizational culture: the metaphor of "layers" underscores just how fluid and changing cultural practices can be—yet another reason for technical communicators to document observations in some kind of working paper.

As newcomers to this organizational culture, Kate and her teammates thus provided an invaluable review that gave the long-term culture members something akin to a usability test. Meanwhile, though, as they grew to know the culture more and more, the technical communication team realized that even if they lacked all the cultural knowledge to be able to represent WFPC as completely as they would have liked in this vital publication for an *external* audience, they did possess knowledge as researchers that would enable them to produce a technical document for *internal* consumption. Their cultural analysis revealed a strong emphasis on women's health issues above and beyond family planning, bringing out the cultural assumption that family planning and women's health are inextricable issues. The technical writing team also knew that the clinical services counseled women on many topics. One such topic was cervical cancer, yet WFPC lacked documentation on the most recent research on this topic, a fact that Kate and her team entered into the working paper. When they proposed to Sylvia that they put their skills in composing annotated bibliographies to work for WFPC, Sylvia immediately approved the project. By the semester's end, when the technical communication team had concluded its work with WFPC, they had produced a ten-page report on cervical cancer research and treatment that lives on to this day in WFPC's culture.

In sum, Kate's technical communication team used several methods to conduct their cultural analysis: engaging in participant observation, capturing fieldnotes, conducting interviews, analyzing artifacts (from the physical décor of the offices to the organizational chart to previous publications), and composing a working paper. In just seven short weeks, they had moved from total cultural outsiders to quasi-insiders who now understood the organization's ethos in much detail. They were able to supply WFPC with a good draft of an annual report, and just as importantly, they were able to help WFPC gauge its own image as represented in that report. As cultural analysts, Kate and her team were also able to perceive a need in the culture that they could supply—the annotated bibliographic report on cervical cancer.

CONCLUSION: SURVIVAL OF THE FITTEST

The phrase "survival of the fittest," often attributed to Darwin, was actually coined by Herbert Spencer (2009, 444), who used the term "fittest" *not* as synonymous with "strongest" (as is often implied in popular usage) but rather with "being most nearly in equilibrium" with one's environment. And so it is with technical communicators in contemporary organizations: those who can analyze their environment effectively and insightfully are most likely the ones who will become the fittest to and for that culture. At the same time, most contemporary organizations witness high turnover in their workforce, which means that technical communicators will be faced ceaselessly with new organizational cultures—whether they remain a good fit in their current culture or move on to another. One technical communicator with whom I have previously collaborated found the turnover at her workplace so daunting that she likened colleagues to "Tasmanian devils" who rushed in and out of the culture (Henry 2006, 199), and in the current globalized economy, workers at all layers of the culture are likely to inhabit somewhat precarious positions. Under such conditions, it may seem daunting to entertain goals beyond survival as an employee, let alone aspire to exercising significant agency in the workplace culture.

Yet Kate's team illustrates how technical communicators can do just that when they incorporate cultural analysis into their everyday work. When I queried Sylvia as to the status of the annotated bibliography produced by Kate and her collaborators, she replied by e-mail on March 11, 2008 with this observation:

> I've referred to it and used some of the research in a grant proposal submitted to the Chamber of Commerce Public Health Fund. I was seeking a second round of funding to support our cervical cancer screening and treatment program, which has helped us secure funding for these services that are not always covered otherwise by the Federal Title X

family planning funding or even all insurances. I plan to get back to it again in the near future as it is time to start writing grants again; the government has just released a report that one in four teenage women has some form of STD, and some of these STDs can prompt cervical cancer. WFPC is strongly committed to educating and counseling such young women. So, yes, it does live on in our culture, in very significant ways.

Were Kate and her team members active agents? It would appear so, and women who visit WFPC in the future will benefit from their efforts. Did cultural analysis help them contribute? Their working paper's observation about the culture's lack of a bibliography on cervical cancer and their proposal to write one answers this question resoundingly. Can your own cultural analysis, achieved by following the heuristic of this chapter, help you shape an organization's culture even as you strive to fit? Let Sylvia's response serve as inspiration.

DISCUSSION QUESTIONS

1. Read Horace Miner's "Body Ritual among the Nacirema" (1996; you can find a reprinted version online at https://www.msu.edu/~jdowell/miner.html). Notice the paragraph numbers in the margin. Beginning with paragraph 3, jot down the main practices you notice. How do these practices compare with those in contemporary U.S. culture?
2. Using Schein's definition of organizational culture from the section that explains approaches to analyzing workplace cultures, divide the class into six teams and assign one part of the definition to each team. Then have each team analyze how you as students have been socialized into your campus culture with respect to your part of the definition. When teams report to the whole group, see if you can add to or challenge the claims of other groups.
3. During one of your instructor's lectures, take notes following the format described in the section "Compiling Fieldnotes." After the lecture, comment to yourself on selected notes in the left-hand margin the way that Kate has commented on hers in figure 3.4. If possible, use a special bullet or symbol in the margins to link one of your instructor's comments with others, indicating the relationship (e.g., "see p. 3 where she gives another example"; "does this comment relate to the work we did on readability?"). At your next class meeting, share your notes with a classmate and note similarities and differences.
4. Interview a classmate to learn as much as you can about her or his composing processes. Begin by elaborating a list of questions that

cannot be answered with a simple yes or no, then add some possible "probe" prompts to them. Take notes furiously as he or she responds, or better yet, use an audio recorder and then index the recording as explained in the section "Interviewing Culture Members." From this set of notes, produce a one-page summary of your classmate's composing processes. As part of the summary, note those processes that you suspect are widely shared in your classroom culture and those you suspect are not. Discuss your summaries as a class.

5. As preparation for your on-site ethnographic research within an organizational culture, visit a department or locale on your campus and take fieldnotes. Share them in class.
6. Fieldworkers sometimes categorize the comments in their fieldnotes as either observational, procedural (notes about steps in the process), or theoretical (hunches or hypotheses about what has been observed). Studying figure 3.4, classify Kate's comments into these categories. Do any of them not fit into a category? What do these comments tell you about Kate's fieldworker mind-set?
7. Study the organizational chart in figure 3.5. What kinds of day-to-day activities would you expect to take place for the employees in each box? How would you expect them to share information across the organizational culture? Considering Kate's comments on the organizational chart and her designation of her team's point of contact, which employees were likely to have contributed what kinds of information to support the annual report? If you are a practicing technical communicator, bring a copy of your organization's org chart to class and see how close your classmates can come to answering similar questions about your organization.
8. If you are a practicing technical communicator currently, find some document in your organization that you can talk about not only for what it *says*, but also for what it *does*. What kind of cultural work does this artifact accomplish, how does it accomplish it, and why?
9. In Sylvia's e-mail of March 10, 2008, she states that the work of Kate's team indicates that the organization's message was "communicated effectively across the different layers" of the organization. Given what you now know about organizational culture, write a paragraph or two probing the metaphor of "layers" and what kinds of cultural dynamics this metaphor might represent.

NOTES

1. Following protocol to assure the anonymity of participants in research involving human subjects, I have assigned a pseudonym to this organization and to the point of contact,

who has granted permission to publish this chapter and who has reviewed it to assure faithful representation of the organization and the work of these three technical writers. The only real name is that of Kate Millen, who graciously agreed to revisit her technical writing project at WFPC and provide excerpts from her working documents to illustrate the approach presented in this chapter. This research was reviewed and approved by an officer of the university institutional review board.

2. Although one of our *W*'s is an *H*, the mnemonic of the five *W's* is commonly used in investigative reporting to assure a comprehensive gathering of basic facts. (Think of the *W* in this word as its *ending* rather than its beginning.) Such reporting may include a sixth *W*, for *why*, yet as the discussion that follows illustrates, the why of organizational practices can be quite complex. Part of your work as a cultural analyst entails uncovering this complexity through your ongoing fieldwork.

3. If you seek to publish your ethnographic analysis, be sure to consult your college or university's institutional review board, which oversees the use of human subjects as part of academic research. You will likely need to obtain informed consent from any subjects who figure in your report and from a representative of the organization.

WORKS CITED

Anderson, Donald. 2004. "The Textualizing Function of Writing for Organizational Change." *Journal of Business and Technical Communication* 18:141–164.

Boyle, Maree, and Ken Parry. 2007. "Telling the Whole Story: The Case for Organizational Autoethnography." *Culture and Organization* 13:185–190.

Cross, Geoffrey. 2001. *Forming the Collective Mind: A Contextual Exploration of Large-Scale Group Writing*. Cresskill, NJ: Hampton Press.

Doheny-Farina, Stephen. 1986. "Writing in an Emerging Organization: An Ethnographic Study." *Written Communication* 3:158–185.

Emerson, Robert M., Rachel I. Fretz, and Linda L. Shaw. 1995. *Writing Ethnographic Fieldnotes*. Chicago: University of Chicago Press.

Fetterman, David. 1998. *Ethnography: Step by Step*. Thousand Oaks, CA: Sage.

Garay, Mary Sue, and Stephen A. Bernhardt, eds. 1998. *Expanding Literacies: English Teaching and the New Workplace*. Albany: State University of New York Press.

Geertz, Clifford. 1973. *The Interpretation of Cultures*. New York: Basic Books.

Hammersley, Martyn, and Paul Atkinson. 1995. *Ethnography: Principles in Practice*. London: Routledge.

Hayano, David M. 1979. "Auto-ethnography: Paradigms, Problems, and Prospects." *Human Organization* 38:99–104.

Henry, Jim. 1995. "Teaching Technical Authorship." *Technical Communication Quarterly* 4:245–259.

———. 2000. *Writing Workplace Cultures: An Archaeology of Professional Writing*. Carbondale: Southern Illinois University Press.

———. 2001. "Writing Workplace Cultures." *College Composition and Communication* 53. http://www.ncte.org/cccc/ccc/issues/v53-2.

———. 2006. "Writing Workplace Cultures—Technically Speaking." In *Critical Power Tools: Technical Communication and Cultural Studies*, ed. J. Blake Scott, Bernadette Longo, and Katherine Wills, 199–218. Albany: State University of New York Press.

Henry, Jim, and "George." 1995. "Workplace Ghostwriting." *Journal of Business and Technical Communication* 9:425–445.

Herndl, Carl G., and Adela C. Licona. 2007. "Shifting Agency: Agency, Kairos, and the Possibilities of Social Action." In *Communicative Practices in Workplaces and the Professions: Cultural Perspectives on the Regulation of Discourse and Organizations*, ed. Mark Zachry and Charlotte Thralls, 133–153. Amityville, NY: Baywood.

Ives, Edward D. 1995. *The Tape-Recorded Interview: A Manual for Fieldworkers in Folklore and Oral History*. 2nd edition. Knoxville: University of Tennessee Press.

Kroeber, Alfred, and Clyde Kluckhohn. 1963. *Culture: A Critical Review of Concepts and Definitions*. New York: Vintage Books.

Matalene, Carolyn, ed. 1985. *Worlds of Writing: Teaching and Learning in Discourse Communities of Work*. New York: Random House.

Miner, Horace. 1956. "Body Ritual among the Nacirema." *American Anthropologist* 58: 503–507.

Odell, Lee, and Dixie Goswami, eds. 1985. *Writing in Nonacademic Settings*. New York: Guilford.

Racine, Sam J. 1999. "Using Corporate Lore to Create Boundaries in the Workplace." *Journal of Technical Writing and Communication* 29:167–183.

Schein, Edgar H. 1990. "Organizational Culture." *American Psychologist* 45:109–119.

Scott, J. Blake, Bernadette Longo, and Katherine V. Wills, eds. 2006. *Critical Power Tools: Technical Communication and Cultural Studies*. Albany: State University of New York Press.

Spencer, Herbert. 2009. *Principles of Biology*. Ithaca, NY: Cornell University Library.

Spinuzzi, Clay. 2003. *Tracing Genres through Organizations: A Sociocultural Approach to Information Design*. Cambridge, MA: MIT Press.

Sunstein, Bonnie Stone, and Elizabeth Chiseri-Strater. 1997. *Fieldworking: Reading and Writing Research*. Boston: Bedford / St. Martin's.

Zachry, Mark, and Charlotte Thralls, eds. 2007. *Communicative Practices in Workplaces and the Professions: Cultural Perspectives on the Regulation of Discourse and Organizations*. Amityville, NY: Baywood.

KELLI CARGILE COOK, EMILY COOK,
BEN MINSON, & STEPHANIE WILSON

4 How Can Technical Communicators Develop as Both Students and Professionals?

SUMMARY

You are likely reading this book because you are considering a career as a technical communicator. In a sense, every chapter you read in this book, every topic you discuss in this course, and every project you complete are the first steps you'll take to develop yourself professionally. Professional development is generally defined as actions an individual takes to establish, strengthen, or maintain knowledge and skills that are necessary to perform a job well or to advance one's position or status in an organization. But what are these actions, specifically, and when should they be taken? When does this work end? This chapter is designed to answer these questions. To do so, it maps the status of technical communication as a profession and considers how technical communicators' own professional growth is connected to the discipline's development as a profession. It provides you with snapshots of the careers of three technical communicators. Using them as examples, it suggests a heuristic that will assist you as you develop professionally, both now as a technical communication student and later as a novice technical communicator.

PORTRAITS OF THREE PROFESSIONALS

To illustrate how successful technical communicators have worked to develop themselves as professionals, this section introduces you to three graduates of the technical writing program at Utah State University (USU). All three students earned their bachelor's degrees in this program, and one of them has also earned her master's degree in USU's online technical writing program. These professionals all started their work as technical communicators just as you are starting yours, by enrolling in an introductory technical communication course and progressing from there. This section, however, showcases their current positions, describing their roles and explaining their work. Later in this chapter, you'll learn the steps they took to arrive at these positions. In both sections, these three professionals will tell you their stories in their own words.

MEET STEPHANIE WILSON

Of the three of us, I am the most recent graduate of USU's technical writing undergraduate program. I graduated in May 2008. Before graduation, I applied for a position at USU with the Programming and Design (PAD) team, a unit within the office of the vice president for information technology (IT). After a few months of waiting, interviewing, and negotiating, I accepted a position on the PAD team as technical support coordinator. My job responsibilities include writing end-user documentation, training users, and managing work flow and resources within the team. I work daily with intelligent and talented IT professionals, as well as many subject-matter experts in most of the departments and research centers at USU.

My job as technical support coordinator requires me to train and support users of the USU content management system, ezPlug. To do this work, I write documentation for ezPlug, provide hands-on training as needed, and answer questions over the telephone about how ezPlug works. End-user documentation is not the only type of writing I do; my supervisor sometimes asks me to plan, author, edit, and publish other texts independently. For example, I write tool tips, error messages, and other text prompts. I sometimes help with technical public relations writing for IT, especially on our team website and other IT websites. One writing project I especially enjoy is creating posts for the blog our team publishes, which is geared toward the professors, students, and administrative assistants responsible for the content of various websites at USU. I also manage projects and work flow for a team of programmers, designers, and student employees. I supervise interns from the technical writing program, who support my work by writing documentation drafts and helping me acquire visuals for our documentation projects. All of these duties require me to write, edit, collaborate, and manage people almost every weekday.

MEET BEN MINSON

I graduated from USU's technical writing program in May 2005. I currently work as a technical writer for the Church of Jesus Christ of Latter-Day Saints on some of the many software applications built internally to support our large organization. I began as an intern shortly after graduation and later became a full-time employee. As a member of a small technical writing and training team, I create online help systems, quick-reference guides, user guides, tutorials, and release notes.

Part of developing content for these projects is interviewing interaction designers, business analysts, developers, and testers. I document tasks, functionality, work flows, and downstream effects of user actions in these applications. I manage most of my projects completely, from writing drafts and creating graphics to delivering the finished documents. When needed, I collaborate with product managers in providing live training. I also work as a reviewer of official department communications to help the department maintain a professional image. In any given day, I take on several different roles, and writing is a small piece of the pie.

MEET EMILY COOK

I have worked as a technical communicator for over ten years now, and I've earned two degrees in technical communication, both from USU. I earned my undergraduate degree in May 2001 and my master's degree in December 2005. My current work is very fast-paced. I am a proposal manager within the marketing communications arm of GC Services' Marketing Department. I oversee the department's workload as well as complete my own projects, including proposals, presentations, brochures, trade-show materials, and our internal company newsletter. I follow up and ensure that my counterparts complete their work and have the resources in order to do so. I also coordinate and manage tasks related to personnel issues, including hiring, interviewing, and training new candidates to our department.

Arriving at this position took some time. Leading up to it, I worked as a technical writer for training curricula, a proposal analyst, a proposal coordinator, and a senior proposal coordinator. You might also find it interesting to know that I've held many of these jobs from a distance; that is, I telecommute to work. Of course, this was not always the case. I had to prove myself in the office first, but now I work over two thousand miles from my company's home office in Houston. I sometimes travel to meet with other members of our staff, but mostly I work independently and communicate with my colleagues electronically.

TECHNICAL COMMUNICATION'S PROFESSIONAL DEVELOPMENT

Although their job duties and workplaces are different, these graduates share common educational experiences and have taken similar paths to reach their current professional goals. In many ways, the road to professional success seems easy to understand and navigate. Almost everyone has at least a basic understanding of how to find a job and keep it. Defining the terms *profession*, *professionalism*, and *professional development*, given this basic understanding, seems relatively simple. Simple definitions, however, can be troublesome. For example, people tend to blend all of these terms into a single, oversimplified definition so that *professionalism* means "become good at what you do." But that definition ignores important historical and cultural meanings of *profession*. Considered in light of these more complicated meanings, the terms *profession*, *professionalism*, and *professional development* are concepts less easily defined or enacted.

For this reason, technical communication scholars have hotly debated whether our discipline is a profession at all. When you first read this statement, you may react negatively to it. You would not be the first person to do so. But before you reject this idea completely, read this section. Its purpose is to help you understand what a profession is; explain why scholars have debated technical communication's status as a profession; provide you with a historical perspective about technical communication's development as a discipline, if not a profession; and suggest that, while technical communication may not currently qualify as a profession, it is a maturing discipline on its way to becoming one.

WHAT IS TECHNICAL COMMUNICATION?

It isn't easy to define technical communication. Defining our discipline, however, is a crucial step to becoming a profession. As Brenton Faber (2002, 307) notes, before we can claim to belong to a profession, "we need to carefully define what is professional about professional communication and how professional communication is distinguished from other forms of workplace writing." Similarly, in his introduction to *Power and Legitimacy in Technical Communication* (2003, 1), a two-volume collection that grapples with the history, identity, and future of technical communication, Gerald J. Savage argues that it will be impossible to call technical communication a profession until we are able to define exactly what technical communication is: "Unless we are able to define our field, we are unlikely to be recognized as a profession—we cannot be recognized by others if we can't even recognize ourselves." The definition problem is compounded by the fact that individuals who call themselves technical communicators work across diverse fields, disciplines, and organizations, making it particularly challenging to find a definition capable of encompassing such difference. In fact, some scholars, such as Allen (1990), have argued that we should not attempt to define technical communication because its definition is specific to organizations and locales, not across them. One anthology, *Defining Technical Communication* (Jones 1996), nevertheless, collects twenty-three essays that work toward a disciplinary definition. In the introduction, Jones notes that while the anthology "brings together many of the best efforts" at defining the field, these efforts do little more than provide us with a list of qualities and skills: "qualities and skills," he acknowledges, "do not, of course, provide us with a satisfactory definition" (v).

Interestingly, in 2005, two years after *Power and Legitimacy* was published, two scholars, Kathy Pringle and Sean Williams, claimed to have arrived at a definition by studying a group of diverse technical communicators to discover how they defined themselves and the work they do. The definition Pringle and Williams derive from their study was not unlike many of the definitions collected in *Defining Technical Communication*: "Technical communicators work at the intersection of technology and people, migrating back and forth between technology and communication as they design products for specific audiences. . . . We approach technology from a human perspective and believe that technology should adapt to people, not the other way around. We design communication products accordingly, using whatever media, software, technology, or tool is most appropriate to achieve this end. People are the ultimate end, we would argue, not the technology" (2005, 369). This work and the values we place on it, they contend, describe not only what technical communicators have done in the past, but also what we are doing now and will be doing in the future (368). If we accept Pringle and Williams's definition, then we have, as Savage suggests, taken the first step toward becoming a profession.

ARE OTHER STEPS NECESSARY TO BECOME A PROFESSION?

Developing and agreeing upon a common identity or definition is the first step toward professionalizing technical communication, but what steps follow? Most scholars agree that a standardized code of professional responsibilities or behavior (more commonly called a code of ethics) is an early step toward professionalization, as is agreeing upon a common set or coherent body of knowledge (Rainey 2005; Faber 2002; Hayhoe 2005; Campbell 2008). Next, the field must identify core competencies or skills, standardize processes, and develop criteria for evaluating products (Rainey, Turner, and Dayton 2005; Hayhoe 2005). Such steps, Rainey (2005, 679) argues, result in a "mechanism for identifying and validating the work that [technical communication professionals] do."

Historically, professions are promoted through organized efforts of those who seek to belong, and professional organizations are generally where this work begins: "Professions will commonly form professional organizations, unions, or guilds which work to unify the practice, represent the field to the public, lobby government officials, advise government and organizations, monitor and promote education of members, promote communication and socialization among practitioners, and maintain codes and standards" (Savage 1999, 358). In the last decade, the Society for Technical Communication (STC), our discipline's largest professional organization, has taken the lead in this work, setting concrete goals and endeavoring to achieve them. According to STC's website, its members include "technical writers, editors, graphic designers, multimedia artists, web and intranet page information designers, translators and others whose work involves making technical information understandable and available to those who need it" (STC 2009–2010). As the largest professional organization for technical communicators, STC is poised to move the discipline toward professionalization. The organization's 2008 strategic plan (STC 2008) outlines four specific tasks necessary to achieve these goals.

- Establish a body of knowledge for the profession.
- Define and promote ethical standards for the profession.
- Identify the skills and aptitudes that are associated with successful technical communicators.
- Adopt, enhance, or create international standards that positively impact the profession.

In response to these tasks, STC already has an established code of ethics, "Ethical Principles for Technical Communicators" (STC 1998), and the organization is investing significant resources toward the development of the "Technical Communication Body of Knowledge" (TCBOK), a draft of which its members are now able to access.

The TCBOK draws heavily upon the scholarship of Kenneth T. Rainey, who dedicated the latter part of his career to studying technical communication competencies and skills and working with STC as well as other national and international technical communication organizations to promote consensus about these competencies. Researching alone or with collaborators, Rainey spent years identifying what he called "core competencies" and then testing his findings through workplace research. Rainey and his collaborators Turner and Dayton encapsulated this research in their award-winning article, "Do Curricula Correspond to Managerial Expectations? Core Competencies for Technical Communicators" (2005, 323), in which they identify three sets of competencies that technical communicators should possess:

- *Primary competencies*: the abilities to collaborate, write clearly, assess and learn technology, take initiative, and evaluate their own and others' work
- *Secondary competencies*: the abilities to use technology to accomplish work in various media and to "write, edit, and test various technical communication documents"

- *Tertiary competencies*: the abilities to conduct research and usability tests, single-source, design instructional materials, budget, present, and respond to cultural differences

This research, derived from academic curricula as well as managerial surveys and interviews, also led to identification of four key skills that technical communicators should possess: excellent communication skills, collaborative skills, interviewing skills, and an affinity for technologies.

Other researchers have conducted workplace research with similar results. For example, Whiteside (2003) examined programmatic curricula and interviewed technical communication undergraduates and managers to identify key job-entry skills. While she identified many of the same competencies as Rainey, Turner, and Dayton, her research participants included business and organizational knowledge in their lists of competencies, and they recommended that soon-to-be graduates also learn project management, computer languages, and software tools (311). Managers valued collaborative skills, as well as time management, in their new employees; they also recommended that students take advantage of internship opportunities and learn to work with subject-matter experts (311). Both graduates and managers in Whiteside's study recommended that students gain knowledge of a technical area to adapt more easily to a workplace setting. Kim and Tolley's (2004) study resulted in comparable findings; the five graduates surveyed in their study recommended that students gain competency in problem-solving, analytical, and rhetorical skills (381). They also recommended that students engage in multiple internships and "familiarize [themselves] with industry needs" (385). Included in these needs is industry-specific knowledge: "Domain knowledge or experience may be especially important when a student wants to specialize in a particular area, such as healthcare, scientific research, business, or technology" (382). These skills and competencies, Kim and Tolley's participants said, would serve the technical communicator well not only when on the job but also when looking for one.

In all of these studies, technology skills and abilities were mentioned, but the need for specific tool or software instruction was inconclusive. While some graduates recommended that students learn specific software tools, managers were less emphatic. One manager in Rainey, Dayton, and Turner's (2005, 333) study noted: "Basically, if I can find a superb writer who understands technology and works well with others, I am willing to provide training for all the other skills I need the person to have." Like this manager, Kim and Tolley's (2004, 382) research led to a similar conclusion: "Although the presence or lack of computer skills was not the only factor in our graduates' job search processes, these days students need to master some level of technology as an expected standard literacy. However, the specific tools they need to know may vary depending on the particular job or area of technical communications." These findings suggest that knowing specific technologies is less important than knowing how to learn technology.

The benefit of these studies is that they not only inform STC members creating the TCBOK, but they also can help current students in technical communication programs identify what skills and competencies they need to be successful in their workplaces. In other words, they work in two directions: both supporting the discipline's development as a profession and identifying key competencies and skills that newcomers to the discipline should possess. It is important to note, however, that the TCBOK is still in an early stage, and STC has yet to achieve buy-in from national and international technical communication organizations, such as the Association of Teachers of Technical Writing, the IEEE Professional Communication Society, and the Council for Programs in Technical and Scientific Communication. While representative members of these organizations have been active in TCBOK development, interorganizational buy-in has stagnated, possibly because many academics who belong to these organizations have yet to be persuaded that professionalization and the standardization of programs and practices are beneficial to them and their students. Like Allen (1990), they argue that their programs are specialized by local needs and competencies; therefore, standardizing curriculum and practices may detrimentally affect their programs' abilities to meet these needs. What Savage wrote in *Power and Legitimacy* (2003, 6) remains true today: "Because of the field's lack of consensus about our knowledge and professional identity, as academic programs proliferate, so do iterations of what constitutes the knowledge of the field." To achieve buy-in for the TCBOK and any other professionalization steps that follow, STC will need to persuade members of these diverse organizations that its professionalization interests are also their own.

WHERE DO WE GO NEXT?

If these early steps are eventually successful, scholars speculate that the next step toward professionalization is the development of certification or licensure standards that determine who may or may not become a member (Savage 1999; Hayhoe 2005), but much work has yet to be done before this can happen. Professions, historically and culturally, connect closely with a specific, identified audience, develop a sense of social responsibility toward this audience and others, and have an ethical awareness that guides and directs their professional identity (Faber 2002, 316). As an example, think of the profession of medical doctors: doctors are recognized because they work with specific individuals (sick people, for instance), because they have a social responsibility to heal, and because they have an ethical system that guides them as they do this work. While technical communication is currently taking many of the early steps toward professionalization, it is far from developing a cultural or historical identity, such as those we recognize for doctors, lawyers, and even engineers. Obviously, technical communicators are far from this type of recognizable professional status, nor can we assume that all technical communicators or even their national organizations agree on what it means for technical communication to be a profession. For these reasons, this latter step is still on the distant horizon, but it is possible that, within your career, this final step toward professionalization will occur.

HOW DO THESE DEVELOPMENTS AFFECT TECHNICAL COMMUNICATION STUDENTS?

Like STC and its members, technical communication students are positioned to strengthen and support our discipline as it matures. In a sense, your professional development directly affects the discipline's development as a profession. As your knowledge, skills, and competencies grow, you will find yourself better situated to respond to the challenges our discipline faces as it matures. To support this maturing process, you should focus your work and your professional development in three areas: by learning everything you can about technical communication, by engaging with others who do work similar to yours, and by leading if you are called upon to do so. The heuristic for the professional development life cycle outlines preliminary steps you can take as a student as well as those you can take as a graduate and a practitioner to continue your personal professional development as well as contribute to that of the discipline's.

HEURISTIC FOR THE PROFESSIONAL DEVELOPMENT LIFE CYCLE

Technical communicators trained in academic programs have at least three stages in their career development. At first, these stages are chronological: you develop your knowledge and gain basic skills or competencies as a student, you apply what you know as a job seeker, and you maintain or expand your basic skills as an employed practitioner. You'll find, however, that you may repeat these stages one or more times as you gain experience or change your goals.

In fact, you may find that you cycle through these stages multiple times, as you change positions or accept promotions. In many cases, you'll need

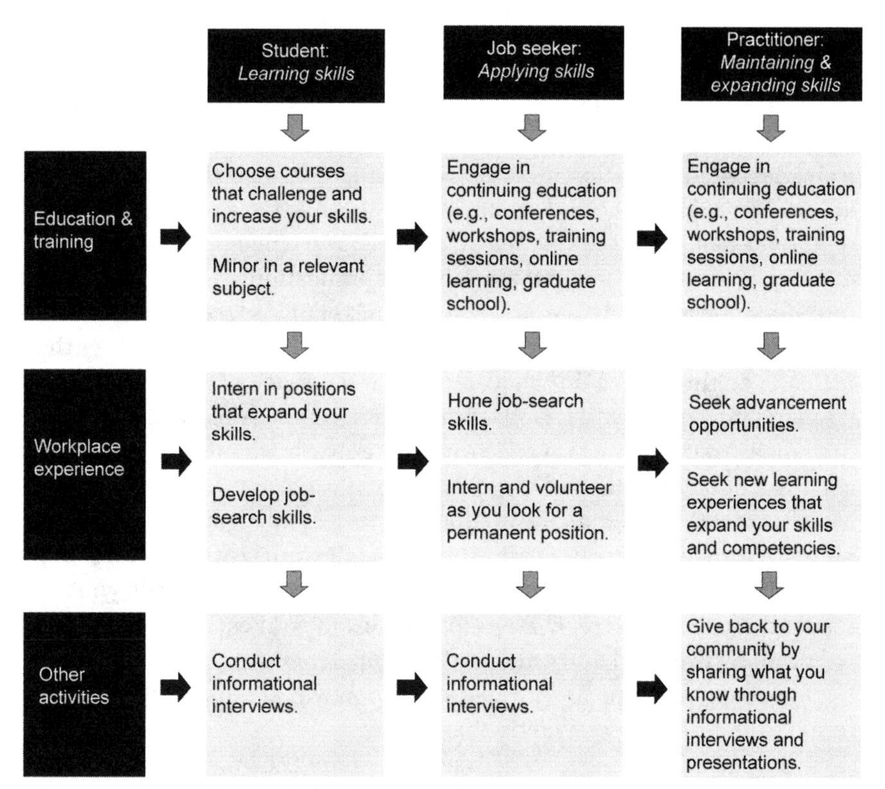

Figure 4.1. The professional development life cycle: stages and activities

to retool with each change. The U.S. Department of Labor, according to Jablonski (2005), reported in 2002 that the average length of time employees spent with the same employer is 3.7 years. "Since 1983, when the U.S. Department of Labor first began keeping such statistics, the average tenure has ranged from 3.4 to 3.8 years" (17). If these averages remain stable, you can expect to change jobs as many as ten times during your career. This section is designed to help you at whatever stage you find yourself. As illustrated in figure 4.1, the professional development life cycle is divided into three stages—student, job seeker, and practitioner—and each stage includes three sets of professional development strategies: education and training, workplace experience, and other activities.

STUDENT PROFESSIONAL DEVELOPMENT STRATEGIES

Professional development strategies you apply as a student are keys to your future goals, whether you are enrolled in undergraduate or graduate studies. First, you should carefully select courses that challenge your abilities and increase your skills. Look for courses that expand your knowledge of technical communication practice or that require you to think about developing or designing information in new ways. As you consider courses to take, remember the core competencies and skills outlined in recent technical communication literature, and choose courses that develop your knowledge of audience and other rhetorical skills, require you to conduct research (including usability tests), allow you to collaborate with others, and strengthen your oral and written communication skills. Courses that introduce you to business practices can build your understanding of what it's like to work and communicate outside the classroom, and courses that focus on culture—both in the workplace and outside of it—help you understand how to work in a global market. Finally, be sure to take a variety of courses that require you to learn different software tools—word processing, content management, graphic, and design applications—and remember that the specific software you learn is less important than learning how to learn new software. Software changes so quickly that you'll always be learning to use something new. Establish these skills while you're a student, and you will apply them for the rest of your career. By matching course objectives and outcomes to the competencies and skills necessary for success as a technical communicator, you will graduate with a strong foundation for work you will do thereafter.

Technical communication course work is not the only part of this picture for students. You should also consider a minor in a relevant subject, depending on the kind of industry or workplace you find interesting. If your program of study does not require you to take a minor or you do not have time for one, consider electives from this list of possibilities: business, marketing, instructional technology, computer science, foreign languages, graphic design, communication, journalism, public relations, corporate communications, the natural sciences (e.g., biology, chemistry, geology, zoology), natural resources, civil engineering, political science, nutrition and food sciences. Courses in preprofessional programs, such as premed and prelaw, can provide you with an excellent introduction to the discourse of these fields. Taking courses in a relevant minor provides you with knowledge of the language that specialists use in these fields, which will make writing with and for them easier if you take a job in one of these fields.

Even as a student, gaining workplace experience is necessary if you want to compete for the best entry-level jobs after graduation. Internships, whether paid or unpaid, give you the opportunity to learn from practitioners and allow you to take the basic skills you learn in your courses and apply them in specific workplace settings. As an intern, you will likely find that your coworkers understand that you are a novice, and they are generally very willing to help you learn. This kind of advantage can greatly increase your understanding of workplace practices while expanding your professional network. Many programs have internship officers or faculty members designated to help students find technical communication internships. Ask your instructor about the opportunities available in your program. Whether your internship is paid or voluntary, it is invaluable experience because it will provide you with many opportunities to put your knowledge into action, and it will give you an understanding of technical communication practice that is hard to replicate in the classroom.

Internships and other workplace experiences also allow you to put your job-search skills into action. You will probably learn how to write résumés and cover letters in your technical communication courses, and you may even create a professional portfolio to use when you interview, but this work is not fully tested until you use it to look for your first job as a technical communicator. Internship interviews can provide you with opportunities to test and hone your skills. Applying for internships, you will quickly learn what works and what does not, and this knowledge will be invaluable to you when you enter the next state of your career as a job seeker.

Finally, as a student you can engage in several other activities that develop your skills for future application. These activities are key to extending your professional network. They include conducting informational interviews, joining professional organizations, and using professional social-networking sites.

Informational interviews are a great way to learn from individuals doing jobs that you desire after graduation. An informational interview is different from a job interview in several ways:

- You are asking the questions, not answering them.
- Your interview's purpose is not to get a job but to learn about technical communication jobs.
- Your secondary aim is to get to know another person who does technical communication work that interests you and, thereby, establish a personal connection that will extend your professional network.

Another opportunity for extending your professional network is to join a professional organization. As you know already, STC is the largest professional organization for technical communicators, and it has many student and local chapters. You may find student chapters of other professional organizations on your campus, especially if you have minored or taken electives in another subject. Any of these organizations can help you connect with others with interests like yours, and these connections may provide you with many opportunities for employment before and after graduation. According to a 2003 survey, almost 62% of interviews for managerial positions were obtained through networking (Bloch 2003, 12). Given these numbers, it is easy to see how important developing your personal network can be.

Finally, you might be surprised to learn that social-networking sites on the Internet are also great ways to connect with others professionally. Professional-networking sites like LinkedIn, Ecademy, and Xing are all dedicated to helping individuals keep track of colleagues and connections. Of course, social-networking sites like Facebook and Twitter are also available to you, but use any Internet site (or archive) with caution. These more popular sites are more often used for disseminating personal information, and your connections may learn more about you than you'd like. In fact, anything you post on the Internet (including the automatic archives generated by many e-mail discussion lists) can be located using search engines. For this reason, you should always be thoughtful and cautious when you post on the Internet. Searching the Internet for an applicant's name is often one of the first steps employers take when deciding whom to interview. The wrong kind of personal information, too much personal information, or indiscreet images can diminish your chances for getting the job of your dreams after graduation.

JOB-SEEKER PROFESSIONAL DEVELOPMENT STRATEGIES

Many of the skills you learn as a student come directly into play when you graduate and start your first job search. Depending on the economic conditions when you graduate, the time between graduation and employment could be negligible or it could last for months. In either case, to find the best jobs, you'll want to apply skills from both your course work and internships.

Even though you have graduated, you may still find that you need to add to your skill set as you apply for jobs. In these situations, continuing your education through self-study and professional workshops is a great way to get the additional knowledge you need. For example, you may find that the jobs available to you require technology skills that you do not have. Fortunately, most software packages have inexpensive or free demonstration copies that you can download to use for a limited time, and the creators often provide tutorials and other documentation. Take advantage of spare time during your job search to learn new technologies, so you can honestly say that you have worked with them when you have interviews.

You might also discover that your professional organizations offer workshops on job searching, technical communication skills, or technologies that will strengthen your knowledge and competencies. Attend your local professional-organization meetings to take advantage of such workshops or investigate the possibility of online training. Many online courses are affordable and reputable. Like workshops, they are often a quick and easy way to gain or sharpen your marketable skills. Finally, when job markets are particularly tight, graduate school enrollments tend to increase. If work is not an option, then the time may be perfect to earn a graduate degree. Consider advanced degrees in technical communication or in related areas that will enhance your marketability upon graduation.

In the meantime, you can add to your workplace experience through additional internships or volunteer opportunities. The career services department at your university or college can often point you to possibilities for employment—both paid and unpaid—after graduation. You can also find many opportunities at nonprofit organizations that are always in need of communication specialists. If you have time, consider volunteering for such organizations to gain additional experience and to improve your community through service. Whether you write a proposal, create marketing brochures, or simply design a form, you will be gaining experience that can help you in your later job search, and you will be doing good for your community at the same time. Additionally, continue the professional development activities you began as a student. Conduct more informational interviews, extend your personal network, and maintain your memberships in professional organizations. These activities are even more important now that you find yourself on the market. Opportunities can materialize through these connections that would not be possible working on your own.

PRACTITIONER PROFESSIONAL DEVELOPMENT STRATEGIES

Once you are employed as a technical communicator, your professional development strategies may change slightly, but they do not end. You should continue to think about education and training, workplace, and other professional development activities to enhance your work and advance your position within the organization. If you have not already done so, then now is the time to continue your education by attending conferences, workshops, and training sessions to extend and sharpen your tool set. You can learn new techniques and technologies easily through these interactions, and if your schedule allows, consider graduate school. You will find like-minded professionals in graduate school, and you will deepen your understanding of technical communication practices by studying the theories behind them and their applications across different industries and organizations.

Doing your job well and continuing your education through opportunities like these will also provide you with new challenges and opportunities to advance your career. These opportunities may become available to you within your current organization or when you move to a new one. According to recent career theorists, one of the defining characteristics of twenty-first-century workers is their mobility: "They often voluntarily move to new companies, new careers, and new geographic regions" (Jablonski 2005, 16). Regardless of whether you find growth opportunities in your organization or outside of it, seek out new experiences that expand your skills and competencies, and be willing to share what you know with others.

One of the best methods of sharing this information is through other professional development activities, which can be found in your community and professional organizations. As you develop your competencies and skills, you will likely find that you are among those individuals asked to provide training for others or that you are asked to lead organizational activities. Doing so places you in a position opposite to the one you had as a student: you are now the mentor, not the mentee. Acting as a mentor, training junior employees or interns, or serving as a leader in a community or professional organization is rewarding and allows you to give back to the community that first supported you. In exchange you will find that you continue to learn as you interact, engage, and lead others in their professional development. By talking to students who request informational interviews, you will learn what is being taught in undergraduate classrooms and be able to provide practical experience to sharpen their understanding of technical communication workplaces. Through training others, you learn to articulate clearly what you know, and often you find that you learn as much through these interactions as you teach. Through leading others, you shape the future of your organization, providing more and better support for those who follow you. Cycling through these stages, you will find many opportunities to grow, learn, and share your knowledge with others.

THE PROFESSIONAL DEVELOPMENT LIFE CYCLE IN ACTION

The three technical communicators you met earlier in this chapter—Stephanie Wilson, Ben Minson, and Emily Cook—have all fully engaged in the professional development life cycle. In this section, they will tell you in their own words how they have put the recommendations and strategies in this heuristic into practice and will provide you with examples for how you might do the same.

STEPHANIE WILSON

My experiences most clearly illustrate the professional development process that students can engage: preparing for a career in technical communication by taking well-chosen courses, completing internships, learning successful job-search skills, completing informational interviews that prepared me for my first job interviews, and engaging with professionals by joining STC.

When I initially approached graduation and the prospect of finding a job, I was nervous. I wondered if I was ready to publish documentation for end users and if I could carry myself with confidence as I met with subject-matter experts. As I agonized over these and other questions, I had an epiphany: I'd already performed many professional duties as an undergraduate, and I possessed the skills necessary to fulfill future job functions. I prepared myself throughout my undergraduate experience by taking relevant course work and consulting with my professors. I took classes in rhetorical theory, document design, writing technology, and the publication process, and I translated my theoretical experience into professional experience by seeking internships and conducting informational interviews with industry professionals. I conducted an informational interview with one individual who just started her first professional job. I asked her questions about the hiring and interviewing process, which put many of my worries in perspective as I approached my own job search.

My first internships accurately prepared me to apply my classroom experiences to the workplace. From January 2007 to my graduation in May 2008, I worked as a tutor in the USU Writing Center and also as the sole technical writer for Utah's Local Technical Assistance Program (LTAP). These consecutive jobs gave me a two-sided perspective: as a tutor, teaching the writing and reading theory that I'd learned in class, and as a writer, trying to employ that theory to communicate complex ideas in clear, accessible language. I began to understand the writing process on a much deeper, more practical level, and my writing, editing, and communication skills improved.

Closer to graduation, I found additional internships, and after working as a public relations assistant, a computer lab consultant, a marketing intern, and a web designer/content manager, I began to trust my new abilities. My internships fostered my sense of mastery as a writer, developed my understanding of the way people think, and introduced me to the conventions of professional workplaces. I began to understand that it was not necessarily a particular skill set that would make me successful in the workplace, but rather the confidence and acuity to use my skills, and the aptitude to learn new skills to apply to future situations.

As I contemplated the number of skills I could develop, I began to consider the various career paths available to me. I joined a few professional organizations, including STC and the Public Relations Society of America, on the recommendations of my professors, and I started participating in their communities, through both online groups and in-person interviews. I expanded my understanding of topics that were interesting to me, such as web content strategy and user experience, by subscribing to professional blogs and participating in local technical communication communities like the USU Content Managers group and the Intermountain Chapter of STC.

As the president of our student chapter of STC, I invited a group of local technical communicators to speak to our students. They discussed their work, including their own career paths, their interns and student employees, and the role that academic theory played in their industry experiences. As I spoke with them, I collected suggestions ranging from how to format a skills-based résumé to how to work with an ineffective team member. Their depth and breadth of industry experience answered many of the questions generated by my lack of practice. With my education nearly complete, my internships at a close, and the advice of industry professionals in mind, I began applying for positions that matched my areas of interest. I have worked in my current position for approximately eighteen months now, and I am applying to graduate programs in technical communication. I am looking forward to this new learning experience and the opportunities it will soon provide me.

BEN MINSON

Like Stephanie, my professional development began as a student in a technical communication program. I also served as an officer in the student STC chapter to which I belonged. My internship helped place me in my first full-time position. As a full-time practitioner, I've continued to serve as an active member of STC; to further engage with other professionals, I also write a blog.

In particular, two courses from my technical writing undergraduate program started me on the professional development path: a class on publishing documents in print format and another on interactive media. In the first, I completed a project in which I learned the basics of Adobe FrameMaker and, at the same time, wrote a beginner's guide to the program. This guide was basic, beginner-level, task-based software documentation, and I believe having this document in my portfolio played a key part in my getting my current job. In the second class, I worked with another student to plan an interactive educational experience, focusing on the audience and on meeting the needs of a client.

As a student, I looked for ways to complement course work (and of course, to add to my résumé and portfolio). Internships were a large part of this effort. I accepted a web administrator internship offered at the USU Writing Center, where I was working at the time as a tutor. In this position, I designed and built a website that presented information about the center to university faculty, staff, and students. This gave me experience creating a plan and design for a project, obtaining approval from a client, and then executing the plan. I also became a technical editing intern, working with engineering students to polish their project documents. This gave me experience collaborating and giving feedback to others. As vice president of the student chapter of STC, I further developed skills in working as a team to make and execute plans. I also gained software training experience by providing workshops for STC members in the English department's computer lab. Being an undergraduate student can seem like a scramble to get as many bits of experience as possible and rack up points on some scoreboard. But due to the generally varied work of technical communicators, I found direct application for this range of professional development experiences I had while I was a student.

Now, as a practitioner, I continue to look for ways to develop my skills. One of the most important ways to do this is to connect with other professionals and learn from them. Social media provide ways to do this quickly. For example, I have a professional blog where I write mostly about technical communication and related topics. My blog demonstrates my interest in my field and the things I learn about it; it also gives me an opportunity to practice my writing and analytical skills and sometimes instructional writing skills. In return, readers comment on my posts and give me additional ideas and perspectives. Interestingly, one of my current colleagues was hired largely because of the expertise and enthusiasm his technical writing blog demonstrated. I read other technical writers' blogs in my RSS aggregator. I also have a Twitter account and follow other technical writers. Some provide useful links to articles and blog posts about professional communication that I likely would not find myself—I can surf the web only so much.

The technical writing and training team I belong to in my job holds regular reviews of our projects and designs. We give each other feedback and talk about best practices. This helps us arrive at a unified presentation of documentation and training beyond merely following a style guide, and it expands our view of what is possible and effective. To connect with other professionals, I continue my membership in STC, where I associate with other technical communicators and learn about what is happening in the field. I have volunteered in situations such as helping the Instructional Design and Learning special-interest group with its website design and serving as webmaster and then president of the Intermountain Chapter. The skills I use to guide chapter activities to completion are comparable to the skills I need for managing documentation projects. I have attended STC annual conferences, which help me see what is current in the field and how other professionals are improving their day-to-day work. Professional development is crucial to keeping my skills and deliverables at the level they need to be for me to be an essential asset to my organization, as well as to be desirable in a competitive field. I can learn some things by working alone, but learning from others takes me farther faster.

EMILY COOK

Like Stephanie and Ben, I first prepared myself professionally by majoring in technical communication, but I extended this knowledge with a minor. While this course work and my internships were essential in getting my first full-time position, I decided within a few years to advance my knowledge with a master's degree, which I earned online while working full-time. In addition to earning my graduate degree, I have also continued my professional development through participation in workshops and other workplace training opportunities.

As an undergraduate, I majored in technical communication, but I also took a minor in marketing. I found that my business course work (and even fellow students) were completely different than in my major. Minoring in business flexed me out of my comfort zone, taking classes in a differently structured and competitive environment. My minor also allowed me to see that I really love business communication. I had to think critically and be able to deduce information a little differently than when I was reading a book for a literature course or writing a paper for a technical writing deadline. This course work was instrumental in landing my first internship and job. It was in one of my business courses that the instructor called me aside after a test and asked if I was interested in doing technical writing in a business environment. Wow, what an opportunity! These courses definitely set me on a path to finding my niche in business communications.

As an undergraduate I also gained invaluable experience as an intern. I worked as a technical writing intern at a small start-up in Logan, Utah, called Ingeo Systems Inc. When I began, the company had approximately fifteen employees. The small size enabled me to work with everyone, ask questions, and develop relationships. At Ingeo, I learned to have a thick skin. I was viewed as "young" or "the newbie," which was frustrating and somewhat empowering at the same time. I worked closely with the full-time technical writer to review drafts; to write FAQs, white papers, and press releases; and to interview subject-matter experts. In the beginning, the diversity of everyday tasks is what really excited me about the field of technical writing as a whole.

After working for a few years, I enrolled in USU's online master's program and began my graduate studies. In both my undergraduate and graduate courses, the work that tested me the most was the kind of class that left me feeling accomplished and motivated to move on to the next big thing. My master's course work was very diverse, but I loved the online group environment. I felt very much in control and accountable for my work, but I also had the flexibility to do it on my own time and hold down a full-time job. I found that I enjoyed classes that would have intimidated me in the past, including usability, designing an online course, and rhetoric. Having completed two degrees, I can honestly say that all of my course work translated in some way into my career as a professional communicator. My business is extremely dynamic, and most of my experiences as a student were as well, which lent itself nicely to some of the experiences I have had in the professional workforce.

As I mentioned earlier, over the years, I've held many positions as a technical communicator. All of these positions revolved around technical writing, but each was geared toward a different industry. These positions were heavily oriented toward documentation coordination and editing. All of them were usually driven by strict deadlines and expectations. To get these jobs, I used techniques I learned in my course work, and I drew upon my personal network to find opportunities. I am glad I had the opportunity to be flexible, learn, and develop the skills to do each of these jobs. As a practitioner, I continue to keep in touch with past colleagues and peers through e-mail, Listservs, blogs, book clubs, and networking sites (e.g., LinkedIn). Attending conferences and workshops allows me to see that learning should never stop. I am social by nature, and I have found that this type of environment gives me a place to network and learn, all while meeting new people and learning the latest trends. I believe that as technical communicators, we can get so busy with our daily tasks that we forget how fulfilling it is to have an opportunity to talk to and be with other professionals within our industry.

In addition to completing my master's degree, I dream of working on my PhD someday. Until then, I have taken an active interest in learning new software when work warranted it. As an example, I have learned Photoshop and InDesign through online help and self-instruction while working full-time. I would like to get more involved with professional organizations again. I just renewed my membership in the Association of Proposal Management Professionals, and always feel I benefit from the knowledge base and camaraderie that such professional organizations have to offer. I find that my professional organization memberships tend to ebb and flow, depending on my workload and the demands of work. When I am not able to commit time to my professional organizations, I focus on other forms of professional development. For example, I recently enjoyed speaking to undergraduates at USU about my experiences. All of these professional activities stretch my creative muscle a bit, requiring me to answer questions and think about how I can improve my skills. When I think about my professional development, I see that I often get so busy that I forget to take time for myself, including my professional advancement. I would love to have a more structured way to incorporate professional development into my everyday life, but, in my case, the deadline-driven work has to come first.

YOUR PROFESSIONAL LIFE CYCLE

All three of these practitioners began their careers just as you are beginning yours. They studied hard, applied what they learned, and engaged with and led others as they matured as practitioners. As you begin your career, remember that at every stage of the professional development life cycle, you should focus on learning, engaging, and leading.

Technical communication is a relatively young but thriving discipline. Its growth as a recognized profession depends, in many ways, on you, the next generation of technical communicators. Your focus and engagement in the professional development life cycle not only will enhance your professional development, but will also enhance the strength, identity, and viability of technical communication as a career.

Although the path ahead for you and for our discipline is not fully visible, technical communication scholars are clear in this message: the future is in our hands. To become professionals, we must engage in the work necessary to develop our discipline of technical communication as a profession, and we should do so with our eyes open to the political, social, and economic challenges and changes this development will require. The development of its identity and status is linked directly to your successes and those of your fellow students. To begin this process, start your professional development today. This chapter, and the rest of this book, shows you how.

DISCUSSION QUESTIONS

1. Find a recent copy of your résumé. Does it demonstrate the professional development steps you've taken? If so, where? If not, what can you add, and where would you add this information?
2. In addition to the medical profession, what other professions do you recognize? Do all these professions meet the characteristics described in this chapter?
3. What professional organizations have student chapters on your campus? What are the requirements to join? What kinds of activities do they sponsor?
4. Does your campus have a career center? What services do they provide to students and graduates of your school?
5. Visit the STC TCBOK (http://stcbok.editme.com/) and analyze its contents. What are the key competencies you find there?
6. What internship opportunities are available on your campus? How do you find out about them? How do you apply for them?
7. Think about recent graduates of your technical communication program. Where are they employed? What courses and minors did they take with their degrees? Were these courses and minors beneficial to them? Discuss how you might connect with these graduates to build your own personal network.
8. As a class, conduct an Internet search for technical writing blogs. Choose an article that interests you and discuss its contents with your classmates. How might the information you gathered from the article help you develop professionally?

WORKS CITED

Allen, Jo. 1990. "The Case against Defining Technical Communication." *Journal of Business and Technical Communication* 4:68–77.

Bloch, Janel. 2003. "Online Job Searching: Clicking Your Way to Employment." *Intercom*, October, 11–14.

Campbell, Alexa. 2008. "Teaching Technical Writing: Teaching Professionalism in the Classroom." *Intercom*, January, 38–40.

Faber, Brenton. 2002. "Professional Identities: What Is Professional about Professional Communication?" *Journal of Business and Technical Communication* 16, no. 3: 306–337.

Hayhoe, George. 2005. "The Future of Technical Communication." *Technical Communication* 52, no. 3: 265–266.

Jablonski, Jeffrey. 2005. "Seeing Technical Communication from a Career Perspective: The Implications of Career Theory for Technical Communication Theory, Practice, and Curriculum Design." *Journal of Business and Technical Communication* 19, no. 1: 5–41.

Jones, Dan. 1996. *Defining Technical Communication*. Arlington, VA: Society for Technical Communication.

Kim, Loel, and Christie Tolley. 2004. "Fitting Academic Programs to Workplace

Marketability: Career Paths of Five Technical Communicators. *Technical Communication* 51, no. 3: 376–386.
Pringle, Kathy, and Sean Williams. 2005. "The Future Is the Past: Has Technical Communication Arrived as a Profession?" *Technical Communication* 52, no. 3: 361–370.
Rainey, Kenneth T. 2005. "Approaches to Professionalism—A Codified Body of Knowledge." In *2005 IEEE International Professional Communication Conference Proceedings*, ed. G. Hayhoe, 679–688. Limerick, Ireland: IEEE.
Rainey, Kenneth T., Roy K. Turner, and David Dayton. 2005. "Do Curricula Correspond to Managerial Expectations? Core Competencies for Technical Communicators." *Technical Communication* 52, no. 3: 323–352.
Savage, Gerald J. 1999. "The Process and Prospects for Professionalizing Technical Communication." *Journal of Technical Writing and Communication* 29, no. 4: 335–381.
———. 2003. "Introduction: Toward Professional Status in Technical Communication." In *Power and Legitimacy in Technical Communication*, vol. 1, *The Historical and Contemporary Struggle for Professional Status*, ed. T. Kynell-Hunt and G. J. Savage, 1–12. Amityville, NY: Baywood.
Society for Technical Communication. 1998. "STC's Ethical Principles for Technical Communicators." http://www.stc.org/about/ethical-principles-for-technical-communicators.asp. Accessed January 30, 2010.
———. 2008. "STC's Strategic Plan: Adopted by the Board of Directors May 2008." http://www.stc.org/about/strategicPlan.asp. Accessed January 30, 2010.
———. 2009–2010. "For the Press." http://stc.org/about/press01.asp. Accessed January 30, 2010.
Whiteside, Aimee. 2003. "The Skills That Technical Communicators Need: An Investigation of Technical Communication Graduates, Managers, and Curricula." *Journal of Technical Writing and Communication* 33, no. 4: 303–318.

PART 2 SITUATING THE FIELD

While part 1 asks relatively specific questions about the nature of technical communication, part 2 steps back slightly to consider a broader set of problems (figure P2.1). In this portion of the book, you will start thinking about ways to situate the field in the context of larger social concerns, including theoretical and historical concerns. Learning how to situate technical communication as one field among many is crucial because no profession exists in isolation from the rest of the world: professions and professionals themselves serve larger constituencies, draw on and contribute to projects in other fields, and generally try to belong in productive ways to both local and global communities. The emergence of networked communication devices has opened up new opportunities for work not

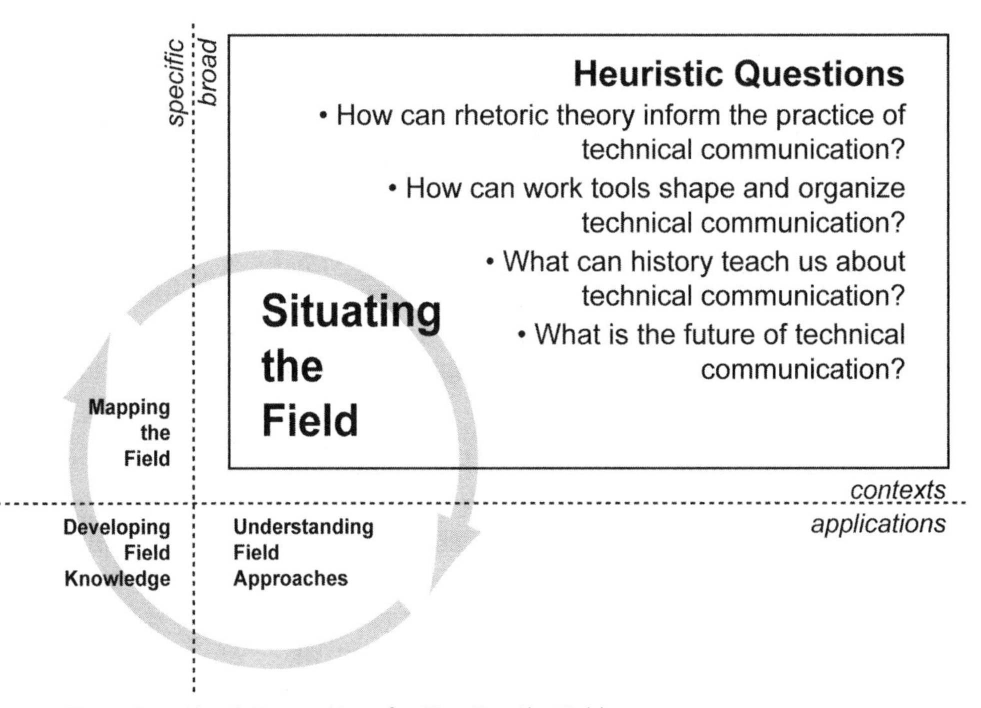

Figure P2.1. Heuristic questions for Situating the Field

only in interface design, but in social networking, microfinance in developing countries, and political activism, to name just a few new areas. Taking advantage of these opportunities involves adopting a broader, more complex perspective on how technical communication is situated as one field among many that address user activities in technological contexts.

Part 2 begins with a discussion of certain key aspects of rhetoric, one of the oldest and still most powerful sets of theories about everyday communication exchanges. James Porter's "How Can Rhetoric Theory Inform the Practice of Technical Communication?" will help you understand and respond to a range of complex problems in workplace situations. As Porter notes, rhetoric is often seen as an ivory-tower issue that concerns only academics. In reality, however, rhetoric theory can provide you with approaches to new communication situations and can help you figure out how to act and work strategically. Porter focuses his efforts on an expanded view of writing and on a key and complex element in all technical communication projects: rhetorical purpose.

Jason Swarts addresses the flip side of the theory/practice binary in "How Can Work Tools Shape and Organize Technical Communication?" Although Swarts focuses on a concrete tool for doing technical communication (cascading style sheets used to structure, among other things, advanced websites), he simultaneously demonstrates the importance of thinking about both concrete and conceptual matters, understanding how the use of a specific tool can affect and be affected by larger issues. His chapter emphasizes the need to always view technology as a nonneutral element, as something that both contributes to and reflects social values, helping to shape but not determining the contexts for technical communication.

In "What Can History Teach Us about Technical Communication?," Bernadette Longo and Kenny Fountain move back and forth between broad and local histories. Beginning with a discussion of several key publications about the history of technical communication as a field, Longo and Fountain construct a set of heuristics that can be used to generate local histories of technical communication within specific organizations. Connecting the larger historical contexts of technical communication with the actual practices adopted at your own organization can provide you with powerful ways to understand the forces affecting how you work, including ways to challenge yourself and change things to make yourself a more effective technical communicator.

Finally, Brad Mehlenbacher closes this part of the book by providing responses to the question, "What Is the Future of Technical Communication?" Mehlenbacher suggests that we answer this question by touching base with our theories, histories, and practices before looking forward: where we have been and where we are now place us on trajectories pointed in specific directions. Technical communicators, as Mehlenbacher points out, will need to deal with increasingly complex communication problems, including users who rely on multitasking as a way of life, increasing amounts of information to filter and sort, and workers and users whose job functions are extraordinarily complex. Although we often think that new technologies can solve these problems all on their own, the heuristics in this chapter demonstrate that emerging technical communication problems are best understood as *communication* problems.

JAMES E. PORTER

5 How Can Rhetoric Theory Inform the Practice of Technical Communication?

SUMMARY

This chapter examines the role of rhetoric theory in technical communication. What does rhetoric theory *do*? How does it help technical communicators do their work, what value does it add, what is its usefulness? While theory is often viewed as an academic pursuit without practical value in the real world of work, this chapter argues that theory is an indispensable tool for the technical communicator. However, the usefulness of rhetoric theory depends on *how it is applied*. This chapter defines *theory*; then considers the role of rhetoric theory in technical communication; and finally presents one example of a rhetoric theory (highlighting different views of writing), showing how that theory can be usefully applied to the work of technical communicators.

INTRODUCTION

Max is a technical communicator working on a development team for a software company developing applications for web-based social media. His job is to write and test online user documentation. So he wears two functional hats: he serves as both documentation writer and usability specialist.[1]

Max is having trouble with a piece of documentation he has written and is currently testing, aimed at helping users register and then set their preferences for a new application. Using an iterative usability testing model, he has tested and fixed the documentation several times—but each time he tests it, the new users continue to have problems. He dutifully fixes the errors in the documentation, takes the revised documentation to the next set of users, and then encounters more problems (new ones). Through each round of testing and document revision, he adds more language and diagrams and provides more thorough explanations to help users understand the various functions and features. Nothing seems to work. Max gets frustrated, and decides that users "just don't get it."

We could characterize Max's problem here as the problem of instrumentalism; that is, he relies on a linear theory, a transmission model of communication, that views documents as message transmitters: documents are supposed to transmit a reality—in this case, a procedural "reality," an ideal narrative about how new users should set up their accounts. The users must first comprehend that reality and then enact the steps to successfully complete a task. When the transmission fails to work, the answer, Max feels, lies in adding more language or more clarification of existing language—"eliminating noise" from the viewpoint of the Shannon-Weaver model.[2] So Max focuses on the document, on getting the language just right and making sure the language accurately and completely reflects the reality of the procedure. That still doesn't work. This model or theory of communication is so deeply embedded in Max's thinking that he cannot think himself out of this box.[3]

Would theory help solve Max's problem? Well, he already has a theory, but the problem is that he doesn't see it; it is, for Max, simply reality, "the way things are," rather than one framework of many for explaining communication. What Max needs is an explicit self-consciousness of theory (his own), an ability to critique theories (to understand their strengths and limitations), and some understanding of alternative theories and of how to apply them to solve problems. He needs to do some *theorizing*, in other words, to examine alternative models of communication, multiple methods for usability, and, surely, to develop a more robust notion of audience. For example, if he were to explore alternative methods of usability, he might become aware of the document-centeredness of his approach. He should think about the document less and the audience more. Methodologically, he might scrap the document altogether and try a different approach focused more on understanding users (e.g., contextual inquiry, starting out by learning more about the users themselves and what help they need to do their work). Such an approach might help product development: maybe the issue is the design of the interface or the nature of the procedures? Rather than focusing so intently on the document as the answer to this problem, Max needs to think more broadly about the entire rhetorical context: about the relationship between the technology, the interface, the user help, and real users' needs.

In other words, Max needs a better theory. Now, the word *theory* carries immense negative baggage outside the academic world: it means abstruse speculation, pie-in-the-sky philosophizing, impractical political critique, jargon-ridden academic blather. It is abstract, idealistic, speculative—the antithesis of what the pragmatic, scientific, empirically minded, precise technical communicator-in-the-workplace wants to be doing. Theorizing is not for real people working on practical issues in the dog-eat-dog world of business, engineering, and IT development. Imagine Rodin's statue *The Thinker*: theorists sit and ponder. They waste time. They don't perform useful or productive action in the world.

In technical communication, theory is often portrayed as counter to practice, and "the theory-practice debate" is one that the field frequently engages in (see Sullivan and Porter 1993, 221). The debate pops up repeatedly in technical communication publications. (See discussion question 3 for some examples.) The theory versus practice question has been an ongoing, perhaps even defining, tension in the field. What value does theory, particularly rhetoric theory, add to technical communication? That has long been a serious question for technical communication.

In this chapter, I consider how one type of theory, rhetoric theory, significantly informs the practice of technical communication—so you can see where I stand in this debate. My discussion examines how theory works to build knowledge and to solve problems in technical communication. This chapter will not be summarizing or covering all the rhetoric theories that you need to know, but rather my purpose is

- to provide a brief literature review explaining what theory is and discussing how rhetoric theory can be useful for technical communication;
- to offer one sample heuristic that highlights different views of writing; and
- to provide an extended example of how this heuristic might work in a particular technical communication context.

WHAT IS THEORY?

Theory is often understood as abstract philosophical musing. Jacques Derrida, Michel Foucault, Judith Butler, Donna Haraway—these are all well-known academic theorists whose writing is abstract and speculative, barely comprehensible to most, unnecessarily verbose it seems, without much apparent reference to actions, events, or phenomena in the world.

But what exactly is *theory*? Conversationally, a theory is simply a speculative idea that someone has: "I have a theory about why that garbage is on the kitchen floor . . . aha, the dog!" This sense of theory aligns with the notion of theory as a research question or hypothesis to be tested: "My theory about this blog interface is that the instructions for creating new pages are unclear: users will have problems creating new pages in their blog space." *Theory* can mean a yet-to-be-confirmed question, hypothesis, or possibility.

This meaning is, ironically, the opposite of another definition of theory: the scientific notion of theory as a well-established, even proven axiom that is "purported to be universal" (Culler 1997, 7). Karl Popper (1963) discusses these two senses of the word: *theory*, first, as a scientifically true proposition, true in the sense of verifiable through inductive observation (e.g., Einstein's theory of gravitation), versus a less empirically verifiable notion of *theory* as an overall system for understanding human behavior (e.g., Marxist theory or Freudian theory). In *A Brief History of Time* (1996, 10), Stephen Hawking further describes the scientific notion of theory: "A theory is a good theory if it satisfies two requirements. It must accurately describe a large class of observations on the basis of a model that contains only a few arbitrary elements, and it must make definite predictions about the results of future observations." *Theory* in this sense means a principle that has universal or nearly universal explanatory and predictive power; it operates as a truth.

But there is yet a third sense of theory—more a rhetorical and humanistic notion of theory as a conceptual framework, a map, or a lens that provides us a way of looking at the world. We all have these conceptual maps in our heads, and they are useful insofar as they help us navigate our lives and solve problems. However, these conceptual maps can also become ruts—as the Max scenario illustrates. These conceptual maps can blind us if *one way* of seeing the world becomes *the only way* of seeing the world. Hence the usefulness of theorizing as an activity: theorizing, or reflecting critically about the strengths and weaknesses of various theories, can expand the way we think by challenging our existing frameworks and giving us new ways of seeing.

This metaphor of theory as "framework" recalls Kenneth Burke's (1966) notion of a "terministic screen," which he describes in photographic terms as a kind of "color filter" that influences one's perception (45–56). To exemplify, he cites three different descriptions of infant behavior, one by a behavioral psychologist, one by a social psychologist, and one by a medieval theologian (St. Augustine). All three analysts are observing and describing the same infant behaviors or "instinctual responses" (crying, smiling, sucking, clinging, following), yet because each analyst has a different methodological frame, each derives different kinds of knowledge about infant behavior. The behaviorist focuses on causation: how certain stimuli produced predictable effects in infants. The social psychologist sees the baby's behavior fundamentally from a social frame of reference, focusing on the behaviors as guided by the relationship between mother and child. St. Augustine sees the infant behavior as fundamentally expressive of a relationship between child and God (48–49). According to Burke, "All three terminologies [terministic screens] . . . direct the attention differently, and thus lead to a correspondingly different quality of observations" (49).

You can see how different theoretical frameworks also influence perceptions of the Internet. Some see the Internet primarily as a *space*, as a communication medium or a publishing space akin to a library. Others see the Internet as a *place*, a community or home or culture, an environment in which people live and move and interact. That metaphoric distinction—a theory distinction—matters in terms of issues such as privacy, research ethics, and intellectual property. If we view the Internet as a space for storing published work, like a library or a vast database, then we might suppose we have the right to treat whatever is posted there as publicly available for research purposes. If, however, we see the Internet as a place where people interact socially—more like a street café (which is public in one sense, but where people could be having a private conversation at a table)—then we might be inclined toward a different ethic, viewing some material on the Internet as "private" even if it is possible to access it and download it.

Think of theory, then, as a kind color filter that you use, that you must use, in order to comprehend and analyze human behaviors, social events, or texts. The ability to self-reflect and to articulate your particular theoretical filter—to identify it, to critique it, to understand its strengths and limitations, and to imagine alternatives—is the activity of theorizing. Don't underestimate the value of theory and of theorizing to your work: theory itself is essential to technical communicators, and theoretical self-awareness, an ability to theorize, is a considerable strength.

As Jonathan Culler (1997, 14) describes theory, it has several distinctive characteristics:

1. Theory is interdisciplinary—discourse with effects outside an original discipline.
2. Theory is analytical and speculative—an attempt to work out what is involved in what we call sex or language or writing or meaning or the subject.
3. Theory is a critique of common sense, of concepts taken as natural.
4. Theory is reflexive, thinking about thinking, inquiry into the categories we use in making sense of things, in literature and in other discursive practices.

In Culler's third point, theory as the critique of common sense, theory almost always makes our lives more difficult by complicating things. When Burke (1966, 44–45) notices that there are two different views of language use—scientistic and dramatistic—he is complicating our notions of language, literacy, and reading. He is making two things out of what we thought was one thing. Is this a violation of Occam's razor, a philosophical principle that insists that we should keep things as clear and as simple as possible? Well, sometimes simplicity causes problems rather than solves them. Complicate only when doing so serves the purpose of providing a better tool for analysis and problem solving. The test of any theory is its usefulness and productivity, whether the complexity helps us in some way: increases our understanding of a situation, helps us solve a problem, or allows us to generate new knowledge.

What does theory *do*? It attempts to arrive at knowledge or understanding by questioning, critiquing, and problematizing something that we normally think or do—"the critique of common sense," in Culler's terminology. It uncovers something hidden or at least underappreciated, calls attention to it, and says, This is important to thinking about this topic or question. It provides a distinctive point of view, a lens or filter, that helps us understand people, actions, texts, events. Theory should have heuristic and explanatory power: that is, it should help you to *do* things, it should help you to *see* things in a different way, it should enable you to *produce* things . . . findings, conclusions, recommendations that have a real effect—an action—in the world.

You can't do very much useful work—basically none—as a technical communicator without theory. If you are running a usability test, you need a theory of usability method to guide how you design your study, select subjects, and manage your testing procedures. If you are working collaboratively on a team to design a content management system (CMS) for your company, you need a theory of collaboration and teamwork that guides how you work with others, and you need a theory about the design of interfaces and systems that guides your development of the CMS product. It also helps to understand the power and usefulness of countertheories and multiple theories in solving problems. *Theory* in this sense is not abstract and useless, but rather it is the fundamental and necessary principles guiding your work as a technical communicator.

HOW DOES THEORY WORK IN TECHNICAL COMMUNICATION?

Let us take a detailed look at one particular theoretical statement in the field of technical communication, a well-known and highly regarded piece of theory: Patricia Sullivan's article "Beyond a Narrow Conception of Usability," published in 1989 in the journal *IEEE Transactions on Professional Communication*.[4]

In this article, Sullivan is not reporting on a usability study she herself conducted. Rather, she is speculating more broadly on the overall methodology of usability. She is doing what is called meta-analysis—that is, analyzing others' usability studies according to a particular framework for organizing and evaluating those studies. It is also important to note that underlying Sullivan's theory of usability is a considerable amount of practical experiential knowledge about usability; Sullivan has consulted with corporate usability departments, and she has conducted her own usability studies—although the primary purpose of her 1989 article is not to share or describe her own research, and that research experience is not explicitly evident in the article itself. She has also read a lot of usability research—and that extensive reading is evident in the article.

Sullivan notices that different disciplinary groups use different approaches, or filters, for conducting usability research: "Psychologists and engineers in human-computer interaction typically use experiments and case studies to study the usability of interfaces and systems; sociologists and anthropologists use ethnography and field methods when they study the computing of organizations; marketers typically use interviews and surveys to study consumer preferences; document designers, educational psychologists, and writers in technical communication use various exploratory and text-based methods to study the usability of educational materials" (257–258).

Sullivan starts by observing that different disciplines favor different methods. She next proposes that researchers do not restrict themselves to methods from their own disciplines, but rather look outside their disciplines to methods that are suited to the research questions and contexts they are examining. Table 5.1 shows available usability methods taken from Sullivan's 1989 article. This table is a heuristic, that is, a generative framework listing the various methods used in usability research and evaluating those methods in terms of the quality of information they provide about various usability metrics, such as accuracy and satisfaction.

This table is a piece of theory—a tabular presentation of a verbal narrative that aligns usability methods (listed vertically in the left-hand column) with usability metrics (listed horizontally across the top of the table). With this table, Sullivan is bringing rhetoric theory to usability studies by emphasizing the importance of context to usability research design. This table presents in digested format a good deal of information, but it also makes an argument based on Sullivan's practical experience. The table presents theoretical and abstract information, but it also references particular usability studies that are discussed elsewhere in the article (the numbers in brackets represent references in Sullivan's article). The purpose of this theory is to encourage usability researchers to think beyond

Table 5.1. *Various evaluation methods' strengths for yielding information about documentation quality*

Method	Placement (of text on page)	Accuracy (of information)	Level of explanation (whether explains what users need)	Style (appropri-ateness)	Satisfaction (user response)
Direct questioning					
Surveys			may		good
Interviews		may	may		good
Comprehension tests [21]		may	good		
Observation					
Informal observation [22, 23]	good	may	good		may
Lab observation [24, 25]	good	good	may		
User protocols [26]	good	may	good		may
Reading protocols [26]			good	good	may
Keystroke records		good			
More traditional evaluation					
Computer text analysis				good	
Editorial review [28]	may	may	may	good	
Technical review [29]		good			

Source: Sullivan 1989, 260.
Note: Numbers in brackets (Methods column) represent references in the original publication.

their disciplinary preferences and employ a broader repertoire of methods useful to solving various usability problems.

Sullivan's analysis is also a critique of how usability studies, in 1989, had drifted into a kind of monomethodological mind-set, seeing the work of usability as consisting of one primary type of method: testing finished documentation, in a lab setting, at the end of the documentation design cycle. (This is of course Max's approach to usability.) According to Johnson, Salvo, and Zoetewey (2007, 320), "Sullivan was among the first to acknowledge that usability testing is weakened when confined to validating all-but-finished documentation" and that "end-of-development usability fails to capture important user input that can be of great value earlier in the design cycle." Sullivan is suggesting that usability has drifted into defining itself too much in terms of methods and not enough in terms of inquiry purposes: that is, focusing too much on the *what* (the method), not enough on the *why* (why are we conducting a usability study?). So she is critiquing the field, but not in a harsh sense. Rather, she is circumspectly pointing out a limitation in the field's approach and recommending a different approach, visually represented in the heuristic table she provides.

Would this kind of usability theory ever show up in a corporate usability report? It is unlikely that extensive theorizing would show up—that's just not tolerated in corporate reports (too much abstraction and verbiage). But I can imagine a report including a citation to Sullivan's work, perhaps even a brief summary or quotation. I can imagine usability specialists using Sullivan's heuristic table to guide their decisions about research design for a planned usability study, or to teach usability methods to new hires, or to argue to their clients for the validity of a certain methodological approach to a usability study. If your clients are program developers who think that lab tests are the only valid method for usability research, then you will need to make the argument that lab tests are good for answering some research questions, but not so good for others.

In their 2007 article, Johnson, Salvo, and Zoetewey provide a retrospective on Sullivan's 1989 article. From a vantage point twenty years later, they see the same problems that Sullivan identified persisting in usability research. They provide yet another theoretical lens, theorizing the problem as being that usability occupies an uneasy space between two cultures—usability as science versus usability as art (Johnson, Salvo, and Zoetewey 2007, 323). The art part is messier, as it involves the complex rhetorical variable of users, audiences, people—and, as we all know, the variability of audiences poses quite a challenge for rhetoric. To win recognition and respect in the workplace, usability needs to be scientific, to an extent even positivistic—that is the scientific paradigm that persuades engineers and computer scientists. But to be valid as a methodology, and to be true to its primary mission, usability needs to function also as an art. This binary tension—a theoretical framework—helps explain many of the challenges that usability specialists face. Usability itself, as a field, has to address two different kinds of audiences (at least) and, often, bridge the gap between those audiences (Johnson, Salvo, and Zoetewey 2007, 328).

Such theoretical discussion—Sullivan's (1989) original piece and Johnson, Salvo, and Zoetewey's (2007) response to it—is abstract to be sure, but it has practical implications for usability education and workplace practice. Taking "a broad view" toward usability means that you shouldn't get comfortable with one method but that you need to develop a broad repertoire of usability methods and understand how different methods apply to different research questions, different products, different audiences. A "one size fits all" approach limits choices and problem-solving potential. Even though Sullivan's article is now over thirty years old, it still does good theory work, is still relevant to usability studies.

Sometimes researchers build a new theory rather than work with existing theories, particularly when existing theories are not adequate to the task at hand. In a research article published in *Technical Communication Quarterly* in 2006, Huatong Sun does precisely this.[5] On one level, Sun's article is a simple research report. She conducted a comparative case study of users using mobile messaging technology in two different locations: Albany, New York, in the United States, and the Hangzhou region of Zhejiang province, China. The purpose of the study was to determine, first, if users in different cultures would employ text messaging in different ways—and, if so, what would be the implications of this finding for design of IT products, particularly mobile technology interfaces.

Before collecting her data, however, Sun recognized that she had to develop a framework for understanding the scene for her users' actions; she needed a robust definition of culture that would take into account both the sense of "dominant cultural values in a national culture" and the sense of "complexities of local contexts" and "subgroup culture (e.g., age group, gender, and organizational affiliation)" (460). In other words, Sun recognized the need to develop a complex notion of culture that would avoid gross generalities and that would, rather, take into account the many ways and levels on which culture operates.

A significant portion of Sun's article (459–466) is allocated to building a theory of cultural usability that would serve as the lens or perspective through which she observed her users' behaviors. She built her theory by synthesizing and critiquing others' notions of culture. Her "new framework integrates key concepts and methods from activity theory, genre theory, and British cultural studies" (461). What resulted from this analysis was a table and an accompanying diagram (Sun 2006, 465) that constitutes a model for cultural usability, a model that Sun then used as the framework guiding data collection and analysis for her study.

This model allows Sun to "see" her data in a richer, more complex way than would be possible without it. She sees, for instance, that her users live in a variety of cultures, not only in a broad national culture (China, the United States) but also in local cultures of work and recreation and social/personal networks of friends and family. One key finding from the analysis is that technology designers need to develop "a deep understanding of concrete use activities in local contexts while considering cultural and structuring factors" (477). In other words, a monolithic and macrounderstanding of culture is not sufficient to explain how users engage and interact with technology. The older notion of culture as an isolated and distinctly different national identity (e.g., "Chinese culture") is not a viable model for understanding the design and use of digital technologies.

These two examples—Sullivan, Sun—also show us the importance of seeing theory as a *conversation*. Rather than seeing theory as the isolated and autonomous statement of a great thinker, see it as a conversation among reflective thinking professionals who are collaboratively and intertextually working together, discussing ideas, sometimes across time and space to develop useful frameworks and systems. That is why it is often difficult to read *one* piece of theory with full comprehension, because to understand a theory usually requires significant awareness of prior context: it helps to be familiar with the previous conversation that has led up to that one piece of theory.

These uses of theory serve a heuristic purpose and are thus a form of rhetorical invention. In classical Roman rhetoric, *inventio* was one of the five key canons of the art of rhetoric (along with arrangement, style, memory, and delivery). To promote invention, classical rhetoricians developed *heuristics*—that is, procedures for prompting and developing ideas, arguments, and content for discourse (see Lauer 2004, 8–9). Heuristics are open-ended questions, prompts, categories, memory devices, and/or visual grids—such as Sullivan's table on usability, Sun's framework for understanding culture, and Stuart Selber's (2004, 25) "conceptual landscape" table that identifies multiple types of computer literacy. Such heuristics aid thinking, discovery, deliberation, research, and design.

A HEURISTIC FOR TECHNICAL COMMUNICATION

Let us examine one very simple piece of rhetoric theory—about views of writing—and its accompanying heuristic to see how viewing theory as a framework or perspective can work heuristically to help technical communicators.

Writing is a deceptively simple term. We all know what it means, we use the term all the time. But its simplicity and ubiquity hide its complexity. Like the terms *culture* and *audience*, the word has a wide variety of meanings. It helps to view writing heuristically from four different perspectives:

- Writing as *product* or *document*: This is the traditional formalist view that sees writing as "words on the page," that views the text as a product, as a formal artifact that you produce following certain genre conventions (a memo, a report, a website, an online user help forum).
- Writing as *self*: This perspective sees a piece of writing as an expres-

sion or extension of an author/writer—either the individual, the team, or the collective group, and often all of the above—what we sometimes call ethos.

- Writing as *process*: This is a dynamic view of writing as the process of composing, as a process that can include inquiry and research as well as collaboration with others. It also includes downstream reception and interaction, the ways in which readers and users interact with a document and, perhaps, provide feedback to developers.
- Writing as *action* conducted for the benefit of some *audience*: This is the performative perspective that focuses on what writing *does*, on the effects the writing has on its intended audience.

Technical communication has traditionally been defined primarily by the product or document perspective, that is, with the emphasis being on the genres and types of documents that technical writers produce. From that perspective, technical communicators have been, primarily, producers of instructional manuals, then computer documentation, then online user help, and finally designers of websites, content management systems, and user help forums. Thanks in part to the influence of the field of rhetoric/composition and to empirical research on workplace practices (aka "workplace research"), the field of technical communication has moved toward thinking more about process, action, and reception. Usability research is of course predominantly focused on questions of reception and interaction: How effectively will documents work with their intended audiences?

Technical communication has never focused all that much on the expressive function. But the field has always acknowledged that all writing creates an ethos or identity, an image of the author. For most organizational writing, the author identity is usually not an individual writer but a collective identity. This identity is often constructed through graphic means—for example, the logo, the website design, the product branding. Companies are flocking now toward an identity of being "green," because that helps develop an ethos of civic responsibility. *Ethos* in this sense is corporate image.

But even internally, all writing has an ethos: for instance, a usability department might need to develop its reputation as "methodologically scientific" in order to convince programmers and designers to heed the results of usability testing—and usability researchers must fashion their usability reports in order to convey their allegiance to quantitative

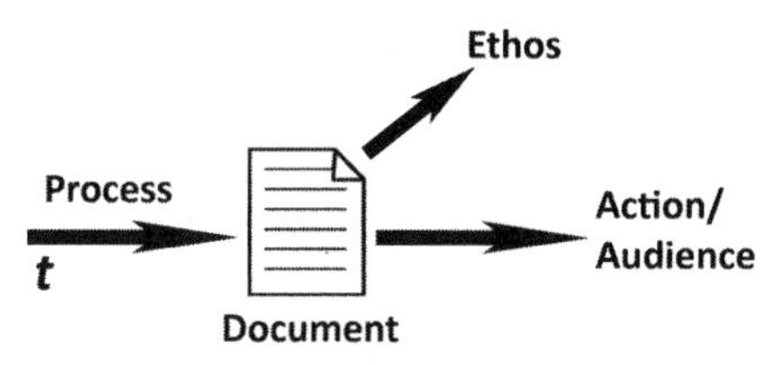

Figure 5.1. Four views of writing (t = time)

measures of validity. If I am a new employee writing memos and reports to my supervisor, then I will take a great deal of care with my first writing on the job, because those writings will play a large part in constructing my new identity in the organization: Am I careful and precise in my language use? Am I attentive to detail? Are my conclusions and recommendations appropriately backed by the data I have collected (or am I prone to exaggeration and overstatement, carelessly imprecise phrasings)? Readers will make judgments about *me*, and about the quality of my thinking and my character, based on the writing I produce. The ethos of a piece of writing—the character that writing displays—says something about the writer or writing team or corporation that produced it.

To simplify this heuristic, let us abbreviate it as DEPAA. (*Depa* is a Sanskrit word meaning "lamp," indeed an appropriate acronym for a heuristic.) See figure 5.1 for a diagram of the heuristic.

How might this simple heuristic apply to the work of the technical communicator? If we see writing primarily from the formalist framework as a text, a document, or a product, that perspective can sometimes be powerful, but it can also lead us to take a limited view of our role in the communication process. Let's say that we are given the assignment to construct a content management system as a way for an organization to produce and distribute information internally in the organization. That's an assignment that asks us to take a product view: produce a document—in this case an online system for document management. If we already think that way, we can happily go off and create a product that we think will do the job. This approach dovetails neatly with the genre-focused way that academic assignments are often pitched: Write an essay. Create a website. Produce a report. Conduct a usability test. These are all assignments that focus on product and genre and hence take a *document* perspective.

If we apply a slightly different theoretical lens to the task, however, we might begin by asking, What is our goal here? What is this system supposed to accomplish? What is the desired outcome for the content management system? Or backing up a bit more: What is the problem we're trying to solve with this content management system? These are *action* or results-oriented questions. What is this system supposed to accomplish? Why—and, very importantly, for whom?

The why and for whom questions are of course the most basic questions of rhetoric theory: Why are we writing (or communicating)? For what audience? And what effect or outcome are we aiming for? Why—or So what? if we want to put a sardonic spin on it—is a key question that should guide all rhetorical events, all writing projects, all document development.

The process of thinking here parallels the kind of thinking applied by Hart-Davidson et al. (2008) in their article "Coming to Content Management." They describe and reflect on two projects involving two clients—the pseudonymous NPO (national professional organization) and Michigan State University Libraries. In their approach to these projects, the team used rhetorical theory heuristically, first to understand the assigned task from a different framework and then to design an approach to the task that would allow them to solve the problem for the client. "Solving the problem for the client" in this circumstance meant, in a sense, questioning or critiquing the assignment as given, challenging the frame in which it was viewed by the client—in a nice way, to be sure—and then constructing a different frame that would be more productive, more helpful to the client in the long run.

When clients come to you and say, "Build us a website," that is a product orientation that requires some critical exploration. You need to back the client up a bit, and ask: Why do you want a website? What purpose are you trying to accomplish, and for whom? Hart-Davidson et al. (2008) apply this shift in perspective to a very practical problem for their client: How do we design our website to provide useful information to our clients and customers? They shift from a typical static, product-oriented notion of *content* to a dynamic notion of *content development*: as they phrase it, "Your website is not something you have, but something you do for, or even with, your clients/members/customers" (12). This shift in perspective allows Hart-Davidson et al. to redefine the location of their client's problem: the locus of the problem is not the writing as document (the website), but rather the writing as process . . . the who and how part (who produces the writing and how).

This shift in the theoretical framework has practical implications for the methodology of solving the problem. Instead of focusing on the content and design of the client's current website, Hart-Davidson et al. ended up studying "the roles and workflows of content developers" (25). In short, they studied the process, not the product, with the aim of developing a content management system that would maximize the efficiency of content development as well as the usefulness of the content for end users. Yes, some of the recommendations for their clients are product based, but many pertain to work flow and communication practices (see pp. 27–29, for example). In short, the solution to the client's problem lies in the relationship between process and product, not simply in a product or document. Here is an example of where a theory about writing—specifically, about writing as composing process—helps solve a real-world client problem.

An ongoing problem in technical communication is the danger of drifting away from audience into "the systems-based approach" that focuses too much attention on the design of products, documents, and systems and forgets the human interaction in systems—that is, the presence and contribution of the audience or user. In his work promoting "user-centered technology," Robert Johnson (1998) most famously reminded us of this problem in regards to the design of computer documentation. George Pullman and Baotong Gu (2008) also caution us about the dangers of systems-based thinking (a kind of product thinking) in regards to the design of content management systems: "Another factor contributing to the difficulty of CMS implementation is that most content management systems take a systems-based approach toward managing content/information/knowledge at the cost of considerations for content and user needs. . . . The very expression content management excludes any idea of writing or communicating and focuses on information independently of the people who produce or consume it" (2). The heuristic perspectives of *process* and *action* can help us think beyond the more static product orientation.

EXTENDED EXAMPLE

So how might this heuristic help Max, our technical communicator with the usability problem? If Max uses the heuristic critically and imaginatively, it should help him get out of the thinking box that is blocking his ability to solve the problem. When Sullivan and Porter (1997) observed Max's behaviors during his usability test sessions, they noticed that he would collect data about his users, but then always turn his attention to the document—adding more clarifying language, fiddling with organization, changing terminology, adding or deleting images, and so forth. In other words, he was focused intently, almost exclusively, on fixing the document, seeing the document as both the source of the usability problem and the means to its solution. His focus was on how accurately and completely the document represented the system that users were trying to learn. So we might also say that he was system-centered as well as document-centered.

Rhetoric theory offers several paths toward solving this problem. First and foremost, a rhetoric-theory perspective insists, *Look at your audience!* But Max *was* looking at his audience, in a way, and that alone wasn't helping him see his task differently. He was looking, but was he really paying attention? What if he turned his perspective away from the document and paid more attention to process and to action, as well as to audience, reframing the writing task in a completely different way? What if he saw the assignment not as "producing a piece of user documentation" but as helping the user understand and learn to use a new system? A dramatic shift in perspective might help Max see that the usability problem maybe isn't "in" the document at all, that maybe the answer lies in a hybrid solution—for example, multiple pieces of documentation; user help forum; instructional video; direct classroom training. This is the kind of perspectival shift that Hart-Davidson et al. (2008) and Pullman and Gu (2008) are talking about in their research articles: instead of focusing on the document and the system, focus more on the process, on the work flow that users are engaged in. What can follow from that shift in perspective is a shift in defining the locus of the problem or need. In Max's case, the need lies in the users—and what is necessary to meeting users' needs is a deeper understanding of those needs in their context of work and use. That perspectival shift should result in Max shifting his focus and his methodology, his entire approach to the task. Instead of fiddling so much with the document, he should fiddle more with the users: he might interview them about their work practices (e.g., contextual inquiry interviews), he might observe their work practices, he might try to differentiate among different types of users. Once he understands his users better, and their work practices and work flows, then he can begin to design a solution to address the problem he is tackling. That solution might or might not involve a document, but it is likely to involve multiple approaches to solving the problem.

CONCLUSION

To be effective, usable, and useful, rhetoric theory must prompt and push deeper thinking; it must be powerful in the respect that it enables you to *see* in a new way—to see events, texts, processes, positions, and people in a way that deepens your understanding and leads to more productive action. Rhetoric heuristics also need to be simple, memorable, and portable—transferable across a wide variety of situations and contexts.

In this chapter, I have offered one simple heuristic based on rhetoric theory that I think can be useful across a wide range of technical communication contexts, ranging from designing websites, to developing content management systems, to setting up virtual worlds for corporate training, to producing environmental impact statements, to creating user help systems, to designing interfaces, to writing business reports. This heuristic can be used as a tool for production of a new product or for critique and evaluation of an existing product or to design a research study. It can be used to diagnose problems or, more strategically and proactively, as a way of thinking about communication products and processes to avoid problems in the first place. Certainly Max needed a heuristic and some theorizing in order to resee his writing situation. Ultimately, the index of a successful theory or heuristic is that it does useful, valuable work in the world. "There is nothing more practical than a good theory" (Lewin 1952).

What I have argued in this chapter is that rhetoric theory is indispensable to technical communication, as it provides the conceptual framework necessary for all work in the field. It guides how technical communicators think about documents and design; it guides their view of the writing process; it influences their approach to research and methodology; it shapes their interactions with others, the audiences they are writing for, the cowriters they are writing with, the clients whom they are representing. Technical communicators who don't acknowledge the significance of rhetoric theory to their practice—or who underestimate its power or who fail to use it productively to deploy its power—are likely caught in a theoretical framework that they can't see and that is therefore likely to limit their ability to adapt to changing circumstances in their work.

DISCUSSION QUESTIONS

1. What different views of writing have you encountered in your academic education so far? Have different instructors provided differing perspectives—or have they conveyed one consistent view? Ask the same question about your own experiences working as a technical communicator or doing client-based projects. Be sure to describe particular moments, events, or cases that provide evidence of differing views of writing and communication.
2. Research the Shannon-Weaver transmission model of communication. What is the model and where does it come from? How has it been critiqued? Have you ever encountered someone who held to this view of writing and communication? In what ways was that model useful or helpful to this person? (It is, after all, a very powerful model, useful in many ways.) In what ways was it not useful? Do you find the model helpful?
3. Examine and report on one of the many instances of a theory-practice debate in the field of technical communication. (See the list below for

some possibilities—or locate another instance of a theory-practice argument.) Be sure to understand the context for the debate. What was the starting point for the debate (if one is identifiable)? What are the issues involved and the main points of contention? Who takes what positions in the debate and why? Evaluate the arguments that are being advanced. What are the claims of various sides? What evidence do advocates provide for their positions?

- The "Cruel Pies" debate in *Technical Communication* over the ethics of visual representation (Dragga and Voss 2001a, 2001b, 2003; Hayhoe 2002; Moore 2008; also see the letters to the editor regarding "Cruel Pies" in *Technical Communication*, vol. 48, no. 4, 2001, and vol. 49, no. 1, 2002).
- The debate in the *Journal of Technical Writing and Communication* over the meaning and relevance of Aristotelian ethics to technical communication (Katz 1992, 1993, 2006; Moore 2004).

4. Using the context provided in question 3, write a short position statement indicating where you stand on this particular argument—and why. Develop your own response contributing to the conversation, imagining that your response will be published as the next installment in the debate. See Warnick 2006 for a model demonstrating how such an analysis might work.
5. Use the DEPAA heuristic as a tool for analyzing a technical communication project that you have worked on. Think about not just the document itself, but the entire project that led to the creation of the document (or documents). Begin by describing a complex technical communication project you have worked on—one that involves working for a client, producing multiple documents over a relatively long period of time (weeks or months), and some collaboration with others. First, identify the stages of the project, and its distinct deliverables. Analyze your experience using the DEPAA model as heuristic. How does applying the heuristic generate different views of the project—from its starting point to its final outcome? How does the DEPAA heuristic provide helpful prompts for thinking about the project and understanding its distinct phases, documents, and, possibly, problems? Are there important aspects of the project that the DEPAA heuristic doesn't bring up or emphasize?
6. Interview a working technical communicator, asking about a problem, issue, or challenge he or she has encountered while producing a report, developing a website, testing the usability of documentation, designing a new interface, and so forth. Listen carefully to the story. Take

detailed notes—perhaps even audiorecord the interview. Make sure to ask questions about the purpose of the document and the process by which it was created. Ask the technical communicator to describe the problem (or problems) involved in the task, and how those problems were addressed or solved. How do differing views of writing create issues or miscommunication? Do the differing views enable technical communicators to resolve problems?

7. Read pages 14–29 of Stuart Selber's book *Multiliteracies for a Digital Age* (2004). In particular, study the chart he provides on page 25, a heuristic chart highlighting differing views toward technology. What characterizes each of these different theories of technology? In what particular ways do these differing views result in different approaches to, for example, teaching technical skills to novice users?
8. Using the context for question 7, indicate where you have seen examples of people holding these differing views of technology—instructors you have had in various courses, seminars, or workshops; team members you have collaborated with; or department members, colleagues, and supervisors you have worked with. Are Selber's terms useful for describing different views of technology—or would you make adjustments in those terms, maybe provide more or different terms? Do Selber's terms account for, match up with, the views of technology you hold or have yourself encountered in academic or professional experiences?

NOTES

1. This scenario of Max is based on an actual case study reported in Sullivan and Porter 1997.

2. The Shannon-Weaver model, a common model in technical communication, has been criticized for being oversimplistic, mechanistic, and unidirectional in its conception of the communication situation.

3. Max is caught in what Robert Johnson (1998) would call a system-centered approach to thinking about technology and its relationship to users.

4. The journal issue in which this article appeared won the 1990 NCTE award Best Collection in Technical and Scientific Communication.

5. This 2006 article is based on Sun's 2004 doctoral dissertation at Rensselaer Polytechnic Institute, which won the CCCC Award for Outstanding Dissertation in Technical Communication for that year.

WORKS CITED

Burke, Kenneth. 1966. *Language as symbolic action: Essays on life, literature, and method.* Berkeley: University of California Press.

Culler, Jonathan. 1997. What is theory? In *Literary theory: A very short introduction*, chapter 1. New York: Oxford University Press.

Dragga, Sam, and Dan Voss. 2001a. The authors respond. *Technical Communication* 48:380–381.
———. 2001b. Cruel pies: The inhumanity of technical illustrations. *Technical Communication* 48:265–274.
———. 2003. Hiding humanity: Verbal and visual ethics in accident reports. *Technical Communication* 50:61–82.
Hart-Davidson, William, Grace Bernhard, Michael McLeod, Martine Rife, and Jeffery T. Grabill. 2008. Coming to content management: Inventing infrastructure for organizational knowledge work. *Technical Communication Quarterly* 17:10–34.
Hawking, Stephen. 1996. *A brief history of time*. Updated and expanded edition. New York: Bantam Books.
Hayhoe, George F. 2002. Back in the classroom. *Technical Communication* 49:273–274.
Johnson, Robert R. 1998. *User-centered technology: A rhetorical theory for computers and other mundane artifacts*. Albany: SUNY Press.
Johnson, Robert R., Michael J. Salvo, and Meredith W. Zoetewey. 2007. User-centered technology in participatory culture: Two decades "Beyond a Narrow Conception of Usability Testing." *IEEE Transactions on Professional Communication* 50:320–332.
Katz, Steven B. 1992. The ethic of expediency: Classical rhetoric, technology, and the Holocaust. *College English* 54:255–275.
———. 1993. Aristotle's rhetoric, Hitler's program, and the ideological problem of praxis, power, and professional discourse as a social construction of knowledge. *Journal of Business and Technical Communication* 7:37–62.
———. 2006. A response to Patrick Moore's "Questioning the Motives of Technical Communication and Rhetoric: Steven Katz's 'Ethic of Expediency.'" *Journal of Technical Writing and Communication* 36:1–8.
Lauer, Janice M. 2004. *Invention in rhetoric and composition*. West Lafayette, IN: Parlor Press.
Lewin, Kurt. 1952. *Field theory in social science: Selected theoretical papers*. London: Tavistock.
Moore, Patrick. 2004. Questioning the motives of technical communication and rhetoric: Steven Katz's "Ethic of Expediency." *Journal of Technical Writing and Communication* 34:5–29.
———. 2008. Cruel theory? The struggle for prestige and its consequences in academic technical communication. *Journal of Technical Writing and Communication* 38:207–240.
Popper, Karl. 1963. Science as falsification. In *Conjectures and refutations*, 33–39. London: Routledge and Kegan Paul. http://www.stephenjaygould.org/ctrl/popper_falsification.html.
Pullman, George, and Baotong Gu. 2008. Rationalizing and rhetoricizing content management. *Technical Communication Quarterly* 17:1–9.
Selber, Stuart A. 2004. *Multiliteracies for a digital age*. Carbondale: Southern Illinois University Press.
Sullivan, Patricia A. 1989. Beyond a narrow conception of usability. *IEEE Transactions on Professional Communication* 32:256–264.
Sullivan, Patricia A., and James E. Porter. 1993. On theory, practice, and method: Toward a heuristic research methodology for professional writing. In *Writing in the workplace: New research perspectives*, ed. Rachel Spilka, 220–237. Carbondale: Southern Illinois University Press.

———. 1997. *Opening spaces: Writing technologies and critical research practices*. Greenwich, CT: Ablex.
Sun, Huatong. 2004. Expanding the scope of localization: A cultural usability perspective on mobile text messaging use in American and Chinese contexts. PhD dissertation, Rensselaer Polytechnic Institute.
———. 2006. The triumph of users: Achieving cultural usability goals with user localization. *Technical Communication Quarterly* 15:457–481.
Warnick, Quinn. 2006. Toward a more productive discussion about instrumental discourse. *Orange* 6. http://orange.eserver.org/issues/6-1/warnick.html.

JASON SWARTS

6 How Can Work Tools Shape and Organize Technical Communication?

SUMMARY

In this chapter, I discuss how tools mediate the work of technical communication, by shaping and organizing how writers conceive of and carry out their projects. Technical communication is a tool-oriented profession. Technical communicators not only write about tools but also use tools of their own. To write procedures for tool use, technical communicators need to understand the user's tasks and the functional capabilities of the tools used. Less frequently, however, do technical communicators consider how their own tools affect their work, by reinforcing a way of thinking about a task until it feels natural. Being aware of how tools shape and organize technical communication will help readers of this chapter become both more effective and critically conscious technical communicators.

INTRODUCTION

Nancy is a technical communicator at a midsize software company. She, like everyone at that company, is responsible for following a style guide. The guide reinforces standards for more than just aesthetic reasons. Information formatted according to approved style rules is more usable for a variety of internal and external audiences. When these audiences take the content they are given, they know that it can be readily adapted for or transformed into other kinds of texts that serve a variety of functions.

Nancy has been tasked with updating her company's style guide, a task occasioned by the company's decision to adopt a content management system (CMS) for use by all divisions responsible for producing various kinds of product documentation, such as user guides and training materials. This CMS relies on cascading style sheets (CSS), sets of instructions for communicating with document readers, Internet browsers, and printers about how documents should be styled when shown or printed (see figure 6.1).

The situation Nancy faces is that other divisions have already adjusted to using CSS and those changes are reflected in their local style guides.

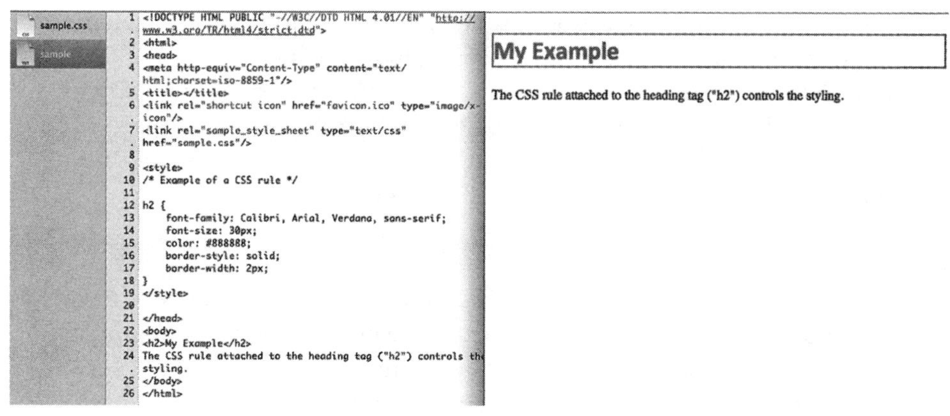

Figure 6.1. Sample CSS. Style information between <style></style> tags on left is rendered in the screen on the right.

Now that the company is making a full-scale change toward the adoption of a CSS-enabled CMS, Nancy must consider how the change will be reflected in the companywide style guide. To do so, she must consider how CSS will shape and organize the practices of technical communication in order to understand the role of the style guide.

The first step is for Nancy to learn about how CSS shapes practices of writing. That knowledge, in turn, will shape her understanding of the kind of shaping and organizing work that the style guide must do in response to the capabilities introduced by CSS. As different tools alter what writers are capable of doing in a text, the company's style rules must change to control the use of those expanded capabilities and to reinforce valued practices of writing. For example, one entry on bulleted lists, in an older version of the company style guide, says "Insert a 0.25″ space between the bullet and the start of the list item. Follow each list item with a semicolon." This rule, Nancy assumes, specifies the look of the list item because the writers at the time acquired tools capable of controlling the spacing to that degree. Changes in tools then prompted changes in the style guide. So in the current revision, the entry must now account for the capabilities of structured writing and CSS. It might now read, "For 'To Do' bulleted lists, use a checkmark instead of a round bullet. Follow each list item with a semicolon." It is also possible that the rule could disappear entirely, depending on how much control the CSS leaves to the writers. This issue and others like it are prompting Nancy's investigation of CSS and related technologies.

Before returning to this scenario, I will briefly discuss the importance of tools in technical communication and present a vocabulary for talking about the influence of those tools. From this literature, I develop a heuristic to assist readers in investigating tools, their histories, and their impacts on actions and interactions in a given context. I then demonstrate the application of this heuristic in order to show how those tools collectively shape the motives driving the style-guide revision and how they organize the approach that Nancy takes to the task.

LITERATURE REVIEW

To examine the shaping and organizing influence of tools, we need to be clear about the nature of that influence. A cause-and-effect structure, in which the introduction of a tool precedes and appears to affect technical communication, is too simple and more problematic than helpful. The problem is that a cause-and-effect structure isolates a tool as the source of a given effect. But technical communication is too complex and comprises so many actions and interactions that are historically, culturally, and technologically situated, that such a simple formulation is not possible. Understanding the influence of tools requires us to examine the larger context of tool use. We will first consider how technical communication is a tool-mediated field, then consider how tools actively shape the work of technical communication, and finally recast that activity as "mediation" and discuss ways to study it.

TECHNICAL COMMUNICATION IS A TOOL-MEDIATED FIELD

From its origins in land grant universities and engineering schools at the turn of the twentieth century (Connors 1982) and its development during the wartime and postwar eras of the twentieth century (Kynell 1999, 148), technical communication has been a tool-centered field. On one hand, technical communicators are users of tools. Steel-tipped pens, typewriters, carbon paper, and photocopiers have, without question, changed the practices of written communication (Yates 1993, 21–64).

On the other hand, tools are the topics of technical communication. At the heart of the field, some argue, is a responsibility to accommodate readers to tools within contexts of use (Dobrin 1983, 242). Yet technical communicators are more than technicians, skilled writers who objectively unveil the workings of a tool. They have a responsibility to understand that tools are not neutral, that tools are built to serve a purpose, an ethic that resides in the social context where they are used (Sullivan 1990). Tools have politics, which is to say that they guide people to interact with one another and with their environments in ways that serve particular social values (Winner 1989, 22). Sometimes these politics manifest as blatant value statements, such as when a city chooses a park-bench design that discourages the homeless from congregating in city centers. At other times, the statements are morally ambiguous, such as computer-aided design software that requires us to seek the assistance of someone skilled at reading the outputs.

TOOLS SHAPE ACTIONS AND INTERACTIONS

In sum, we can say that a tool shapes both the practice of technical communication and the social interactions that technical documents foster. It may appear strange to attribute such influence to tools, until we consider a broader definition of the term, one that shows both the plenitude of tools around us and the subtle ways that they alter our perception of the world.

Lev Vygotsky offers a broad definition of tools that reveals a connection to writing. To Vygotsky (1978, 7), tools are a broad category of objects that includes texts, signs, and written language, in addition to more traditional tools like keys and chisels. His work suggests that tools influence mental processes, by shaping the activities in which tools are employed. For example, a set of sticks helps children learn to count, but over time, those children learn to substitute words and signs to serve the same function. "The sign acts as an instrument of psychological activity in a manner analogous to the role of a tool in labor" (52). It has an "organizing" function (24) by which it shapes and extends thinking beyond what a person might otherwise be capable of without the tool (39). This definition gives us one way to understand the kind of shaping and organizing that tools do.

To say that a tool shapes an activity means that with a tool, one sees and approaches that activity differently than without. For example, when creating a website, one might approach the task differently with a simple text editor than with a graphical HTML editor. Although the result may be the same, a text editor draws the designer's attention to marked-up lines of text, while the HTML editor draws attention to the rendered design, which is a different object of work. The tools support different views that foreground one activity (visual design) over another (coding).

A more common way to describe how tools shape and organize activities is to say that they "mediate" those activities by imposing a structure on them (Hutchins 1997, 338). Hutchins offers the example of a checklist that mediates the act of shopping. The checklist presents shopping as a series of simple activities (i.e., items to purchase) combined with the conventional checklist form (e.g., checking off completed items). Items to purchase are the activities to be checked off, and in this way, the checklist mediates, shapes, and organizes shopping. The same list also potentially constrains the act of shopping by limiting purchases to only those items appearing on the list.

While tools may have the potential to mediate, people must still decide to use them. They must recognize that the tool could provide some useful mediation, and when people recognize a tool's potential, we say that they have perceived an affordance (Norman 2002, 1999). James J. Gibson (1986) originally used the term *affordance* to describe a tool's "action possibilities" or the potential to be used in service of a given task, in a given setting. For example, a maul may afford splitting wood, but only if the user is able to perceive that use and is motivated to take advantage of that affordance. A component of both Gibson's and Norman's definitions is that the ability to perceive affordances is socially situated, shaped by a desire to participate in an activity that is particular to a given community.

Tools shape our actions in ways that enable us to enact our affiliation with those communities. In other words, people perceive and take advantage of a tool's affordances within a social activity, often with the aim of participating in it. Insofar as a tool mediates the actions of an individual, it also mediates that individual's interactions with other people (Ihde 1991, 96, 102). Returning to our example of the HTML editor, before using the tool the technical communicator might have relied on colleagues to assist in the development of the website. Now those work relationships may not be necessary. Or the relationships may be altered, creating different kinds of professional dependencies or conflicts, given how information is presented and shared.

Tools shape not only the appearance of the information but also the ways in which others can use it. Users with different backgrounds and specializations use the same information for different purposes (Winsor 2001), enabling complex, distributed, multidisciplinary teams to work together. Even the material form that information takes can enable social interactions to take shape or to degenerate, as Sellen and Harper (2002, 110) discovered when observing the importance of paper slips to air traffic controllers as literal reminders of aircraft movements in and out of different airspaces. What begin as mediating interactions between people and their tools become mediating interactions between other texts and people the farther out we trace the effects.

STUDYING MEDIATION

Analyzing mediation is not easy because there are few direct causal relationships to track. What is needed, instead, is a broader scope of analysis that accounts for the mediating influence of tools on individuals as well as the ripple effects throughout a social setting.

One useful approach comes from activity theory, which advocates understanding the mediating influence of a tool in a larger historical and cultural context. The activity theoretic approach considers a community and its rules, structures, and divisions of labor (Cole and Engeström 1993, 8) when trying to understand mediation. How does a person formulate an objective and recognize the value of a tool for meeting that objective? The process both influences and is influenced by community norms, rules, and relationships. Engeström's (1993) use of activity theory for describing how tools like digitized medical records mediate the operations of a health clinic provides an engaging example.

Engeström begins by describing a changing economic reality for health care in Finland, where one particular clinic recognized a need to see more patients, spend less time per patient, and still maintain a high level of overall care. Digitized medical records were thought to be a way to respond to the rule of seeing more patients for shorter amounts of time by systematizing both the input of patient information and the access of it. There was an impact, however, on the community norms. Because the clinic started to see more patients, it became harder for them to schedule appointments, resulting in more urgent-care requests, to which doctors responded by doing less listening to patients' nonmedical explanations of their conditions and relying instead on the standardized medical information about the patient captured in the digital record. Further, the increased demand for physicians meant a need to break down the typical doctor-patient relationships (division of labor) by distributing the patient burden across more of the medical staff, some of whom had little to no history with the patients and so needed to rely even more on the medical records. One of Engeström's points is that a change in tools does result in a different kind of mediation, one that might stand in unexpected conflict with the values of the community where it is introduced.

The activity theoretic position that mediation be understood in the context of a social setting of action could just as easily explain other mediated interactions, such as how scientists use shared data sets to participate in distributed research projects, or how people signal affiliation with a particular group through distinctive and conspicuous use of mobile phones.

While activity theory is helpful for analyzing a social setting and its influence, this perspective can be enriched by another that considers how the social comes together, holds together, and takes on the appearance of reality (Jonassen 2004, 110). From actor network theory (see Latour 2005), I borrow the idea that tools develop historically, by connecting with and incorporating other tools. At different moments in a tool's development, it reflects the needs of the people who enacted earlier tool designs. Tools have histories, and by reading a tool's history one can understand how that tool has shaped an activity over time and how those mediating influences persist in the accumulated design. Summarizing both kinds of mediation discussed in this section: "The use of tools is an evolutionary accumulation and transmission of social knowledge, which influences the nature of not only external behavior but also the mental functioning of individuals" (Kaptelinin, Nardi, and MacCaulay 1999, 32).

Through their designs, tools accumulate knowledge and perspectives that have situated value. For example, the units of measurement etched on some navigational instruments reflect valued ways of measuring the world, marking locations, and plotting courses that developed historically and gradually became default frames for the world when they became crystallized in maps, alidades, software, and other instruments used in modern navigation (see Hutchins 1995). In this case and others, the perspectives afforded by tools reflect some values to the exclusion of others. The next section considers how these ideas can be investigated with a heuristic.

A HEURISTIC FOR ANALYZING TOOL MEDIATION

There are many questions one could ask about tools and how they mediate, but to develop the social and historical account described here, we can look to activity theory and actor network theory for inspiration.

Activity theory suggests that all tools are situated in a social activity, a goal-oriented task, through which one participates in a community. A person uses all manner of tools as well as mundane artifacts like clothing, to do something within the context of a community, even if all that person is attempting to do is fit in. Given the factors that motivate a person to use a tool, certain operations and functions will be more apparent because of their immediate applicability. Sometimes these operations and functions are perceived informally, and at other times they are directly identifiable in the tool's interface. How to understand these operations is the first question in our heuristic.

How do the concepts and operations associated with a tool mediate a user's understanding of a given task? Visually, we might present the point as in figure 6.2.

Tool use, seen from the perspective of the user, is only part of the picture, however. Activity theory reminds us that tools also mediate how users relate to other people. In other words, the tools shape not only the individual's work but also that person's relationships with others. From this expectation, we get our second question.

How do the concepts and operations associated with a tool mediate social

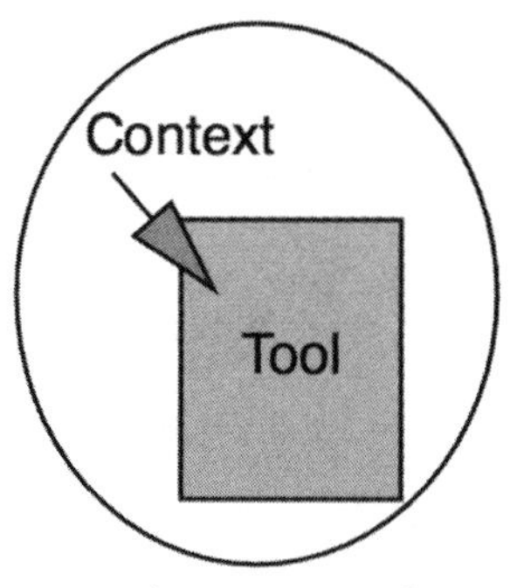

Figure 6.2. Tools are always situated in contexts of activity.

relationships occurring around it? This question leads us to understand more about the people who are in the context surrounding the tool being examined (i.e., the circle encasing the tool in figure 6.2). What motivates them to interact as they do?

Turning attention to the tool itself, activity theory and actor network theory both tell us to look for design histories. Activity theory tells us that tools change in significance and meaning as they become associated and disassociated with particular contexts and tasks. Actor network theory tells us that some tools grow by linking together tools (and contexts of activity) over time. For example, mirrors are tools, and lenses are tools, and the microscope, which followed in a temporal sense, links together both under the operation of the microscope. This observation leads to our third question.

What is the tool's design history, and how does that history influence current uses of the tool? With this question, we engage in a little historical detective work, trying to determine the origins of the tool under examination. What other tools does it link together? Are there particular skills or expertise associated with those underlying tools that are now folded into the interface of the new tool? Are there operational or functional assumptions carried forward through the tool's design? For example, as we will see, the CMS carries forward the functional assumption that texts are made up of modular chunks of text that can be manipulated in isolation. We can add design history to our visualization (figure 6.3). Early on, the tool started as separate tools that gradually linked together into more complex tools. The spheres of activity associated with those tools also started to merge.

The last two questions are closely related. They attempt to uncover the broad, distributed network of other tools and people through which the mediating effects of any particular tool will be seen. For example, by using a tool like an MRI, one produces radiographic data about a patient, in a

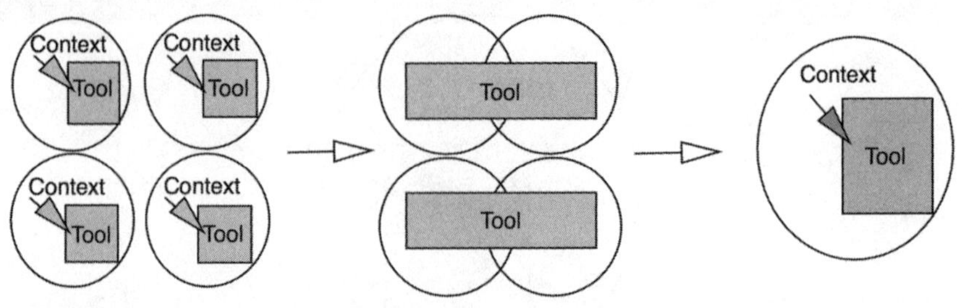

Figure 6.3. Over time, tools and their contexts of activity start to merge.

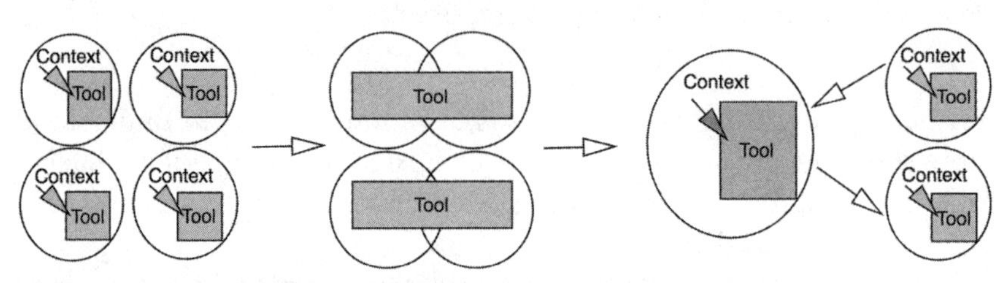

Figure 6.4. A tool's scope of mediating effect includes other people and tools within that context of work.

form that is required for use in certain charts or for display in a particular piece of software. The data also allow medical personnel of different specializations to talk about a patient and coordinate the care they will provide.

To what other tools is this one connected in practice or function? By what means are the mediating effects of a tool formalized and made more stable? To complete the visual that now includes a picture of the tool across time, we can illustrate the distribution of that tool's mediating effects in space—its scope of mediational influence (figure 6.4). In other words, by tracing out the connections to other tools (and to other people and the relationships among them), we can see how a particular tool both shapes and organizes those activities even while its own operation and significance is held steady.

These questions are the focus of the next section, in which our hypothetical technical communicator, Nancy, uses them to understand the mediating influence of CSS and style guides.

APPLYING THE HEURISTIC

By applying the heuristic, Nancy began to explore how CSS mediates practices of technical communication and to trace those effects as they impact the development of a style guide.

How do the concepts and operations associated with a tool mediate the user's understanding of a given task? Nancy's investigations revealed that, early on, the company style guide was a single, comprehensive document used to enforce standards on any text. As the company became more distributed and diversified and the range of texts increased, there was a growing realization that one style did not fit all needs and that some divisional styles might need to take priority, while inheriting a look and feel from a common set of style standards. Through conversations with colleagues responsible for promoting content management in their divisions, Nancy came to realize that CSS had two important operational qualities that filtered into those divisions' style guides.

One concept was the cascade, the ability for a single document to have different styles attached to it, prioritized depending on the output. For example, the author of a document may assign a style sheet that specifies the look of <ul> (i.e., an unordered or bulleted list) and <li> elements (i.e., bulleted points), but the user may apply a different, local, style sheet that overrides some style elements. Originally, prioritizing the application of different style guides was accomplished informally as different divisions decided when to apply the company style guide or their own local style guides. With the adoption of CSS, such cascading capabilities invited technical communicators to see their texts as having a greater range of potential uses and styling considerations.

Another concept that appeared to have influenced technical communication was that styles are inherited, referring to how style characteristics are passed on from parent content elements to child content elements. For example, an unordered list <ul> contains list items <li>. The unordered list is the parent element, and the bullet points are the child elements. Styles associated with the <ul> parent element, such as font, are inherited by the <li> child elements.

More than just concepts, cascades and inheritance represented operations that these division heads valued and that Nancy's company now wanted to ingrain as habits of thought about how writing is practiced. These technological assumptions were what Nancy needed to uncover in order to build them into the company style guide.

Nancy also recognized that problems associated with enforcing style rules would be circumvented in CSS, which would substitute the need to compel writers to follow the rules with the very functionality of the tool itself. Writers wouldn't need to be reminded to follow the style guide. They would need only use their writing tools.

How do the concepts and operations associated with a tool mediate social relationships occurring around it? Delving into the company's historical records, Nancy learned that as the company expanded, diversified, and spun off subsidiaries, the desire for consistency intensified. Consistency helped manage distributed units of the company by facilitating the flow of documents and information. Later, concerns about consistency grew, concurrent with rapid development and change in writing tools that made it easier to manipulate visual design and layout. Nancy knew that style guides fostered consistency by passing down style rules, but now that work would be delegated to forms, templates, macros, and CSS rules. In a large company like hers, these delegations of style rules to technologies would become necessary for enforcement that is more consistent and reliable than editors might provide (see Hart 2000, 12, 14).

Before computers could check consistency, the responsibility fell to writers and editors. As the company produced more documents, the cost to vet them rose. Around this time, the company style guide started referring to forms and templates, which lowered the cost of enforcing stylistic consistency by reducing the number of style decisions writers could make. Responsibility for adhering to style rules started to fall back on the writers and their templates, relieving editors of some responsibility. With this shift in responsibilities, Nancy recognized, the editors and writers could develop different working relationships.

In addition to being costly to enforce, style guides were also costly to produce. Early versions were printed, copied, and bound in limited numbers. Conceivably, the limited availability of the style guide effectively consolidated the distribution of that knowledge within the company. Those with access to the style guide were those who could speak about the rules most authoritatively. So responsibility for interpreting and applying the rules fell to them. While the ability to produce and distribute style guides electronically helped lower production costs, CSS will improve cost savings by embedding style rules in the writing tool's interface. A downside, however, is that this move will divorce writers from a consideration of stylistic concerns, which are a necessary part of a document's rhetorical effectiveness.

What is the tool's design history, and how does that history influence current uses of the tool? Nancy's initial consideration of how CSS-enabled content management affects social relationships led to her observation that while the company had long used style guides, at some point in the past the function of those guides started to merge with CSS itself, which had merged with structured authoring tools. Puzzling out this complex mediation required Nancy to investigate three parallel histories: structured writing, CSS, and a local history of style guides at her company. Only by seeing how each came about could Nancy see what they had become in sum.

A brief investigation of the history of CSS revealed that tagged and structured content could be traced to the development of standard general markup language (SGML) in the 1970s, which facilitated the development of HTML in 1989 (Turner 2008). HTML developed out of a desire to create structured content for the web; in doing so, it conflated structure with style, a problem addressed by the development of CSS in 1993 (Bersvendsen 2005). CSS merged back with SGML just before the adoption of CSS1 standards in 1996. At the time, it was proposed that CSS could be applied to SGML, meaning that the styling rules could be applied to specific content objects or overridden, depending on the output. CSS2 standards introduced the ability to reuse structured content and to style content dynamically, based on its destination (Lie and Bos 1998; Bos et al. 2008).

This abbreviated technical history clearly showed how characteristics of older technologies were preserved and enacted in CSS. In particular, CSS works with and anticipates structured content. It assumes a need to keep content separate from style. This observation alone would impact the work that Nancy's style guide would need to accomplish. Companies found SGML and HTML worth adopting because they served their values at the time. Once adopted, use of those tools uncovered problems or shortcomings that helped spur the development of new tools like CSS, which improved on the existing tools but still worked within the framework of mediation established by them. What this technological history further suggests is that by adjusting a style guide to account for the influence of CSS, Nancy would also need to keep in mind the mediating framework, values, and assumptions the style guide itself has inherited from its precursors.

The earliest style guides Nancy could find date to the 1970s. In them, there was little explicit mention of document design and layout. Instead, the overwhelming focus was on creating consistency by controlling expression (see Blair 1970, 1). The style guides contained little discussion of layout, and when it was addressed, its importance was downplayed or treated as falling outside the realm of a writer's concern (see Lee 1970, 3). The focus on expression over layout was partly due, Nancy surmised, to the limited range of layout choices available to writers. Much of that work would have gone to specialists.

Throughout the 1980s, the company style guide expanded to include guidelines for controlling layout and visual design, while retaining a strong focus on controlled expression. Around this time, the company started to expand, spin off subsidiaries, and open international offices. Documents now had national and international audiences. The resulting translation and localization needs required increased control over variation in expression and layout (see Blakely and Travis 1987, MPD-63).

The style guides also reflected changes in tools. As word processing and document design technologies grew in sophistication and availability, technical communicators at the company had control over a larger range of stylistic features (see Lalla 1988, WE-176).

A style guide from the 1990s revealed significant shifts in emphasis, away from specifying detailed rules of usage and style to focusing on templates (see Caernarven-Smith 1991, 140–142) and style rules for genred sections of common documents—for example, overview sections, feature lists, examples, and screen captures (see Washington 1991, 554). In other words, the focus shifted to content types to which different style rules might be applied. The significance of the shift, Nancy recognized, is that it reflected a motivation to see documents as structured content, a perspective afforded by structured writing tools that would come to be associated with CSS, developed around the same time.

Later in that decade, style guides started to show tighter integration with CMSs by referencing document type definitions (DTDs), or rules specifying how structured content was to be assembled for different document types. Increasingly, Nancy noted how the style guides explicitly and implicitly referenced structured writing tools, even deferring to them as some of the enforcement of style rules was rolled into their overall functionality.

As content migrated to the computer screen and to the Internet, the style guides started to reflect the importance of localized styles (see Dalton 2002, 529; Rude 2002, 139), and guidelines for controlling the appearance of text started to address lingering problems with poor on-screen legibility (see Nichols 1994, 436).

By early 1998, the documentation division started using structured writing tools with CSS capabilities. A revision to their local style guide later that year revealed reliance on CSS for applying style rules automatically (see Perkins and Maloney 1998, 25). Style enforcement continued to move into the interface as CSS1 standards allowed writers to attach style sheets to content types.

By 2000, the integration of style guides with tool interfaces appeared well established, as visualized in figure 6.5. Submerged in the interface, style rules implicitly guided writers, while shaping the product that emerged (see Hart 2000, 12, 14). The separation of style from content also appears to have facilitated company adoption of single-sourcing, outputting the same

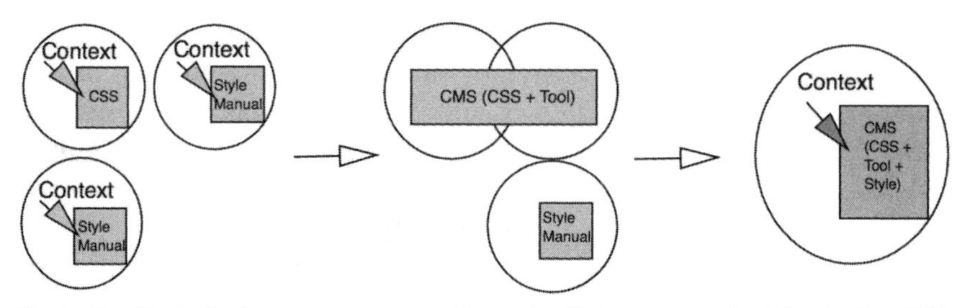

Figure 6.5. Tools that were once separate gradually became embedded in the CMS interface.

content into multiple document types, each with its own style sheet (see O'Neill 2002; Quesenbery 2001, 3).

To what other tools is this one connected in practice or function? Implicit throughout the historical record of changes in the company's style was the changing technological landscape in which the writers worked. First was the increase in personal computing, which ensured the ready availability of sophisticated word processing and document design tools, and which likely led to a general increase in the number of texts that people produced and in the number of interactions happening around those texts. Personal computers were also convenient points for enforcing style rules. Second was the development of networking technologies (e.g., e-mail, video conferencing) and network infrastructures, which made it easier for the company to diversify and globalize. As the effects of both were being felt, the values of consistency and cost savings were highlighted in ways that encouraged adoption of technologies like CSS.

Each of these technological developments connects to another in some way. While some tools influenced how technical communication is practiced at the company today, other tools had a broader mediating effect on the context in which technical communication now occurs. The increase in personal computing power and sophistication gave Nancy's colleagues the ability to make more stylistic choices. Corporate diversification and globalization created new opportunities for distributing the labor of technical communication. Both factors increased the threat of inconsistency that could then be addressed via CSS, instead of through an ever-expanding company style guide. Tools work within networks of tools and humans that are temporarily aligned toward a given end. Mediation is an effect of the network rather than of a single tool, and Nancy was starting to see clearly how technical communication at her company is practiced across this assemblage of tools.

By what means are the mediating effects of a tool formalized and made more stable? Mediation is situated within a community and its values, rules, and divisions of labor, which are continually changing. Often this means that mediation is a temporary influence, changing as the community changes. The values, rules, and divisions of labor at the company in our example, however, were reinforced by the adoption of structured writing and CSS. Nancy recognized that the widespread adoption of these tools and the tight integration of the tools and their products into various company work practices (e.g., single-sourcing of documentation and marketing literature) stabilized the mediating effects of these tools. The question Nancy needed to ask, then, was who or what are the actors stabilizing those effects? One such actor is the style guide.

A style guide is a tool that mediates writing. Changes in style guides reflect changes in what companies value in their texts as well as changes in the contexts where a style guide might be employed. In some ways, a style guide articulates and reinforces values derived from and reinforced by the tools technical communicators may use. The style guide elevates guidelines on writing, document design, and usage to the level of policy.

A second formalizing effect comes through the ways that precursor tools and their mediating frameworks become enmeshed in the software and other tools that technical communicators use. Today, many technical communicators use some kind of structured-writing tool that either relies on SGML or XML to run DTDs, sets of rules that govern how structured content should be assembled into different text types. CSS is integrated into those tools to mediate the outputs, and those tools become inextricable parts of that interface.

WHY MEDIATION MATTERS

After a bit of investigation and reflection on the role and function of CSS at her company, Nancy learned a great deal that helped explain the function that the company style guide now serves. At the very least, this heuristic evaluation revealed

- why style guides at her company look like they do;
- how those style guides mediated the way technical communicators looked at texts and shared them with their colleagues;
- why tools were adopted, based on changes in the company, and how the features of new writing tools (e.g., embedded and largely invisible CSS rules) served distinct economic exigencies and corporate values; and

- how making changes to a style guide requires an understanding of the kinds of work and social relationships that different writing tools afford.

What Nancy discovered is that the style guide now reinforces the importance of the CMS because of the extent of their integration. Given the company-specific values that derive from producing content in the CMS (e.g., automatic application of style rules, version control, single-sourcing), there are compelling reasons why writers should be encouraged to use the tool. The style guide does its part by deferring authority to the CMS and the embedded CSS, and in this way reveals how the guide is shaped by CSS.

After considering the complicated work that the style guide accomplishes, Nancy can formulate a clear plan for revision. The plan starts with a mapping of the various objectives that the style guide is meant to accomplish. These include standardization of content and form, flexible reuse of content, cost reduction, and improved coordination between divisions in the company. In the process, Nancy will determine what writers, editors, and managers are attempting to accomplish with their documents, what norms govern those practices, and what social relationships are supported through those documents. Further, through interviews with some of these stakeholders, Nancy will determine the role of the style guide as a tool supporting that work and how it does so alongside other technologies like the CMS and its integrated style sheets. By understanding how the style guide works with and through other technologies, finally, Nancy can determine the scope of the rules to be covered, technological concepts that can be referenced, and assumptions that can be made about stylistic choices that will be in the writer's control and those that will not be. The result will be a style guide that is designed to accomplish its various objectives while reflecting the social and technological contexts in which it does that work.

DISCUSSION QUESTIONS

1. Choose a function from a writing tool (e.g., columns or page layouts) and trace a brief history of it. Where did this technology or function come from? What does a person have to understand to know how to apply that function?
2. Using the same example, think of concepts or operations associated with that specialized knowledge. When applied to writing, how do those concepts and operations influence the way you think about a text?

3. Think about a tool that you use in class or at work (e.g., e-mail, a course management system, an inventory control system). With whom do you interact through that tool? How does the tool mediate that relationship?
4. Using the same example, consider what kinds of identities are associated with the users of that tool (e.g., a course management system assumes teachers and students). What kinds of values are associated with those interactions?
5. Based on your reading of the scenario explored in this chapter, does it seem accurate to state that changes in tools caused the changes that Nancy saw in the style guides? Why or why not?
6. What could a technical communicator do, if anything, to be more critically conscious of the mediating effects of tools?
7. What information could one gather to illustrate the influence of tools on technical communication practice?
8. Where might one look for evidence that a particular tool's mediating function is becoming formalized?

WORKS CITED

Bersvendsen, Arve. 2005. Who created CSS? CSS early history. Arve Bersvendsen Blog. http://virtuelvis.com/2005/01/who-created-css-css-early-history/.

Blair, Eugenie. 1970. Development of the Bell Laboratories editorial style guide. In *Proceedings of the 17th International Technical Communication Conference*, 1–5. Fairfax, VA: Society for Technical Communication.

Blakely, J. Paul, and Anne S. Travis. 1987. The nine-year gestation of a unified technical style guide. In *Proceedings of the 34th International Technical Communication Conference*, MPD63–MPD64. Fairfax, VA: Society for Technical Communication.

Bos, Bert, Tantek Celik, Ian Hickson, and Hakon Wium Lie. 2008. Cascading Style Sheets level 2 revision 1 (CSS 2.1) specification. World Wide Web Consortium. http://www.w3.org/TR/CSS2/.

Caernarven-Smith, Patricia D. 1991. Aren't you glad you have a style guide? Don't you wish everyone else did? *Technical Communication* 39:140–142.

Cole, Michael, and Yrjö Engeström. 1993. A cultural-historical approach to distributed cognition. In *Distributed Cognitions: Psychological and Educational Considerations*, ed. Gavriel Salomon, 1–46. Cambridge: Cambridge University Press.

Connors, Robert J. 1982. The rise of writing instruction in America. *Journal of Technical Writing and Communication* 12 (4): 329–352.

Dalton, Tracy. 2002. Style sheets: The abbreviated answer. In *Proceedings of the 49th Annual Society for Technical Communication Conference*, 529–532. Fairfax, VA: Society for Technical Communication.

Dobrin, David N. 1983. What's technical about technical writing? In *New Essays in Technical and Scientific Communication: Research, Theory, Practice*, ed. Paul V. Anderson, R. J. Brockmann, and Carolyn R. Miller, 227–250. Farmingdale, NY: Baywood Publishing.

Engeström, Yrjö. 1993. Developmental studies of work as a testbench of activity theory: Analyzing the work of general practitioners. In *Understanding Practice: Perspectives on*

Activity and Context, ed. Seth Chaiklin and Jean Lave, 64–103. New York: Cambridge University Press.
Gibson, James J. 1986. *The Ecological Approach to Visual Perception*. Hillsdale, NJ: Lawrence Erlbaum Associates.
Hart, Jeffrey. 2000. The style guide is dead: Long live the dynamic style guide. *Intercom*, March, 12–17.
Hutchins, Edwin. 1995. *Cognition in the Wild*. Cambridge, MA: MIT Press.
———. 1997. Mediation and automatization. In *Mind, Culture, and Activity*, ed. Michael Cole, Yrjö Engeström, and Olga Vasquez, 338–353. New York: Cambridge University Press.
Ihde, Don. 1991. *Instrumental Realism: The Interface between Philosophy of Science and Philosophy of Technology*. Bloomington: Indiana University Press.
Jonassen, David H. 2004. *Handbook on Research on Educational Communications and Technology*. Hillsdale, NJ: Lawrence Erlbaum Associates.
Kaptelinin, Victor, Bonnie A. Nardi, and Catriona MacCaulay. 1999. The activity checklist: A tool for representing the "space" of context. *Interactions* 6 (4): 27–39.
Kynell, Teresa. 1999. Technical communication from 1850–1950: Where have we been? *Technical Communication Quarterly* 8 (2): 143–151.
Lalla, Sharon Trujillo. 1988. The state-of-the-art style guide development. In *Proceedings of the 35th International Technical Communication Conference*, WE176–WE179. Fairfax, VA: Society for Technical Communication.
Latour, Bruno. 2005. *Reassembling the Social: An Introduction to Actor-Network-Theory*. New York: Oxford University Press.
Lee, Joseph. 1970. Writing the company style guide. In *Proceedings of the 17th International Technical Communication Conference*, 1–6. New York: Association of Computing Machinery.
Lie, Hakon Wium, and Bert Bos. 1998. Cascading Style Sheets, level 1. World Wide Web Consortium. http://www.w3.org/TR/REC-CSS1-961217.html.
Nichols, Michelle. 1994. Using style guidelines to create consistent online information. *Technical Communication* 41:432–438.
Norman, Donald. 1999. Affordance, conventions, and design. *Interactions* 6 (3): 38–43.
———. 2002. *The Design of Everyday Things*. New York: Basic Books.
O'Neill, Jennifer. 2002. A global style guide: Working together around the world. In *Proceedings of the 49th International Technical Communication Conference*, 357–358. New York: Association of Computing Machinery.
Perkins, Jane, and Cassandra Maloney. 1998. Today's style guide: Trusted tool with added potential. *IEEE Transactions on Professional Communication* 41:24–32.
Quesenbery, Whitney. 2001. *Building a Better Style Guide*. Whitney Quesenbery Usability. http://www.wqusability.com/articles/better-style-guide-slides.pdf.
Rude, Carolyn. 2002. *Technical Editing*. 3rd ed. New York: Allyn and Bacon.
Sellen, Abigail J., and Richard H. R. Harper. 2002. *The Myth of the Paperless Office*. Cambridge, MA: MIT Press.
Sullivan, D. L. 1990. Political-ethical implications of defining technical communication as a practice. *Journal of Advanced Composition* 10 (2): 375–386.
Turner, Linda. 2008. SGMLUG SGML history. Cover Pages. http://xml.coverpages.org/sgmlhisto.html.
Vygotsky, Lev S. 1978. *Mind in Society: The Development of Higher Psychological Processes*. 14th ed. Cambridge, MA: Harvard University Press.

Washington, Durthy. 1991. Developing a corporate style guide: Pitfalls and panaceas. *Technical Communication* 38:553–555.
Winner, Langdon. 1989. *The Whale and the Reactor: A Search for Limits in an Age of High Technology*. Chicago: University of Chicago Press.
Winsor, Dorothy. 2001. Learning to do knowledge work in systems of distributed cognition. *Journal of Business and Technical Communication* 15 (1): 5–28.
Yates, JoAnne. 1993. *Control through Communication: The Rise of System in American Management*. Baltimore, MD: Johns Hopkins University Press.

7

BERNADETTE LONGO & T. KENNY FOUNTAIN

What Can History Teach Us about Technical Communication?

SUMMARY

This chapter explores how taking a historical perspective can be helpful for understanding the contexts in which technical communicators make decisions about document development. Connections between history and organizational culture are introduced as one approach for understanding why some things are considered appropriate or inappropriate in a particular workplace, and how these values can change over time. In order to honor the rich, diverse, and unfinished history of our profession, we need to know not only the history of a document, the people who created it, and its current usage, but also the role we play as technical communicators in the social practices of science, technology, and business.

INTRODUCTION

If you have ever visited someone in the hospital or been in the hospital yourself, you have seen all the paperwork involved in keeping complete medical records. This record keeping is extensive in regular clinical units and even more so in intensive care units (ICUs), where medical care is more complicated. Nurses, doctors, and other hospital staff complete medical forms as a routine part of their jobs. But where do these forms and other medical paperwork come from? Who develops routine medical paperwork and the systems that regulate it? What roles do organizational history and institutional culture play in the creation and use of these documents?

Rita, a new technical communicator, finds herself asking these questions and more. A few weeks ago, she was hired by a hospital to assist with their technical communication tasks, and now she has been assigned to work with staff in the ICU to develop a protocol document for both monitoring patients' blood glucose levels and treating patients when these levels get too high. Specifically, Rita must revise the existing ICU protocol and clarify the steps nurses should take to monitor glucose levels and decide when to administer insulin. While considering how she might improve the document, Rita realizes she needs to ask a number of questions about the existing protocol, like how it came into being and how it is used. Who initially decided that the ICU needed an insulin protocol, and when did they? Who was involved in developing the document? What important decisions were made in that development, and how were they made? How do ICU nurses use the protocol, and what improvements might they suggest? These are just some of the questions Rita must ask in order to understand the existing protocol and how she can improve it. As you can see, these questions focus on the history of the existing document in order to revise it and improve how people use it to care for patients.

Technical communicators create scientific, technical, and business documents that not only convey information but also create systems of order that influence the routine practices of hospitals and clinics. These technical documents help medical professionals do their job by helping them understand how to handle the tasks in front of them. In large part, technical communicators compose and design these documents based on their understanding of what worked well in the past—on their knowledge of institutional and organizational history. In fact, whether you're an engineer writing a user manual or a technical communication professional creating that same manual, you will rely largely on your historical knowledge of your organization and your specific situation to decide what to include and exclude.

Assembling a history, even an organizational history, is more complex than merely a chronological cataloging of important events and dates, because history involves difficult questions: Which events are deemed important and thus worthy of remembering? Who decides this, and how will these events be recorded? From whose perspective will these events—these stories—be told? Your telling of a historical event would not be the same as another person's, even if you were both participants. Though even a simple recounting of past occurrences can be complicated, people still write histories, because historical accounts provide valuable information for understanding present and future situations. Making use of history involves more than knowing what happened: one must also consider the cultural, social, and individual forces that shaped those events—the forces that have given us one story instead of another.

In this chapter, we explore what history, specifically organizational history, can teach us about technical communication practice by illustrating important ways history influences the daily workplace practices of technical communicators. First, we review some of the significant research into the history of our field to discuss technical communication as a practice that orders scientific, technical, and business knowledge. Second, we turn to the research literature in organizational acculturation to examine how organizational history operates as a contemporary tool of socialization, shaping writers' and designers' communication practices. Then, we introduce a three-part heuristic designed to help new and experienced technical communicators explore the role history plays in workplace communication. The questions in our heuristic ask technical communicators to reflect on the ways historical practices shape our ideas about our work, our values involving the profession, and our identities as communicators. Lastly, we offer an extended example of one new technical communicator's work to create a very specialized technical document—a medical protocol instructing health-care workers on the use of insulin in intensive care settings.

LITERATURE REVIEW

Though much emphasis is placed on the contemporary workplace, the field of technical communication has a rich and vital history that can provide productive insights into the way communication works today. In particular, understanding past practices offers a better awareness of the ways technical communicators display, shape, and, more importantly, order scientific and technical knowledge. Our literature review is organized in two sections to correspond to our chapter's two main contributions. First, we explore how technical communication can be understood as a process of ordering scientific and technical knowledge and practice. To do so, we turn to research in technical communication history that recovers the discipline's lost historical figures or emphasizes the importance of history for understanding current practices. Second, we show how technical communication practices and documents operate as a socializing force in organizational cultures. That is, technical documents shape our practices, actions, and viewpoints.

HISTORY, SYSTEMS OF ORDER, AND TECHNICAL COMMUNICATION

In *The Order of Things* (1994), philosopher Michel Foucault understands history, particularly the history of Western knowledge, as a complex process of ordering. Foucault seeks to better understand how "our [Western] culture has made manifest the existence of order" (xxi), which involves two aspects: (1) a process of creation, of making categories and systems of order; and (2) a state of being ordered. Though we often understand this ordering as a logical process, we are actually the ones who create the definitions and systems of classification that seem so inevitable. Think, for example, of how you save files to your computer. Think of all the decisions you make about which documents go in which folder, the names of those folders, the way those folders are stored, and the number of folders to use. All of this represents a process of ordering, of creating order and finding what (at least to you) seems an intuitive system. In the work we do to create order, we often fail to acknowledge the historical and cultural influences that shape our systems of order—the forces that have made the world what it is today.

Many historical studies of technical communication describe the field as practices of ordering, of naming and shaping scientific and technical knowledge. This historical research often emphasizes the discipline's European and North American emergence in relation to science, engineering, and business. Hoping to uncover a more complete picture of the past, these researchers show how technical communication works hand in hand with science and technology. In fact, they contend that technical communication is the means through which scientific and technological ideas and processes are communicated to both general and expert publics.

For example, Teresa Kynell and Michael Moran's *Three Keys to the Past* (1999), one of the first collections on the history of technical communication, initiated this historical research, which often aims at one of three goals: (1) to recover the discipline's lost historical figures, (2) to recontextualize the significance of existing figures within the field, or (3) to stress the importance of history for understanding contemporary practices. In this collection, James Zappen's (1999) study of Francis Bacon's contribution to the history of technical communication, R. John Brockmann's (1999) essay on early-nineteenth-century American inventor Oliver Evans, and Charles Bazerman's (1999) work on Joseph Priestley make plain the role engineers and scientists played in the development of technical communication, specifically their use of sophisticated rhetorical techniques to accomplish their professional goals. This spotlight on overlooked moments in our field's history demonstrates how technical communication pervades all aspects of science and industry. For example, Tolbert's (2001) work recognizes Nicolas-Claude Fabri de Peiresc's attempts to "determine terrestrial longitude" while keeping the Church's blessing; Di Renzo's (2000) study of Marcus Tullius Tiro's roles as Cicero's confidante, secretary, and archivist demonstrates how our history is embroiled in the controversies of the past. Peter Hager and Ronald Nelson's (1993) study of Chaucer's *Treatise on the Astrolabe*, and Bernadette Longo's (2000a) look at Vitruvius's *De Architectura* demonstrate the textual elegance and rhetorical savvy of these works as examples of technical communication. Such research unearths the ways technical communication's past has been shaped by the contribution of historical figures commonly appreciated as scientists, engineers, literary artists, and politicians. By emphasizing their work as communicators of scientific knowledge, these studies underscore technical communication's role in shaping scientific information and influencing Western science and technology.

Others have shown how technical communication functions as a unique discourse of science, engineering, and industry. For example, Elizabeth Tebeaux's *Emergence of a Tradition* (1997) focuses on English Renaissance technical texts and genres and traces the emergence of technical communication (in the English-speaking world) as a collection of texts, namely manuals and how-to books, and as a set of conventions, such as the use of page layout, audience awareness, and visual illustrations. Tebeaux explains the origins of verbal, textual, and rhetorical elements that have come to dominate technical communication practice and pedagogy. Arguably, the work of Tebeaux and Michael Moran, beginning with his rhetorical analysis of the 1590 map of Virginia, allowed others to take seriously the supposedly mundane history of technical documents by proving their relevance to contemporary practice (Moran 1999, 2002, 2003). These studies strengthen connections between our field and the history of ideas, illustrating technical communication's involvement in the creation, promotion, and deployment of scientific and technical knowledge.[1]

Because of this connection to the history of business, science, medicine, and technology, technical communication has played a powerful role in shaping scientific knowledge. Bernadette Longo's (2000b) cultural history, *Spurious Coin*, focuses on the ways this powerful role has gone unnoticed. For example, technical communicators, scientists, and engineers have mistaken technical documents for uncomplicated texts used only to contain and transport technical knowledge. As a result, they have often failed to appreciate the way technical documents shape knowledge and practice. Working as a control mechanism within organizations, technical communication functions to authorize whose knowledge is, and is not, valued (Longo 2000b).

Technical communicators, then, use documents to order knowledge, shape information, and make implicit and explicit arguments about what is to be valued. In fact, technical communication does not simply influence social practices—encouraging certain ways of being while discouraging others—but actively develops social practices. For example, Miles A. Kimball's (2006, 353) historical analysis of Charles Booth's maps of London poverty from 1889 to 1902 investigates the "cultural basis" of such technical communication concepts as "transparency and clarity in information graphics." By reading Booth's maps back into their original context, Kimball argues that the preference for transparency in visual displays "arises from a broader visual culture," which influences readers' expectations (355). These particular visual texts were created in response to

London poverty and reflect the contemporary social and cultural attitudes that represented poverty as an easily corrected ill. Carol Siri Johnson's (2006) analysis of the letters written by an eighteenth-century Scottish engineer brings to light the "prediscursive" tradition of the iron-making industry. By focusing on the development of the industry from an oral and apprenticeship-based tradition (prediscursive) to a text-based one, Johnson explores how one important area of technical communication flourished through a network of social connections that created and shaped technical knowledge even before this knowledge was set down in writing.

Such work illustrates how technical communication has been implicated in larger cultural contexts and debates for centuries, intimately connecting our field to the historical circumstances that surround it. This is as true today as it was yesterday, as it will be tomorrow. This historical influence can be viewed from the macrolevel of histories of science and technology to the microlevel of organizational and institutional contexts where history still matters and where technical communication still operates as a system of daily ordering.[2]

ORGANIZATIONAL HISTORY, SOCIALIZATION, AND TECHNICAL DOCUMENTS

In *How Institutions Think*, anthropologist Mary Douglas (1986) investigates the role organizations and institutions play in the thoughts and actions of individual members. Douglas asserts that institutions work by offering members the necessary frameworks, systems of thought, even "analogies with which to explore" and "justify the naturalness and reasonableness" of the institution's rules and ways of operating (112). These analogies, which underpin that institutional culture, shape the way members view the larger group, the way they view themselves, and the types of practices they engage in: "[the institution] provides the categories of their thought, sets the terms for self-knowledge, and fixes identities" (12). To become a full-fledged member of an organization, or workplace, we come to accept a certain view of our work and ourselves, a view mediated in large part by institutional history and organizational culture. This history and culture, which is rooted in the past but plays out in the present, determines our decisions on the job. Both the tougher choices and the easier ones we make as technical communicators are constrained by workplace history and culture: "An answer is only seen to be the right one if it sustains the institutional thinking that is already in the minds of individuals as they try to decide" (Douglas 1986, 4). We need only remember the space shuttle *Challenger* disaster, and other failures of science and technology that involved communication, to understand the powerful force of organizational history in shaping current decisions and actions (see Dombroski 1991; Winsor 1990).

But how do we become members of an institution to the extent that it shapes our thinking and actions? How do we learn to "fit into" an organization? And how are the historical practices of that organization still playing out in the present? These questions have been the concern of management research for decades (see Argyris 1964; Cable and Judge 1997). Much of this research has focused on how people learn salient information about an organization's culture in order to know what is acceptable or unacceptable behavior in that culture—in other words, how they become socialized into that organization (see Van Maanen 1976; Trice and Beyer 1993). An assumption underpinning this research is that people new to an organization want to succeed. One way of doing a job well is to know how things are done in that workplace. After working at a job for some time, we seem to *know* the right and wrong way to do things in that setting. But this knowledge of acceptable and unacceptable practices comes through direct and indirect processes of acculturation.

Even after working at a job for a while, situations arise in which the right or acceptable course of action is unclear. How do we know then which actions and practices will be successful? Meryl Reis Louis (1980) argues that individuals learn how to succeed on the job by first learning about the organization through socialization. She describes socialization into an organization as "the process by which an individual comes to appreciate the values, abilities, expected behaviors, and social knowledge essential for assuming an organizational role and for participating as an organizational member" (229). This process of socialization is not merely a process of learning the expected tasks, because each "role change" a worker must make involves an adjustment to that socialization process (230). Thus, a worker's socialization is never complete. As we move into a new position, we find a new culture to learn. And as new information or research changes how a job is done, this new knowledge can change an organization's ideas about acceptable and unacceptable actions.

An organization's culture is a result of its members' previous decisions and actions. The values and background knowledge embedded in that culture are the historical artifacts of those decisions and actions. Some values and knowledge stem from successful decisions people want to emulate; others stem from unsuccessful decisions that provide cautionary tales of action to avoid. Whether as good examples to emulate or bad decisions to avoid, an organization's history lives in its current culture and is communicated through its stories and documents. Lee Odell (1985, 250) finds that this culture "influences 'practically everything' in the life of an organization." On the job, technical communicators get work done by internalizing "values, attitudes, knowledge, and ways of acting that are shared by other members of the organization" (250). In fact, communication is often the tool people use to internalize organizational culture. Jean Ann Lutz (1989, 114) argues that communication is "the primary tool through which members participate in an organization, through which nonmembers learn about an organization, and through which a corporation's climate and image are codified." By understanding that an organization's culture is communicated through various media, including texts, verbal stories, and the design of physical spaces, we can appreciate the role workplace documents play in socializing people on the job and standardizing their actions.

Documents serve to socialize technical communicators because communicators look to prior documents for examples of how to complete current tasks; we look to earlier documents to learn how to develop future ones. Lee Clark Johns (1989, 153) describes this colorfully by saying that the "file cabinet has a sex life; it reproduces itself." The file cabinet stores the historical information that *we* reproduce when we design those future technical documents. In the example of the ICU insulin protocol, when we look at a past document to understand a current writing situation, we are using historical knowledge to make decisions about our current practices. This use of past documents transmits existing information and organizational culture into current documents and practices. It is as if technical communication is one way of sending information from a past sender to a future receiver. But the situation is far more complex.

David Dobrin (1983, 242) famously defined technical communication as "writing that accommodates technology to the user." One interpretation of this model suggests that the technical communicator conveys only appropriate technical information to a user, without the ability to use historical understanding to impact the information or the culture of the communication. As Jennifer Daryl Slack, David James Miller, and Jeffrey Doak (2003, 178) argue, this simple understanding leaves the technical communicator only the ability to "transmit the sender's meaning as a perfectly executed message," with no power to influence or change the meaning or cultural context of the message. If the technical communicator is merely a conduit—a neutral channel through which information passes—then the communicator becomes an impotent worker, whose agency is denied except in cases of failure. Slack, Miller, and Doak argue that a more empowered way to think about the technical communicator's role is as someone who articulates meaning, using an understanding of historical and social circumstances to craft documents that participate in knowledge-making processes.

In this view, technical communicators must know what came before and use that information to either recreate the past or perform different actions in response to new information and circumstances. Being successful as an articulator of meaning relies on the communicator first being successfully socialized into the organization—understanding what is currently acceptable and unacceptable based on historical knowledge, and exercising good judgment regarding what needs to be changed and how to effect that change within the organization's current culture. This socialization into an organization's culture is communicated largely through actual documents and practices of the technical communication workplace. These technical documents themselves often exert an influence on individuals' communication decisions and actions.

Again, an organization's history lives in its current culture and is communicated through its documents, which most often resemble earlier documents. A memorandum written today will have similar format and content structure as earlier memos—even though most memos today will be sent as e-mail. Documents constituting a particular genre embody conventional structural elements that help readers anticipate the kinds of information a document will contain and its location and format. Charles Bazerman (1988, 62) describes a genre as "a social construct that regularizes communication, interaction, and regulations"; the "formal features" of a genre are the "solution to a problem in social interaction." That is, a document's regularized verbal, visual, and textual features help that document solve a social problem; this document can be understood as having a kind of agency in a social interaction. JoAnne Yates and Wanda Orlikowski (1992, 302) find that "genres are enacted through rules, which associate appropriate elements of form and substance with certain recurrent situations," thus giving these documents the ability to shape an individual's actions within an organization. A genre is especially influential when considered in relation to other professional genres that shape workplace communication. For example, Amy Devitt (1991) shows how various genres of accounting documents (memoranda, letters, tax reviews, research reports, etc.) work together to regulate professional practices in that field. These genre networks of professional documents are based on historical knowledge and social relationships that regulate future document decisions, influencing technical communicators' decisions about what is appropriate in a current situation.

Recent studies of medical technical communication have investigated the genre of medical forms as instruments that reflect larger institutional cultures and shape the practices of members. Roger Munger's (2000) study of the evolution of emergency medical services run reports finds that (1) changes in the forms themselves reflect changes in the ways practitioners used the forms and (2) medical forms created for multiple users offer insights into the status, authority, and power of various users. Susan Popham's (2005) examination of medical forms analyzes common paperwork of medicine (such as diagnosis, insurance, and billing forms), in order to understand how the fields of science and business are constructed and represented through these documents that influence patient care. These studies illustrate how social practice and disciplinary knowledge are represented in technical documents. In a concrete way, historical knowledge and judgments are reflected in a document's visual and verbal conventions, which serve to address social needs and relationships. These documents are artifacts shaped by historical circumstances; thus, they direct human action and offer a window into that action.

In sum, the history of technical communication is a history of ordering knowledge, of socializing communicators into certain cultures of writing, of creating technical documents that reveal the past and shape the present. To prove this point, we have made several claims about technical communication. First, technical communication operates as a system of ordering that shapes and produces technological, scientific, and often business knowledge through the creation and use of texts. Because all forms of communication are influenced by their contexts and circumstances, technical communication practices are embedded in historical and cultural processes that influence what *counts* as technical communication, as well as the *consequences* of that communication. Second, these processes of ordering often take the form of organizational culture or institutional history, which continue to influence the contemporary workplace environment. Third, through this continued influence of the past on the present, technical communication practices and documents play a socializing role; they encourage or value certain practices while discouraging or complicating others. And finally, the traces of this history, and this socializing function, can be witnessed in the actual documents of a workplace.

LEARNING FROM TECHNICAL COMMUNICATION HISTORY: A HEURISTIC

How exactly do we learn from our organization's past? What lessons can our local historical context teach us about technical communication more broadly? At this point, we will answer these questions by providing a three-part heuristic for exploring the role history plays in workplace communication. This heuristic translates our previous discussion into a practical framework for guiding technical communication practice. In particular, these questions will be helpful to technical communicators, novices and experts alike, in two major ways: (1) helping communicators reflect on the greater, present-day significance of their workplace's cultural and organizational history and (2) providing communicators with the "big picture" view of the tasks they perform and the documents they create.

STEP 1: WHAT SYSTEMS OF ORDER DOES MY WORKPLACE CREATE?

When we find ourselves in a workplace, our first question should be, What systems of order does my workplace create? We must know and understand what our company produces or provides as well as the consequences of these products and services. Because technical communication practices work to order scientific, technical, and business knowledge, each technical communicator is involved in this ordering process. That is, we are doing more than simply communicating information by creating technical documents and presentations. If we focus only on the task before us, we may forget and neglect the important cultural and historical role our work plays in shaping and making meaning. The first step, then, is to consider our part in the process by understanding what types of order our specific workplace seeks to create. We can do this by stepping back and thinking more conceptually about our workplace, what it produces, and the consequences of those products and services. And we can do this by talking to our coworkers and clients and thinking critically about the work we do.

STEP 2: HOW DOES THE HISTORY AND CULTURE OF MY WORKPLACE INFLUENCE THIS PROCESS OF ORDERING?

The second step is two-part: (1) consider what larger historical forces and movements directly influence the work you do today and (2) consider how the history and culture of your organization encourage certain actions and discourage others. Technical communication does not happen outside of historical circumstances, but instead is connected and responds to them. The systems of ordering that each workplace initiates are linked to larger historical contexts, which in turn influence that workplace and its practices. The second question asks us to consider the larger historical context and how it shapes the work of that organization. We can discover this by asking historically directed research questions: What is the history of this company? When and why did the company begin making a product or providing a particular service? What cultural or historical influences shaped the making of this company as well as its products and service?

What current cultural, political, and economic influences continue to shape this workplace?

We can answer these questions by turning to our coworkers, bosses, and published material about our workplace and profession (from newspapers, books, and articles). Remember, institutional histories are still alive in the organizational culture of a workplace. These histories shape the values, environment, and practices—or culture—of contemporary technical communication. And when we learn to succeed in a particular workplace, we absorb the culture of that place and become socialized in particular workplace practices. This socialization operates as a perceptual lens, allowing us to see the work we do through the lens of that organization's history. In order to operate as ethical communicators, we must not forget how the history and culture of a workplace influence our decisions.

STEP 3: HOW WILL MY DECISIONS IN THIS WORKPLACE SHAPE AND ORDER THE ACTIONS OF OTHERS?

Just as technical communication cannot be understood as standing outside of history, the technical communicator cannot be understood as standing outside of an organization's culture. Our socialization in the culture of a particular workplace shapes the way we understand our work, and also the work itself. The texts we create communicate processes of ordering to others: telling the users what to do, how to do it, and what to avoid doing. In these documents, we pass along this user-centered information as well as the values, assumptions, and beliefs of our workplace. With every document we create and every presentation we give, we communicate the organizational culture of our workplace.

As technical communicators, we are not passive conduits for transmitting technical information, so we must reflect on the consequences our actions have for other people. Specifically, we must consider how our documents shape the practices and ideas of the user. The final question is one we can ask ourselves whenever we tackle a new task, join a new team, or compose a new document. Asking this question means stepping back and considering the larger consequences of our work. What am I composing, designing, or presenting? How will my work influence, constrain, or make possible the actions of others?

MEDICAL TECHNICAL COMMUNICATION AND ICU PROTOCOLS

To illustrate the utility of this heuristic framework for technical communication, we offer an extended example of one communicator at work. Here we return to Rita and her hospital workplace to explore how one document—the insulin protocol used in ICUs—makes previous knowledge tangible in people's actions. When Rita started working at the hospital, the ICU staff was already using an insulin protocol they had developed. They realized that nurses were having questions about how to measure blood glucose and administer insulin to their patients. So the medical director of the ICU determined that they needed to revise the protocol and asked Rita to help. When Rita got this assignment, she did not know what a protocol document was, so she first had to look into the history of this type of document, both in general and in the context of this hospital. To do this, she talked with her coworkers and turned to medical research on the topic.

Protocols are used in medical settings to convey medical information and regulate the patient care delivered by clinical professionals. Medical protocols are forms of scientific and technical communication, specifically instructional documents, which convey patient-care directions for use by medical professionals in clinical contexts. In these documents, medical knowledge is presented through both visual conventions and verbal text, often as flowcharts or step-by-step directives, which nurses, doctors, and other health-care workers use to regulate patient care. Through her research, Rita learned that the purpose of protocols is to "reduce inappropriate variations in clinical care, minimize harm, promote cost effective practice, and produce optimal health outcomes for patients" (Steering Committee 2004, 874). Protocols function more like standing orders than recommendations, because they "authorize nurses and pharmacists to administer [treatments] according to an institution- or physician-approved protocol without a physician's exam" (CDC 2000). A clinical protocol, then, authorizes a specific set of patient-care actions to be carried out by qualified nurses, pharmacists, or technicians without a doctor's orders for that individual patient. Though these protocols take different forms, from "elaborately designed" charts to a list of "general recommendations," they all have two features in common: (1) they are intended to lead medical professionals through a sequence of events, and (2) they are inspired by (if not always specifically conforming to) medical research findings (Berg 1997, 1081).

WHAT SYSTEMS OF ORDER DOES MY WORKPLACE CREATE?

Then Rita began looking into how the current insulin protocol had been developed. She found that the ICU medical director had learned of a research study showing improved patient outcomes when blood glucose was monitored and insulin administered using a different procedure than was standard medical practice for most hospitals. This medical director wanted to implement those research findings in clinical practice by developing a protocol based on those findings, one used to monitor blood glucose and administer insulin in his ICU. The medical director formed a committee of people from different areas of the hospital—pharmacy, nursing, the business office—to look at the research and translate its findings into guidelines for patient care in the form of a protocol. By creating this document, the committee standardized and ordered one area of medical practice with the goal of improving patient care and outcomes. It was important for Rita to know what prompted the creation of the existing document, who had been involved with its development, what they considered important and unimportant as they developed the protocol, and how this document fits into and orders medical practices.

Rita found that after using the existing protocol for almost a year, the staff often had questions about how to carry out the procedures because of the protocol's poor design. They sometimes did not monitor a patient's blood glucose at the right time or administer insulin according to the protocol because the information was not presented in a way that nurses could quickly understand in the hectic ICU setting. Sometimes the sequence of events as set out in the protocol did not happen. So to improve how ICU staff followed these guidelines, the medical director decided to bring in a technical communicator to redesign the protocol.

Once Rita understood the history of the existing protocol, she needed to find out more background on the research the medical director used and the purpose of protocols within the larger medical context. She learned that the standards of practice in the existing protocol were based on research findings of a clinical trial published in 2001 in a widely circulated medical journal. This study showed a 42% reduction in mortality in patients that had hyperglycemia managed with a protocol with a glucose target of 80–110 mg/dl (Van den Berghe et al. 2001). Thus, the intentions and content of the existing protocol were based on the (historical) findings of a research study that took place before the document was developed. Because this past knowledge promised greatly improved outcomes for ICU patients, hospital staff considered it important to create a protocol that would translate these research findings into a document that would regulate clinical practices at the bedside.

When Rita looked into why protocols are used in medicine, she found that they are one component of a larger evidence-based medicine (EBM) movement in contemporary health care, which seeks to "eschew unsystematic and 'intuitive' methods of individual clinical practice in favor of a more scientifically rigorous approach" (Goldenberg 2006, 2621). EBM seeks to make clinical research and deliberate decision making the foundation of medicine. By structuring patient care according to standardized guidelines drawn from clinical research findings and other verifiable evidence, proponents of EBM argue that such practices—based on historical findings—potentially enhance health care by increasing evidence-supported procedures across a number of clinical settings (Berg 1997, 1082). And though EBM has been the subject of ongoing debate, most medical professionals agree that EBM is here to stay (see Berg 1997).

After looking into the history of the existing protocol within a larger context of contemporary health-care practices, Rita understood that the development of the ICU insulin protocol was intended to present research-based medical knowledge both visually and verbally in order to provide step-by-step directions needed to measure blood glucose levels and, if needed, administer insulin to nondiabetic patients. Rita's redesigned protocol will regulate health-care providers' decision-making processes as they ensure that standards of care are achieved. This protocol is one technical document intended to create order by guiding medical practice. As the technical communicator who designs and composes it, Rita is working within that larger system of ordering.

HOW DOES THE HISTORY AND CULTURE OF MY WORKPLACE INFLUENCE THIS PROCESS OF ORDERING?

To understand this particular technical document, the insulin protocol, Rita must understand the larger historical, cultural, and institutional processes that constitute the document. First, as medical paperwork, the existence of the insulin protocol is a result of the history of medical record keeping and documentation. Second, the insulin protocol is part of a larger cultural and professional shift in late-twentieth-century Western medicine, namely EBM. These protocols, then, are created as a response to the EBM movement and represent a manifestation of it. Third, this specific insulin protocol for ICU patients finds its origin in both EBM and the clinical research of Van den Berghe and her team, whose findings became the gold standard of practice for regulating glucose levels in all nondiabetic ICU patients. In creating this technical document, the staff turned to the Van den Berghe et al. (2001) study because it was recognized as authoritative medical knowledge in this area. Here we see the development of a technical document generated from an awareness of the historical context (the Van den Berghe et al. study) and institutional/organization practices (EBM).

In creating this protocol, hospital staff not only translated research findings, but also articulated updated goals for clinical outcomes in the ICU through new practice regulations. In this case, the hospital staff who designed the document performed as technical communicators in order to implement research findings into standardized professional practices within their organization. (For an analysis of the visual design of protocols and that design's effect on practice, see Longo, Weinert, and Fountain 2007.) This particular protocol was created to order clinical practice by guiding the decisions of medical professionals. Protocols are a form of "regulatory objectivity" approached through "the systematic recourse to the collective production of evidence" (Cambrosio et al. 2006, 189). These technical documents function as a type of objectivity that orders medical practice on agreed-upon evidence, such as clinical research, decision-making committees, and other types of verifiable results. Rather than the individualistic deliberation processes showcased on television shows like *Grey's Anatomy*, this insulin protocol—its creation and use—helps manage medical decision making by offering a procedure to follow. As a technical document, Rita's protocol will order human activity by instructing medical professionals how to act in a given situation.

HOW WILL MY DECISIONS IN THIS WORKPLACE SHAPE AND ORDER THE ACTIONS OF OTHERS?

In creating this medical protocol, one fundamental question Rita must ask is this: How will this document influence, shape, and order the practices of medical professionals? How exactly will this document be used in the ICU? The development and use of a protocol has significant consequences; in the case of ICU patients, it means the proper regulation of blood glucose. Nurses and other medical professionals will use this document (as either a printed text or a computer interface) to care for patients. More than just a set of instructions, this document makes possible a specific medical intervention, and thus the promotion of health and the treatment of illness. And one of the most powerful ways Rita will contribute to this system of ordering is to produce a document that (whether embraced or questioned) allows the work of medicine to take place as smoothly and as responsibly as possible. Rita's document must respond to, and even anticipate, the needs and expectations of medical professionals. And to do this, Rita must understand the scientific and technical practices her document will make possible. This means she must observe the work of medical professionals using the current document and keep that social context and action in mind as she composes the new one. After all, the implementation of Rita's protocol will shape the work of these professionals, telling them what to do, how, and when. And they will accomplish this ordering as part of the historical configurations that merge medical science with the business of health care (see Popham 2005).

The development of protocols represents one way that institutional and cultural histories shape contemporary technical communication practices. Yet the existence of a protocol does not translate into a transparent system of use. How closely a medical professional follows this document depends on a number of factors that Rita must consider. How is the ICU protocol used now? How do administrators want it to be used? How do nurses feel about protocolized medicine? Whether or not a worker faithfully administers care using the protocol comes down to a number of factors, such as medical training, position of authority, and the organizational culture of Rita's hospital and the ICU ward. Lara Varpio and her research collaborators (2007), in their study of record-keeping practices at an optometry program, found that students learned to keep accurate records not just by following the visual and verbal cues of the forms, but also by understanding the social context. Here, students learned authorized and official social practices, which often created tension between the task and the paperwork, in the actual social contexts of use (also see Schryer, Lingard, and Spafford 2007). Using an insulin protocol effectively and authentically involves a process of socialization through which a worker comes to share the same ideals and values of a medical ward or institution—ideals and values that encourage some actions while discouraging others. Because of this, Rita must also take time to talk with several users to understand their needs and suggestions. After all, Rita's document will contribute to the socialization process that shapes these medical professionals. By creating a document that tells a health-care professional what to do and not to do, Rita is working to fulfill historical and cultural goals of a particular organization, as well as the medical profession more broadly.

CONCLUSION

So what can history teach us about technical communication? The short answer is a great deal more than we realize. The history of technical communication involves both the past historical events of the field and the historical narratives and research written about those events. Our knowledge of our professional past comes to us through the texts of others, specifically the texts of historians and researchers who investigate what used to be. Though this is an important avenue for uncovering past practices, research into the past is not the only way to understand the roles and responsibilities of yesterday's technical communicators. We must also look to our current workplace—where technical communication is composed, designed, and presented. These contemporary practices and the histories from which they developed shape our ideas about our work, our values involving the profession, and our identities as communicators. In order to honor this rich, diverse, and unfinished history of our profession, we need to know not only the history of a document, the people who created it, and its current usage, but also the role we play in the larger social practices of science, technology, and business.

We can begin to get in touch with the cultural and organizational history of our workplace by asking ourselves these questions: (1) What systems of order does my workplace create? (2) How does the history and culture of my workplace influence this process of ordering? (3) And how will my decisions shape and order the action of others? As we have seen with Rita's work to create an ICU insulin protocol, a rigorous understanding of history can bring that knowledge into greater focus.

DISCUSSION QUESTIONS

1. What does it mean to say that technical communication operates as a system of order and a process of ordering scientific knowledge?
2. How might this chapter's definition of technical communication influence the way we understand the history of science, of technology, of business?
3. Many people are called upon to develop documents that meet the needs of their workplaces, from administrative assistants to software developers. Thus, people who are not technical communicators accomplish a great deal of technical communication. What unique expertise and skills do technical communicators contribute to an organization?
4. What does it mean to say that hospitals and clinics are technical communication workplaces? Does the presence of technical communication in these settings influence your ideas about (1) medicine as a profession and (2) technical communication as a profession? Consider specific examples and scenarios in your response.
5. What does it mean to be socialized into the culture of an organization? How might we understand technical documents as socialization tools?
6. Concerning the decision-making processes of people in workplaces, Mary Douglas (1986, 4) asserts that "an answer is only seen to be the right one if it sustains the institutional thinking that is already in the minds of individuals as they try to decide." What exactly does she mean? Consider the institutions of which you are a member (school or your workplace). How might Douglas's assertion apply to these institutional contexts? Consider scenarios involving school or a workplace where Douglas's ideas ring true. Consider scenarios in which we might challenge her assertion.

7. Find a technical document created at least fifteen years ago (a user manual or set of instructions, for example).

    - How might we understand that document as a historical artifact?
    - How do the features of that document (organization, style, language, and graphics) represent its social and cultural setting?
    - What type of order does this historical document create?
    - What influence does this document have on current technical communication practice?

8. Go online and check out the website for a technical communication company or any company where technical communicators work.

    - How much information is given about the company's history and professional philosophy? Why might they have included this information?
    - What view of that organization's culture and values do we perceive from the company's website? How are the culture and values communicated?
    - How might that company's philosophy and organizational culture manifest itself in the company's technical documents? In what ways do documents on the site represent the company's philosophy and culture?

NOTES

1. Related to these histories of recovery is feminist research into the gendered past of technical communication (see Lay 1994; Durack 1997; Lippincott 1997, 2003; Tebeaux 1998; Kynell 1999).

2. Elizabeth Britt (2006, 135) makes a distinction between organizations and institutions, namely that organizations "exhibit self-consciousness" while institutions "are marked by a certain taken-for-grantedness." Though we agree with her assertion, we do not distinguish the two, because our observations apply to both.

WORKS CITED

Argyris, Chris. 1964. *Integrating the Individual and the Organization*. New York: John Wiley.

Bazerman, Charles. 1988. *Shaping Written Knowledge: The Genre and Activity of the Experimental Article in Science*. Madison: University of Wisconsin Press.

———. 1999. "How Natural Philosophers Can Cooperate: The Literacy Technology of Coordinated Investigation in Joseph Priestley's *History and Present State of Electricity*." In *Three Keys to the Past: The History of Technical Communication*, edited by Teresa C. Kynell and Michael G. Moran, 21–48. Stamford, CT: Ablex.

Berg, Marc. 1997. "Problems and Promises of Protocols." *Social Science & Medicine* 8:1081–1088.

Britt, Elizabeth. 2006. "The Rhetorical Work of Institutions." In *Critical Power Tools:*

Technical Communication and Cultural Studies, edited by J. Blake Scott, Bernadette Longo, and Katherine V. Wills, 133–150. Albany: State University of New York Press.
Brockmann, R. John. 1999. "Oliver Evans and His Antebellum Wrestling with Rhetorical Arrangement." In *Three Keys to the Past: The History of Technical Communication*, edited by Teresa C. Kynell and Michael G. Moran, 63–91. Stamford, CT: Ablex.
Cable, Daniel M., and Timothy A. Judge. 1997. "Interviewers' Perceptions of Person-Organization Fit and Organizational Selection Decisions." *Journal of Applied Psychology* 82 (4): 546–561.
Cambrosio, Alberto, Peter Keating, Thomas Schlich, and George Weisz. 2006. "Regulatory Objectivity and the Generation and Management of Evidence in Medicine." *Social Science & Medicine* 63:189–199.
Centers for Disease Control Advisory Committee on Immunization Practices. 2000. "Use of Standing Orders Programs to Increase Adult Vaccination Rates, Records, and Reports." *Morbidity and Mortality Weekly Report* 49 (March 24). http://www.cdc.gov/MMWR/preview/mmwrhtml/rr4901a2.htm. Accessed November 13, 2006.
Devitt, Amy. 1991. "Intertextuality in Tax Accounting: Generic, Referential, and Functional." In *Textual Dynamics of the Professions: Historical and Contemporary Studies of Writing in Professional Communities*, edited by Charles Bazerman and James Paradis, 336–357. Madison: University of Wisconsin Press.
Di Renzo, Anthony. 2000. "His Master's Voice: Tiro and the Rise of the Roman Secretarial Class." *Journal of Technical Writing and Communication* 30 (2): 155–168.
Dobrin, David. 1983. "What's Technical about Technical Writing." In *New Essays in Technical and Scientific Communication: Research, Theory, Practice*, edited by Paul V. Anderson, R. John Brockmann, and Carolyn R. Miller, 227–250. Farmingdale, NY: Baywood.
Dombroski, Paul. 1991. "The Lessons of the *Challenger* Investigation." *IEEE Transactions on Professional Communication* 34 (4): 211–216.
Douglas, Mary. 1986. *How Institutions Think*. Syracuse, NY: Syracuse University Press.
Durack, Katherine T. 1997. "Gender, Technology, and the History of Technical Communication." *Technical Communication Quarterly* 6 (3): 249–260.
Foucault, Michel. 1994. *The Order of Things: An Archaeology of the Human Sciences*. New York: Vintage Books.
Goldenberg, Maya J. 2006. "On Evidence and Evidence-Based Medicine: Lessons from the Philosophy of Science." *Social Science & Medicine* 62:2621–2632.
Hager, Peter J., and Ronald J. Nelson. 1993. "Chaucer's *A Treatise on the Astrolabe*: A 600-Year-Old Model for Humanizing Technical Documents." *IEEE Transactions on Professional Communication* 36 (2): 87–94.
Johns, Lee Clark. 1989. "The File Cabinet Has a Sex Life: Insights of a Professional Writing Consultant." In *Worlds of Writing: Teaching and Learning in Discourse Communities of Work*, edited by Carolyn B. Matalene, 153–187. New York: Random House.
Johnson, Carol Siri. 2006. "Prediscursive Technical Communication in the Early American Iron Industry." *Technical Communication Quarterly* 15 (2): 171–189.
Kimball, Miles A. 2006. "London through Rose-Colored Graphics: Visual Rhetoric and Information Graphic Design in Charles Booth's Maps of London Poverty." *Journal of Technical Writing and Communication* 36 (4): 353–81.
Kynell, Teresa C. 1999. "Sada A. Harbarger's Contribution to Technical Communication in the 1920s." In *Three Keys to the Past: The History of Technical Communication*, edited by Teresa C. Kynell and Michael G. Moran, 91–103. Stamford, CT: Ablex.

Kynell, Teresa C., and Michael G. Moran, eds. 1999. *Three Keys to the Past: The History of Technical Communication*. Stamford, CT: Ablex.
Lay, Mary M. 1994. "The Value of Gender Studies to Professional Communication Research." *Journal of Business and Technical Communication* 8:58–90.
Lippincott, Gail. 1997. "Experimenting at Home: Writing for the Nineteenth-Century Domestic Workplace." *Technical Communication Quarterly* 6 (4): 365–380.
———. 2003. "Rhetorical Chemistry: Negotiating Gendered Audiences in Nineteenth-Century Nutrition Studies." *Journal of Business and Technical Communication* 17 (1): 10–49.
Longo, Bernadette. 2000a. "(Re)Constructing Arguments: Classical Rhetoric and Roman Engineering Reflected in Vitruvius' *De Architectura*." *Journal of Technical Writing and Communication* 30 (1): 49–55.
———. 2000b. *Spurious Coin: A History of Science, Management and Technical Writing*. Albany: State University of New York Press.
Longo, Bernadette, Craig Weinert, and T. Kenny Fountain. 2007. "Implementation of Medical Research Findings through Insulin Protocols: Initial Findings from an Ongoing Study of Document Design and Visual Display." *Journal of Technical Writing and Communication* 37 (4): 435–452.
Louis, Meryl Reis. 1980. "Surprise and Sense Making: What Newcomers Experience in Entering Unfamiliar Organizational Settings." *Administrative Science Quarterly* 25:226–251.
Lutz, Jean Ann. 1989. "Writers in Organizations and How They Learn the Image: Theory, Research, and Implications." In *Worlds of Writing: Teaching and Learning in Discourse Communities of Work*, edited by Carolyn B. Matalene, 113–135. New York: Random House.
Moran, Michael G. 1999. "Renaissance Surveying Techniques and the 1590 Hariot-White-de Bry Map of Virginia." In *Three Keys to the Past: The History of Technical Communication*, edited by Teresa C. Kynell and Michael G. Moran, 153–171. Stamford, CT: Ablex.
———. 2002. "A Fantasy-Theme Analysis of Arthur Barlowe's 1584 *Discourse* on Virginia: The First English Commercial Report Written about North America from Direct Experience." *Technical Communication Quarterly* 11 (1): 31–59.
———. 2003. "Ralph Lane's 1586 *Discourse on the First Colony*: The Renaissance Commercial Report as Apologia." *Technical Communication Quarterly* 12 (2): 125–154.
Munger, Roger. 2000. "Evolution of the Emergency Medical Services Profession: A Case Study of EMS Run Reports." *Technical Communication Quarterly* 9 (3): 329–346.
Odell, Lee. 1985. "Beyond the Text: Relations between Writing and Social Context." In *Writing in Nonacademic Settings*, edited by Lee Odell and Dixie Goswami, 249–280. New York: Guilford Press.
Popham, Susan L. 2005. "Forms as Boundary Genres in Medicine, Science, and Business." *Journal of Business and Technical Communication* 19 (3): 279–303.
Schryer, Catherine F., Lorelei Lingard, and Marlee M. Spafford. 2007. "Regularized Practices: Genres, Improvisation, and Identity Formation in Health-Care Professions." In *Communicative Practices in Workplaces and the Professions: Cultural Perspectives on the Regulation of Discourse and Organizations*, edited by Mark Zachry and Charlotte Thralls, 21–44. Amityville, NY: Baywood.
Slack, Jennifer Daryl, David James Miller, and Jeffrey Doak. 2003. "The Technical Communicator as Author: Meaning, Power, Authority." In *Power and Legitimacy*

in Technical Communication, vol. 1, *The Historical and Contemporary Struggle for Professional Status*, edited by Teresa Kynell-Hunt and Gerald Savage, 169–192. Amityville, NY: Baywood.

Steering Committee on Quality Improvement and Management. 2004. "Classifying Recommendations for Clinical Practice Guidelines." *Pediatrics* 114:874–877.

Tebeaux, Elizabeth. 1997. *The Emergence of a Tradition: Technical Writing in the English Renaissance, 1475–1640*. Amityville, NY: Baywood.

———. 1998. "The Voices of English Women Technical Writers, 1641–1700: Imprints in the Evolution of Modern English Prose." *Technical Communication Quarterly* 7 (2): 125–152.

Tolbert, Jane T. 2001. "Seventh-Century Technical and Persuasive Communication: A Case Study of Nicolas-Claude Fabri de Peiresc's Work on a Method of Determining Terrestrial Longitude." *Journal of Business and Technical Communication* 15 (1): 29–52.

Trice, Harrison M., and Janice M. Beyer. 1993. *The Cultures of Work Organizations*. Englewood Cliffs, NJ: Prentice Hall.

Van den Berghe, G., P. Wouters, F. Weekers, C. Verwaest, F. Bruyninckx, M. Schetz, D. Vlasselaers, P. Ferdinande, P. Lauwers, and R. Bouillon. 2001. "Intensive Insulin Therapy in Critically Ill Patients." *New England Journal of Medicine* 345 (19): 1359–1367.

Van Maanen, John. 1976. "Breaking In: Socialization to Work." In *Handbook of Work, Organization, and Society*, edited by Robert Dubin, 67–130. Chicago: Rand McNally.

Varpio, Lara, Marlee M. Spafford, Catherine F. Schyrer, and Lorelei Lingard. 2007. "Seeing and Listening: A Visual and Social Analysis of Optometric Record-Keeping Practices." *Journal of Business and Technical Communication* 21 (4): 343–375.

Winsor, Dorothy. 1990. "The Construction of Knowledge in Organizations: Asking the Right Questions about the *Challenger*." *Journal of Business and Technical Communication* 4 (2): 7–20.

Yates, JoAnne, and Wanda J. Orlikowski. 1992. "Genres of Organizational Communication: A Structurational Approach to Studying Communication and Media." *Academy of Management Review* 17 (2): 299–326.

Zappen, James. 1999. "Francis Bacon and the Historiography of Scientific Rhetoric." In *Three Keys to the Past: The History of Technical Communication*, edited by Teresa C. Kynell and Michael G. Moran, 49–62. Stamford, CT: Ablex.

BRAD MEHLENBACHER

8 What Is the Future of Technical Communication?

SUMMARY

Do the same communication principles that worked for offices and industrial workplaces in the twentieth century work in the online and distributed workspaces of the twenty-first? After providing a scenario of technical communication work in a contemporary organizational context, this chapter draws on research describing the geographically and temporally distributed contexts that envelop technical communication activities, and offers a heuristic that helps readers conceptualize some of the twenty-first-century capacities necessary to operate effectively as rhetorically sensitive multicommunicators. Future technical communicators face ill-structured communication-design situations characterized by audiences with limited attention, doing several things at once, attempting to deal with too much incoming information across too many media devices (phone, television, iPads, laptops, etc.).

INTRODUCTION

Janine is part of a virtual team employed by a major university to provide support for faculty and instructors using a web-based open-source learning management system (LMS). Her team is providing a full documentation suite using XML-GL, a graphical version of the XML markup language. XML allows her team to define all the features of the documents that she builds, including their formatting, font, headings and subheadings, and so on. Her full documentation suite includes an overview video, a general tutorial, a getting-started guide, quick reference information, a user's guide and reference information, several "teaching tip" demos, and a comprehensive online help system. Before this project, she served as a student intern assisting a team with an established documentation suite of similar size and complexity, although for a sophisticated spreadsheet application. The on-the-job experience she gained using HTML and then XHTML to design web-ready versions of some of the key documents will reduce her learning curve for XML.

Thus, Janine's work is fragmented and her problems are wicked (that is, unstructured, requiring immediate attention, without easy solutions or solutions that are easily compared to alternative solutions). She shares virtual spaces with coworkers, and she shares physical spaces with them as well. The shared virtual spaces span numerous established and developing technological platforms and applications, spatially and temporally distributed and high speed, collaborative and isolated. She e-mails, organizes conference calls, instant messages, contributes to several technical user forums, and collects, shares, and synthesizes feedback on her draft documentation. She coordinates continuously with members of the primary development team. Occasionally Janine gets the opportunity to interact with students and faculty who have used or are learning to use the university LMS. Team-member responsibilities are self-defined, the (documentation) "problem" she is solving is constrained only by time and resources, and "completeness" of the project is determined by her team's collective goals, changes to the LMS, and definitions of a useful and well-designed documentation suite.

Her current team is multidisciplinary and multilingual—consisting of a project manager, a programmer, an interface specialist, a graphics designer, and Janine; and her organization is nonhierarchical, her team interacting with other product development and product support units. The work flow of her team is intimately connected to the documents that her team creates and maintains. They meet regularly via a web conferencing application, rely heavily on real-time chat for day-to-day exchanges, and use a document-sharing application to keep track of multiple existing and developing versions of the documentation. Members of the team value Janine's collaborative abilities, including her interpersonal and communication skills as well as her experience multitasking and working with tight deadlines. She is constantly in an in-between state, balancing her attention, energy, interpretive capabilities, and cognitive-processing abilities across numerous cues from her work environment.

This scenario offers a realistic view of the contemporary work of technical communicators. Historically, technical writers worked in isolation producing documentation for hardware or software and receiving minor input from subject-matter experts who understood the software or hardware at a technical level; occasionally, technical writers were technical enough to become experienced users of the software or hardware themselves, incorporating this knowledge into their documentation-writing process. Contemporary technical communicators, however, rarely work in isolation and therefore spend a considerable amount of time and energy communicating their contributions for others. While many undergraduate and graduate programs present technical communication students with opportunities to build websites, design manuals, and produce proposals, procedures, and tutorials, most contemporary organizations place technical communicators on teams, managing projects too big for any one person, supporting software and hardware efforts that will be designed, tested, documented, and used by audiences around the world and across time zones.

This chapter presents a somewhat stoic professional role for technical communicators, acknowledging the complex worlds in which they work and the contingent nature of all forms of technological expertise. While technological literacy is valued in the workplace, including familiarity with both specific (e.g., Adobe Acrobat) and general software applications (e.g., word processors and graphics programs), the greatest challenge facing future technical communicators is largely a communication challenge. Technical communicators can benefit considerably from focusing on their problem-solving capacities and learning processes, on their specialized capacities as researchers, organizers, and synthesizers, on their role as sociotechnical mediators and genre specialists, and on the development and cultivation of conceptual artifacts (that is, "texts") that support rather than undermine human understanding and activity. To that end, this chapter explores the relationships between communication abilities, emerging technological interfaces, and heuristics for strategic problem solving. This chapter also describes the relationship between our work as communicators and our work as technical specialists, emphasizing the skills that communication specialists bring to technical situations rather than the abilities that technical specialists bring to communication events.

LITERATURE REVIEW

Future technical communicators will operate in work contexts where their work is not well defined for them, contexts that demand flexible problem-solving abilities, that is, short- and long-term solutions achieved collaboratively. The problems that they encounter in these contexts will require expertise that no single person is likely to have (due to limited time, memory constraints, incomplete access to learning materials, or complex systems) and that necessitate ongoing sensitivity to sociotechnical mediation (to numerous technologies and to the many audiences that participate in contemporary technological developments). These problems also demand learning during an ever-increasing time famine punctuated by increasingly reduced product cycles, interruptions, and accelerated local and international deadlines (Perlow 1999), even while workers enjoy unprecedented freedom from the traditional constraints of space and time in their virtual work lives.

Technical communicators have become experts in document-production technologies and frequently influence not only the design of the documents and user-assistance systems that support software and hardware but also the interfaces and designs of the software and hardware. Technical communicators no longer work solely for the military or for IBM, writing systems documentation as they did in the 1970s (Rigo 2001). Indeed, in Rainey, Turner, and Dayton's (2005, 326) survey of sixty-seven technical communication managers, respondents reported that their technical communication groups work with an extraordinarily diverse range of genres, including, for example, PDF and hardcopy documentation, online help, style guides, reference and training materials, intranet sites, books, newsletters, annual reports, magazines, proposals, company websites, performance evaluations, video scripts, usability reports, and marketing materials.

To anticipate the future of technical communication, then, it is useful to review the present: the work contexts that technical communicators currently inhabit and the problems that they work with and solve. This naturally leads us to review the nature of expertise in our work. In the past, expertise was viewed two-dimensionally, with the subject-matter expert being an expert on the technology or process and the technical writer an expert on writing and document design. Today, expertise is distributed, if it exists at all. The technical communicator's developing position is as sociotechnical mediator, balancing his or her knowledge of technologies and technological processes with the numerous audiences, users, and developers of those environments.

THE PROBLEMS OF TECHNICAL COMMUNICATION

Technical communicators routinely generate documents in ill-structured domains, that is, in environments that are unstable, that demand flexibility and a creative ability to organize across similar but always different problems and to understand, argue, and evaluate both conceptually and pragmatically. Organizing across complex problems requires not only that technical communicators work with information differently but also that they come to understand knowledge in new ways. As Resnick, Lesgold, and Hall (2005, 79) point out, our understanding of what constitutes knowledge has changed dramatically during the last several technology-rich decades. Knowledge is no longer represented in the form of lists, *primary* sources, controlled areas of expertise, or fixed private states of understanding. Instead, knowledge is contingent, framed by higher-order and changing structures, publicly distributed, and drawn from multiple, emergent sources.

As Spinuzzi (2007, 265) describes, our ill-structured work contexts are frequently characterized by "downsizing, automation, flattening of work hierarchies, increasing numbers of relationships between companies, continual reorganization, the breaking down of silos or stovepipes in organizations, and perhaps most importantly, the increase in telecommunications . . . , which has made it possible to connect any one point to any other, within and across organizations." Within these contexts, technical communicators collect, sort, analyze, interpret, design, and report data, and collaborate, communicate, interact, and negotiate with other professional problem solvers. And they do so by creatively acting as "presence allocators," that is, as problem solvers who can "survey the available communication technologies, choose a medium that provides the right cues for each interaction, and divide [their] presence among two or more interlocutors" (Turner and Reinsch 2007, 47).

Technical communicators are routinely confronted with increasingly wicked problems, a term originally employed by Karl Popper (1972) to describe complex problems (Buchanan 1992). Wicked problems can be contrasted with "tame problems." Conklin (2005, 9–10) defines tame problems: they are well-defined, have explicit stopping points and solutions that can be evaluated as correct or incorrect, and belong to a class of similar problems that have similar solutions (or that have a limited set of alternative solutions).

Thus, although playing a game of chess may be complex and require considerable expertise, problems and solutions in chess can be defined as tame versus, for example, the problems and solutions involved in designing documentation for a new vehicle. Problems in chess involve high-level strategies that can be repeated across chess games, learnable goals that can be improved over time (e.g., control of the center or protection of the king), discrete, single moves, fixed turn taking, shared definitions of action and response, parts that behave in consistent ways, set beginnings and endings, definitive closure, clear winners and losers. Wicked problems are not games (although some attempt to navigate them as though they are). Wicked problems frequently have difficult-to-identify beginnings and endings, incomplete information about the rules of play, strategies that can succeed in one setting and fail in another setting that looks identical, unpredictable resources (or pieces), players who do not know the rules or follow them (yet they are shareholders in the outcome of the engagement), and no checkmate—ever—unless we define checkmate as a conclusion defined by running out of time or resources.

Wicked problems invite numerous misconceptions on the part of busy problem solvers. The first misconception is oversimplification; that is, learners either develop incomplete conceptual understandings or generalize features of one problem instance to other instances with different characteristics. Unfortunately, problem solvers often tend to develop lone mental representations consisting of general features and to apply these representations to all future cases (e.g., all situations in which you propose something are the same). The second misconception is the development of inflexible knowledge structures or rigid definitions of the problem that end up being applied as procedures to more complex cases (e.g., all proposing situations require written proposals). And the third misconception is that problem solvers resist or altogether ignore indeterminate or uncertain information in favor of building problem representations that are easy to apply (e.g., all written proposals contain an executive summary, rationale, and budget section) (Spiro et al. 1987). Although an absolute or perfect understanding of complex subject matter may not exist, there can certainly be identifiably incorrect understandings—that is, insightful problem solving is only possible with deep understanding, and deep understanding involves deep involvement with content and with different audiences, situations, and contexts.

Wicked knowledge work demands that technical communicators generate more in less time more efficiently. Stinson (2004, 167) characterizes our knowledge age as a time where employees are expected to have greater competencies, to manage complex projects, to work harder across more hours of the day, and to juggle both long-term goals and day-to-day organizational needs. And these new realities are exacerbated by high unemployment rates, increased competition for less lucrative jobs, and organizational and market uncertainty.

Given the pressures to solve problems quickly while working with complex problems in ill-structured environments, the technical communicator's ability to achieve what Bazerman (1988) describes as "rhetorical self-consciousness" is exceedingly difficult. Rhetorical self-consciousness involves constant application of the following strategies:

- consider your fundamental assumptions, goals, and projects;
- consider the structure of the literature, the structure of the community, and your place in both;
- consider your immediate rhetorical situation and rhetorical task;
- consider your investigative and symbolic tools;
- consider the processes of knowledge production; and
- accept the dialectics of emergent knowledge. (323–329)

Bazerman's (1988, 147) call for rhetorical self-consciousness parallels Selber's (2004) recommendation that "rhetorically literate" learners be versed in persuasion (interpreting and applying both implicit and explicit arguments), deliberation (acknowledging that ill-defined problems demand thoughtful representation and time), reflection (demanding both articulation and critical assessment), and social action (defining all technical action as social action). And Fleming (2003, 105–106), as well, advocates the preparation of rhetorically sensitive professionals, maintaining that such individuals would have an understanding of circumstantial knowledge (people, places, events, history of the situation at hand), verbal formulas (discourse patterns of a particular community and situation), common sense (community truths, norms, and values), models of textual development (patterns of argumentative thinking in the community), and logical norms (knowledge, warrants, argumentative rules).

Before technical communicators can aspire to the role of rhetorically literate, sensitive, and self-conscious contributors to their profession, they must first acknowledge how their wicked contexts will continue to modify historical notions of expertise in emerging sociotechnical settings.

THE DEATH OF EXPERTISE

It is exceedingly difficult to find individuals we can label, with any degree of confidence, as "experts," that is, if we are defining an expert as someone who knows "everything" about a database we are accessing, a similar version of the same software application, a particular corporate policy or procedure for managing an unusual employee situation, or the features of a genre that is uncommon to our corporate setting. Yet much of the early cognitive-science research was organized around the assumption that, if we learn how experts behave and think, novices can learn how to behave and think the way experts do (Chi, Glaser, and Rees 1982).

We have learned, however, that both experienced and inexperienced learners develop rich mental models of learning tasks and concepts that guide them as they apply knowledge to given situations and acquire new knowledge for use in new situations (Johnson-Laird 1983). These rich models make it difficult for them to communicate effectively with each other.

So expertise is intensely contextualized and social (Brown and Duguid 2000). Moreover, expertise is dynamic and socially constructed and often changes from one problem setting to another. It may even be, as Sternberg (2003) notes, that expertise comes in many different forms, including the ability to think critically (analysis, evaluation) or creatively (invention, discovery) or practically (implementation, use) or wisely (social good, humility). And Brown and Duguid (1992, 172–173) point out that it is certainly possible for people to be both experts and novices in different circumstances: expertise operates as a continuum from novice to expertise rather than as a dichotomy, and experts still need to learn no matter how much knowledge they acquire. Progressive myths of technological progress and notions of the "self-made man" encourage us to overlook the many situations in which experts must learn, and unfortunately, ill-structured situations increase how frequently these moments occur. Ill-structured situations demand forms of expertise that emphasize intelligent relationships to things or situations in the world rather than factual or easily statable knowledge.

Given the historical positioning of technical communicators as user advocates who explain or translate technical concepts for nontechnical novices, the distributed nature of expertise has serious implications for the profession. Most importantly, technical communicators need to adapt themselves as facilitators and mediators rather than as instructors or experts. Mediators, as "interface persons," operate at the edges of communities, understanding, communicating, and negotiating solutions for different audiences with different rules for participation and contribution (Créplet, Dupouët, and Vaast 2003). These activities require that we create and contribute to various "intentional networks" where "joint activity is accomplished by the assembling of sets of individuals derived from overlapping constellations of personal networks" (Nardi, Whittaker, and Schwarz 2000).

THE STRATEGIES OF SOCIOTECHNICAL MEDIATORS

Wicked twenty-first-century work forces us to admit that no one person can know everything about his or her area of expertise. Learning how to learn, therefore, needs to be viewed as a chief professional and educational goal (Fischer 2000). Wicked problems tend not to have easy solutions, are ongoing rather than having identifiable closure, cannot be tested for total accuracy, and can have many "causes" rather than just one obvious reason for existing. Our incomplete knowledge of wicked problems means that most technical communicators are not going to fully understand all of the technical processes and products with which they work. Still, at the most profound level—drawing on the disciplines of rhetoric, psychology, linguistics, and communication—technical communicators need to strive to understand and mediate the relationship between complex symbolic systems and human beings.

We live in worlds filled with technology where work, leisure, and learning are blurring and where distinctions between real and representation are increasingly difficult to maintain (Mehlenbacher 2010). If you watch a

YouTube video of a lecture at Yale, how is your experience of the lecture different from the person who attended the lecture itself? Burbules (2004) emphasizes that we rarely have "direct perceptions" of anything. Technologies—either obvious, such as eyeglasses or cameras, or conceptual, such as stereotypes or assumptions—infiltrate many of our most direct interactions with the world. Our direct interactions, then, are but *versions* of the world (165).

Technical communicators operate at the intersection between these technological versions of the world and conceptual ones. So, ultimately, technical communicators must understand and invent the technological realities that we describe and create. Technical communicators must construct and write an audience-friendly description of the web application, rather than being able to capture any particular *true* version of the web application. Our preparation for this role can be drawn from an important historical precedent, the shift from an oral to a textual culture. Walter J. Ong describes writing as a technology in his *Orality and Literacy: The Technologizing of the Word* (1982, 82):

> To say that writing is artificial is not to condemn it but to praise it. Like other artificial creations and indeed more than any other, it is utterly invaluable and indeed essential for the realization of fuller, interior, human potentials. Technologies are not mere exterior aids but also interior transformations of consciousness, and never more than when they affect the word. Such transformations can be uplifting. Writing heightens consciousness. Alienation from a natural milieu can be good for us and indeed is in many ways essential for full human life. To live and to understand fully, we need not only proximity but also distance.

Writing is a technology with given document or genre characteristics such as text, syntax, lexicon, intended purpose, and audience. And writing is also a technology for mediating between technologies and humans, allowing interactions between various audience attributes (e.g., reading level, demographic characteristics) and authorial goals for reader response, text use or purpose, and so on (Redish 1993). The information that technical communicators produce is as much *formed by* our technological contexts as it *forms* our technological contexts. As well, just as the technologies we invent require articulation, so too do those technologies invent us. The information that we produce to conceptualize, explain, support, market, and help us act is not an object, entity, or module that operates apart from our communities or our contexts. In addition to information *about* and *for* reality, Borgmann (2000, 2) forwards "information through the power of technology . . . as a rival of reality."

So the technical accomplishment that allows cell-phone users to see the name of the person calling, in turn, produces numerous unanticipated social behaviors. In contrast to early telephone interactions where callers either left identifying information on voice mail (or, before that, remained anonymous unless we answered our phone), capturing caller data allows receivers to *decide* whether to take a call or not, that is, to decide whether a caller is important enough to warrant interrupting other interactions or activities, or whether returning the call later is a better strategy for handling the caller, given previous interactions. On the front end of the interaction, callers now conclude that, when they phone someone, they *specifically* are being ignored, since the assumption is that all cell-phone users monitor incoming callers' names as they receive them. Modified verbal exchanges arise: answering a phone call might invite "Wow, you never answer my calls" rather than "You never answer *your* calls" (suggesting that your phone behaviors are based on callers rather than other, personal patterns of behavior between you and your phone). New technologies invent, revise, and reassemble new patterns of interaction.

To mediate the relationship between technology and people, technical communicators must understand their scientific and technological contexts, both how to function effectively within them and how to respond to them thoughtfully and critically. They need to understand audiences, their backgrounds, interests, motivations, better and baser emotional and cognitive states. They need to be sensitive and to understand the human actions and activities that surround them. And they need to understand not only human and technological interactions but also the complex communication that occurs between humans.

The core of this understanding is not, as some might suggest, the technologies that we support and explain but, rather, our considerable investment and commitment to effective communication design. DiSessa (2000, 112–113) captures the elemental yet powerful nature of our primary "interface," text: "Text is linear; it is black and white; it doesn't zoom around the page in 3-D; it isn't intelligent by itself; in fact, in terms of immediate reaction, it is quite transparently boring. I can't imagine a single preliterate was ever wowed at the first sight of text, and yet text has been the basis of arguably the most fundamental intellectual transformation of the human species."

It is through these "texts" (whether audio, visual, animated, graphical, or haptic) that technical communicators ultimately exhibit rhetorical self-consciousness (Bazerman 1988) by interpreting, contributing, critiquing, amending, and elaborating on existing and emerging technologies.

THE TIME FOR LEARNING AND REFLECTION

Our individual professional goals always interact with our social commitments, whether to our team, our organization, our shareholders, or our intended audiences. As sociotechnical mediators, technical communicators operate at both the cognitive and social level, as what Schön (1983) describes as "reflective practitioners," that is, "agents of society's reflective conversation with its situation, agents who engage in cooperative inquiry within a framework of institutionalized contention" (352), agents who must contend with "problematic situations characterized by uncertainty, disorder, and indeterminacy" (15–16).

Because we understand that learning requires reflection, technical communicators need to value the *time* they spend developing understandings of new information, collecting, reviewing, and synthesizing existing resources, and coordinating with others who can help us accomplish our communication goals. This process requires that we acknowledge and rise above widespread perceptions of increased "busyness" (Putnam 2000), information overload, and the necessity of multitasking and polychronicity, that is, our preference for working on two or more tasks at a time (Turner and Reinsch 2007). Reflection requires focus.

Indeed, framed as we are by our hurried, ill-structured contexts, it is all the more important that we balance action with reflection. As Verbeek (2005, 113) points out, "The facts that technological artifacts can be conceived as constructions, always exist in a context, and are interpreted by human beings in terms of their specific frameworks of reference do not erase the fact that systematic reflection can be undertaken of the role that these contextual and interpreted constructions play concretely in the experience and behavior of human beings. That 'the things themselves' are accessible only in mediated ways does not interfere with our ability to say something about the roles that they play, thanks to their mediated identities, in their environment." Our hurried contexts are unlikely to become any more hurried when we build reflection into our problem solving and emphasize learning as part of our professional process.

Some of our communication processes will call on our experiential knowledge and can be carried out quickly, but others—social, technical, audience-oriented issues—may be new to us with every new project. Norman (1993) compares experiential cognition to reflective cognition, blurring the distinction between what we traditionally define as subconscious and conscious task processing. Experiential cognition is automatic and well learned, and, Norman (1993, 22) emphasizes, "the appropriate responses [are] generated without apparent effort or delay." In contrast, reflective cognition involves choice and decision making: "Reflective thought requires the ability to store temporary results, to make inferences from stored knowledge, and to follow chains of reasoning backward and forward, sometimes back-tracking when a promising line of thought proves to be unfruitful. This process takes time."

Effective technical communicators design projects to account for the learning each situation will demand. Eraut (2004, 259) describes the relationship between modes of cognition in terms of workplace learning, performance, and time, noting that "references to the pace and pressure of the workplace . . . raise the question of when and how workers find the time to think." Thus, one's mode of cognition can range from reflexive cognitive processes (pattern recognition, instant response, routinized action, and situational awareness) to rapid cognitive processes (intuitive interpretation, routines with decisions, and reactive reflections) through deliberative or analytic cognitive processes (review, discussion, analysis, planning, and monitoring) (260).

Experienced technical communicators will know when to apply one cognitive process and when to apply the other. They will understand how existing and emerging genres help them mediate their work, how alternative technologies support their communication processes, what "ecological niches" need to be filled beyond their documents to meet their audience's needs, and how effective design can improve their products (Spinuzzi and Zachry 2000, 177).

HEURISTIC

It has become a truism that simple skills preparation cannot prepare us for a twenty-first-century workplace made up of wicked problems, accelerated time lines and distributed expertise, and exponential technical and scientific development. Given the challenges and constraints that face future technical communicators, it is all the more important that we thoughtfully reinvest ourselves in our own learning and communication processes. This requires that we take the time to focus on our learning processes and on eight general activities that we engage in any time we work through complex tasks, activities, or problems.

Figure 8.1 should be interpreted as a heuristic overview, with the outside circle operating as a series of eight recursive activities that all rhetorically sensitive problem solvers engage in as they go about focusing, representing, identifying, exploring, analyzing and explaining, solving, communicating, and evaluating solutions to contemporary problems. Many other goals operate when we carry out these activities: examples are listed in the

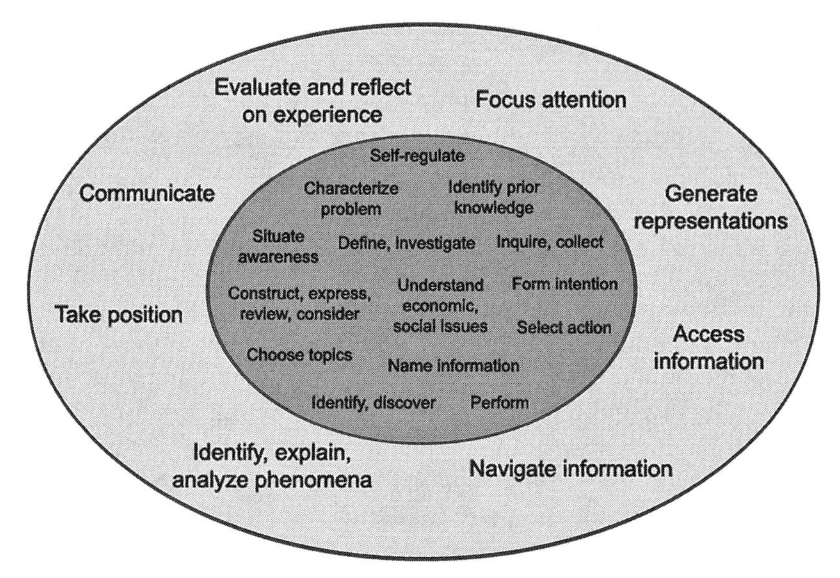

Figure 8.1. Problem-solving activities for the twenty-first century

inside circle. These inside activities are less obvious than the outer eight general activities in that they can change depending on the nuances of the problem situation we are facing. To focus attention and represent our problems, we frequently draw on prior knowledge, characterize our situation or impasse, and develop goals or intentions. But not all our problem situations will be situations that we have encountered before. We may have experience writing hardcopy user manuals, but only some of this experience will translate to the task of building effective help systems. To take a position toward new problems, a considerable amount of investigating, collecting, combining, and comparing is required. We act, learn, reflect, and revise our actions in context. We learn about the differences between creating an effective hardcopy user manual and of creating a usable help system, and we take the time to understand those differences for future problem situations involving different genres or alternative media.

We know these things about learning and problem solving, but that does not mean that most problem solvers attend to the process or its parts when they are actively engaged in problem solving. My interpretation of an effective heuristic is that it encourages active reflection, and my hope is that future technical communicators will improve practice by reflecting intelligently on their own unfamiliar or well-learned communication and problem-solving processes. The ultimate goal is to emphasize our learning processes as technical communicators and to strive through reflection to develop our awareness and capabilities as rhetorically sensitive socio-technical mediators working in complex scientific and technical contexts.

When we focus on our problem-solving activities, on how we learn to learn, on how we know and come to know, and on how we understand our work and our profession, our actions are influenced by reflection rather than reaction. When we complete professional tasks and activities, processes and projects, we attend to what we now know and what we do not know, and we express how we will take that reflection into future projects. Bereiter (2002) describes what it means to understand something and applies his theory of understanding to what it means to understand another person (102–103) and to what it means to understand Newton's theory (109–110). Both examples of understanding turn out to be similar and complex. Indeed, understanding *anything* involves considerable commitment, engagement, practice, feedback, and *time*. His list of eleven ways of understanding something applies as well to what it means to *deeply* understand technical communication.

- Understanding technical communication "depends on your relationship to it" (109). Understanding differs depending on whether you are a programmer, a teacher, a document designer, an engineer, an author of how-to books, an instructional designer, a journalist, or an academic researcher studying communication in the workplace.
- Understanding is critical to acting intelligently in relation to technical communication. What it means to understand technical communication depends on who you are and how intelligently you are able to act in relation to technology, managing technical specialists, deciphering research on technical communication, guiding learners as they become familiar with specific types of technical communication, or supporting technical communication activities.
- Understanding interacts with interest. That is, it is difficult to imagine someone who has no interest in technical communication being able to claim an understanding of it.
- Understanding technical communication requires some understanding of systems theory and logic, the social and cultural forces that have shaped and are shaping technology and literacy, and so on.
- Understanding technical communication does not mean that one

can explain it. Explanation, however, can play an important part in developing and extending understanding.

- Just as "no single correct, complete, or ideal understanding" (110) of technical communication can exist, there can be identifiably incorrect understandings.
- Conversations about technical communication generally emphasize the products or processes of writing, their usefulness, importance, strengths and limitations, and so on.
- Understanding is often conveyed through narratives containing key ideas such as orality and literacy, scientific and technological society, discourse, design, social and cultural influences, and so on. Incomplete or incoherent narratives reveal problems with understanding.
- A deep understanding of technical communication requires knowledge of deeper things related to it, such as state-of-the-art technological developments and historical developments in rhetoric, literacy, communication, and design.
- Insightful problem solving is possible with deep understanding.
- Deep involvement with technical communication, for various audiences, situations, and contexts, is required for deep understanding.

Professional technical communicators need to learn to approach complex situations keeping similar high-level problem-solving goals and strategies in mind. This aim requires paying attention to one's strategies for learning. Communicators are then able to generate rich problem representations, which involves recognizing, finding, identifying, discovering, or framing their problems intelligently. And this activity allows us to form goals and, ultimately, to characterize problems and possible and potential solutions given the constraints of our situation.

Once we have established a working representation of our problems, we can begin the (sometimes extensive) process of accessing and navigating related and relevant information. These are increasingly complex activities, given the proliferation of information resources available to productive professionals. In addition to forming an intention, we can now identify courses of action, begin naming the information types we are accessing and how we intend to use them or to revise our goals if required. Working with information resources allows technical communicators to identify, explain, and analyze phenomena, connecting our efforts to our initial representations, evaluating information critically, redefining our problem, and investigating our subject matter further from available information resources. This process often requires that we employ the rhetorical strategies that Bazerman (1988) and others (Fleming 2003; Selber 2004) have suggested, including reflection, deliberation, purposeful action, and thoughtful understanding of one's circumstances and potential audiences.

Throughout these activities, communication plays an integral role, depending on the problem and specific demands of the situation. Although elaborating on all the activities involved in the communication process is outside the scope of this chapter, we can assume that most communication situations will require the planning, composing or designing, and evaluating necessary to use information effectively to accomplish a specific purpose for a specific audience. Sophisticated communicators will understand the economic, legal, organizational, and social issues surrounding the information need. Moreover, we will use language in context to strategically communicate understanding and to make explicit connections and our representations of particular phenomena for meaningful purposes, well-defined audiences, and different contexts.

These goals demand that we evaluate and reflect on our experience during the problem-solving process. At the most general level, this requires that we consider what worked and what did not work. If we actively engage in reflection, we will also review our experiences in the light of prior experiences for potential improvements, progress, and in terms of intrinsic meaning and effort expended.

Thus, the heuristics for problem solving in the twenty-first-century workplace are meant to outline explicitly the various stages that technical communicators engage in while learning and solving problems. The goal is to pay attention to and reflect on these stages whenever we are engaged in tasks; ultimately, paradoxically, our systematic application of attention and critical reflection will allow us to balance our abilities with the complexity of our tasks (Csikszentmihalyi 1990). This becomes even more important the more our settings appear to expect us to act and react quickly. The heuristic is not to be viewed as a set of prescriptive steps but, rather, to present technical communicators with a telegraphic overview of thirty years of research on how people learn and solve problems. As well, the recursive heuristic process summarized in figure 8.1 is meant to remind professional technical communicators that we move through these physical and mental activities whenever we address communication problems in any context. If our problems are increasing in complexity and our methods for solving them now require both individual and collective problem solving, it becomes even more important that we familiarize ourselves with the fundamental processes that guide and inform our basic problem-solving activities. When we generate representations, and we always do, the activity always requires at some level that we are able to identify and characterize our problem; when we take a position toward a phenomena (product or process), we decide how we will express, review, and articulate a position on the phenomena.

Within our accelerated organizational contexts, technical communicators will increasingly act as problem solvers, attempting to discover or invent—through varying combinations of trial, error, and selectivity—accurate descriptions and explanations of some element of our problem situation and environment (Newell and Simon 1972). Our problem-solving processes are always both cognitive and social, and we therefore act in concert with other problem-solving individuals. Effectively maintaining our personal and professional networks requires that we employ a host of individual abilities (planning, inquiring, choosing, interpreting, arguing), coordination activities (communicating with others), and production activities (acting with others). Our work and learning will increasingly need to incorporate cognitive, social, and design sensibilities (Dietz 2005).

Understanding how we learn and come to understand and the integral role of communication that is part of these processes should, in turn, help us focus on things we can control and contribute to in the future, rather than on the

- increasingly complex and wicked problem situations we face,
- distributed and diffused role of complete expertise in technological and scientific settings,
- exponential development of technically sophisticated devices and genre ecologies that characterize our products and processes, and
- accelerated workspaces, time lines, and learning worlds that make up our everyday professional practice.

EXTENDED EXAMPLE

Although the process that I have described is in line with the productive and critical rhetorical positions that researchers such as Bazerman (1988), Fleming (2003), and Selber (2004) advocate, I draw primarily on learning theory to emphasize that learning plays a critical role in our current work contexts and will play a central role in our future professional lives. Returning to our earlier example, as a professional technical communicator and team member working for a major university providing support for instructors and faculty on the use of an open-source LMS, Janine's knowledge and expertise varies depending on the part of the job she focuses on.

For example, although she has some understanding and experience using HTML and XTML, she will need to extend and refine her knowledge while she learns XML. We have established that she is knowledgeable about standards of writing, genres of support documentation, document-design principles, and the management of editorial collaboration and review. She is less knowledgeable about XML-GL, although she has experience in similar environments that she expects will transfer to this new space. She is practically a novice with the open-source LMS that she is documenting and is therefore aggressively learning about the system from existing online materials, materials developed at other universities using the same LMS, and by comparing the LMS to a prior LMS used at her current university.

Janine's awareness of what she understands and what she does not understand is critical to her strategies for proceeding and for coordinating with the other members of her team. She understands, for example, to turn to her team for recommendations about existing resources and materials that she can draw on to conceptualize parts of the documentation suite she is building; she knows that the project manager can help her construct a reasonable schedule and share it in order to communicate with others on the team and meet the needs of their faculty audience; she knows enough about technologies that support conference calls, forum discussions, and groupware to apply her knowledge to the particular virtual communication technologies that her team uses; the team programmer and interface specialist can help her with XML-GL and LMS questions; and the interface specialist and graphics designer can help her integrate her support documentation into the LMS environment that instructors and faculty will access.

The heuristic complicates prior notions of audience, purpose, and problem solving by anticipating that in the future, technical communicators are no longer expected to act as writers *documenting* as accurately as possible a technology with content provided by a subject-matter expert. Expertise on the team is distributed, and the audience for the documentation suite will bring their own types of expertise to the design challenge (e.g., preferred strategies for instruction). In addition to communicating with her team and with potential users of her documentation suite, she is also aware of various communication-design communities outside her organization that can answer her questions or provide her with helpful solutions to her design problems.

Janine is also expected to work with many technologies, to conceptualize and solve problems at various points in the development process. Her problems will require careful representation because the problems will not have simple solutions (e.g., this particular LMS feature also needs to be supported). Emerging genres and audiences will provide her with many potential solutions, depending on time, resources, and the creativity of the desired solutions. Video, animated assistance, or audio bits may emerge as preferred methods for communicating online instruction. Recommendations for designing textual and graphical information that support each other, for instance, are not only available via a quick search of the web, but can also be gained from published research and trade magazines readily accessible online. Webinars and online professional communities offer additional learning opportunities should she find herself in need of specific design solutions. Where new problems and subproblems emerge, Janine will coordinate where helpful, develop strategies for addressing and proceeding where possible, and seek assistance when necessary.

The team members involved in different parts of the development process will all require communication to facilitate collaboration. And Janine's responsibilities will be simultaneously to her own particular project goals, to her various stakeholders, and to the real and anticipated audiences for her support materials. Her strength will be as a thoughtful communicator who understands the multiple audiences aiming to understand her work, products, and the complex systems that they support. Increased shareholder involvement and communication demands will tighten her already accelerated development cycle. The heuristic, though, stresses the importance of decision making, problem representation, and coordinated action, as well as critical and situational awareness and reflection. Although Janine's professional context and the work contexts of future technical communicators seem at odds with these higher-level individual and social goals, attending to her strategies for learning will help her understand her strengths and the many ways that she can contribute as a sociotechnical mediator.

CONCLUSION

Effective technical communicators understand and reflect on their own problem-solving and learning processes. They understand and invest in their role and knowledge as communicators and are able to contribute sociotechnical designs that mediate technologies and audiences. Future technical communicators will serve as knowledgeable team members, learning, researching, organizing, and synthesizing the many support materials that are required to mediate between communication design, humans, and complex technological processes and products.

Although the future of technical communication is uncertain and indeterminate, our relationship to the study and practice of the multidisciplinary field continues to develop, and our development as rhetorically sensitive sociotechnical mediators continues to hold an important place in our scientific and technical world. Deep involvement and commitment to our role as communication designers will only increase as economic, social, technological, and cultural structures continue to reconfigure themselves in unpredictable and exciting ways.

DISCUSSION QUESTIONS

1. If we are unable to produce perfect solutions to complex, ill-structured problems, what activities can we engage in that produce solutions we can explain and defend?
2. How does our accepting that contemporary knowledge evolves out of our interaction with and between technology and the real world help us understand the role of information overload in our professional and personal lives?
3. What "research" activities do technical communicators frequently have to do on the job?
4. If we are no longer experts who explain technical products and processes for nontechnical audiences, how do we define our emerging identity?
5. How does viewing writing as a technology help us conceptualize what we do as technical communicators?
6. What strategies can we use to engage audiences that are even busier than the audiences of twenty years ago, who also did not read manuals?
7. What role do focus and reflection play in workplace learning and problem solving?
8. Since reflection requires time and time is of the essence in our professional and personal lives, what strategies can technical communicators use to foster this critical part of problem solving?
9. How do recursive heuristics for problem solving differ from recipes or procedural steps provided to help technical communicators generate usable documents?

WORKS CITED

Bazerman, Charles. 1988. *Shaping Written Knowledge: The Genre and Activity of the Experimental Article in Science*. Madison: University of Wisconsin Press.

Bereiter, Carl. 2002. *Education and Mind in the Knowledge Age*. Mahwah, NJ: Lawrence Erlbaum.

Borgmann, Albert. 2000. *Holding On to Reality: The Nature of Information at the Turn of the Millennium*. Chicago: University of Chicago Press.

Brown, John Seely, and Paul Duguid. 1992. "Enacting Design for the Workplace." In

Usability: Turning Technologies into Tools, ed. Paul S. Adler and Terry A. Winograd, 164–197. New York: Oxford University Press.
———. 2000. *The Social Life of Information*. Boston: Harvard University Press.
Buchanan, Richard. 1992. "Wicked Problems in Design Thinking." *Design Issues* 8:5–21.
Burbules, Nicholas C. 2004. "Rethinking the Virtual." *E-Learning* 1:162–183.
Chi, M. T., Robert Glaser, and Ernest Rees. 1982. "Expertise in Problem Solving." In *Advances in the Psychology of Human Intelligence*, volume 1, ed. R. J. Sternberg, 7–75. Hillsdale, NJ: Lawrence Erlbaum.
Conklin, Jeff. 2005. *Dialogue Mapping: Building Shared Understanding of Wicked Problems*. New York: Wiley.
Créplet, Frédéric, Oliver Dupouët, and Emmanuelle Vaast. 2003. "Episteme or Practice? Differentiated Communitarian Structures in a Biology Laboratory." In *Communities and Technologies*, ed. Marleen Huysman, Etienne Wenger, and Volker Wulf, 43–63. Boston: Kluwer Academic.
Csikszentmihalyi, Mihaly. 1990. *Flow: The Psychology of Optimal Experience*. New York: Basic.
Dietz, Jan L. G. 2005. "The Deep Structure of Business Processes." *Communications of the ACM* 49:59–64.
diSessa, Andrea A. 2000. *Changing Minds: Computers, Learning, and Literacy*. Cambridge, MA: MIT Press.
Eraut, Michael. 2004. "Informal Learning in the Workplace." *Studies in Continuing Education* 26:247–273.
Fischer, Gerhard. 2000. "Lifelong Learning—More Than Training." *Journal of Interactive Learning Research* 11:265–294.
Fleming, David. 2003. "Becoming Rhetorical: An Education in the Topics." In *The Realms of Rhetoric: The Prospects for Rhetoric Education*, ed. Joseph Petraglia and Deepika Bahri, 93–116. Albany: SUNY Press.
Johnson-Laird, Philip N. 1983. *Mental Models: Towards a Cognitive Science of Language, Influence, and Consciousness*. Cambridge, MA: Harvard University Press.
Mehlenbacher, Brad. 2010. *Instruction and Technology: Designs for Everyday Learning*. Cambridge, MA: MIT Press.
Nardi, Bonnie A., Steve Whittaker, and Heinrich Schwarz. 2000. "It's Not What You Know, It's Who You Know: Work in the Information Age." *First Monday* 5. http://firstmonday.org/htbin/cgiwrap/bin/ojs/index.php/fm/article/view/741/650.
Newell, Alan, and Herbert A. Simon. 1972. *Human Problem Solving*. Englewood Cliffs, NJ: Prentice Hall.
Norman, Donald A. 1993. *Things That Make Us Smart: Defending Human Attributes in the Age of the Machine*. Reading, MA: Addison-Wesley.
Ong, Walter J. 1982. *Orality and Literacy: The Technologizing of the Word*. New York: Methuen.
Perlow, Leslie A. 1999. "The Time Famine: Toward a Sociology of Work Time." *Administrative Science Quarterly* 44:57–81.
Popper, Karl. 1972. *Objective Knowledge: An Evolutionary Approach*. Cambridge: Oxford University Press.
Putnam, Robert D. 2000. *Bowling Alone: The Collapse and Revival of American Community*. New York: Simon and Schuster.
Rainey, Kenneth. T., Roy K. Turner, and David Dayton. 2005. "Do Curricula Correspond

to Managerial Expectations? Core Competencies for Technical Communicators." *Technical Communication* 52, no. 3: 323–352.

Redish, Janice C. 1993. "Understanding Readers." In *Techniques for Technical Communication*, ed. Carol M. Barnum and Carl Carliner, 14–41. New York: Macmillan.

Resnick, Lauren B., Alan Lesgold, and M. W. Hall. 2005. "Technology and the New Culture of Learning: Tools for Education Professionals." In *Cognition, Education, and Communication Technology*, ed. Peter Gårdenfors and Petter Johansson, 77–107. Mahwah, NJ: Lawrence Erlbaum.

Rigo, Joseph. 2001. "SIGDOC Reminiscences." *ACM Journal of Computer Documentation* 25, no. 2: 31–33.

Schön, Donald A. 1983. *The Reflective Practitioner: How Professionals Think in Action.* New York: Basic.

Selber, Stuart A. 2004. *Multiliteracies for a Digital Age*. Carbondale: Southern Illinois University Press.

Spinuzzi, Clay. 2007. "Technical Communication in the Age of Distributed Work" (introduction to special issue). *Technical Communication Quarterly* 16, no. 3: 265–277.

Spinuzzi, Clay, and Mark Zachry. 2000. "Genre Ecologies: An Open-System Approach to Understanding and Constructing Documentation." *Journal of Computer Documentation* 24, no. 3: 169–181.

Spiro, Rand J., W. P. Vispoel, L. G. Schmitz, Ala Samarapungavan, and A. E. Boerger. 1987. "Knowledge Acquisition for Application: Cognitive Flexibility and Transfer in Complex Content Domains." In *Executive Control Processes in Reading*, ed. B. K. Britton and S. M. Glynn, 177–199. Hillsdale, NJ: Lawrence Erlbaum.

Sternberg, R. J. 2003. "What Is an 'Expert Student'?" *Educational Researcher* 32, no. 8: 5–9.

Stinson, J. 2004. "A Continuing Learning Community for Graduates of an MBA Program: The Experiment at Ohio University." In *Learner-Centered Theory and Practice in Distance Education: Cases from Higher Education*, ed. Thomas M. Duffy and Jamie R. Kirkley, 167–182. Mahwah, NJ: Lawrence Erlbaum.

Turner, Jeanine W., and Lamar L. Reinsch, Jr. 2007. "The Business Communicator as Presence Allocator: Multicommunicating, Equivocality, and Status at Work." *Journal of Business Communication* 44, no. 1: 36–58.

Verbeek, Peter-Paul. 2005. *What Things Do: Philosophical Reflections on Technology, Agency, and Design*. University Park: Penn State University Press.

PART 3 UNDERSTANDING FIELD APPROACHES

This part of the book moves back from theory to practice. However, it does not lose sight of the importance of theory; instead, it begins to look at particular ways to apply theories to practices at a broad level (figure P3.1). You will be connecting broad issues to actions: not just "How do I do this?" but "Knowing what I know about the Big Picture, *how* and *why* might I do this?" In some ways, what you will be learning here is how to use larger frameworks that have long-term rather than (or rather than only) short-term goals. As with questions from part 2, "Understanding Field Approaches" keeps you from becoming so immersed in trees that you lose sight of the forest.

In the first chapter in this part, Blake Scott tackles one of the toughest

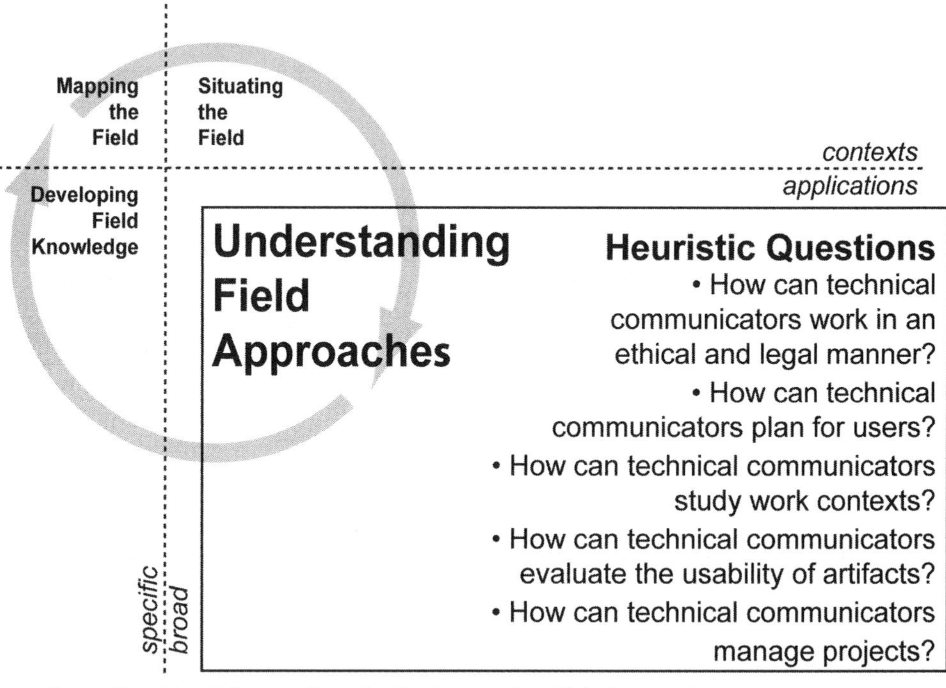

Figure P3.1. Heuristic questions for Understanding Field Approaches

and oldest issues that disciplines face when they begin to step back to consider the broader implications of their applications: "How Can Technical Communicators Work in an Ethical and Legal Manner?" Scott describes the need for carefully considering both ethics and law, but points out that while legal responsibilities are often relatively straightforward, the related issue of ethics can be much more difficult to sort out because it often involves stakeholders with competing interests, ill-defined communication situations, and unfamiliar events or concerns. In exploring these complicated issues, Scott develops a heuristic that will help you engage with stakeholders, identifying their key issues and negotiating solutions that address the needs—legal and ethical—of everyone involved.

In "How Can Technical Communicators Plan for Users?," Antonio Ceraso considers how to more fully involve the most important people in the success of technical communication: users. While technical communicators understand that their texts eventually land in the hands of people who will use them, all too often users are only a vague, back-of-the-mind presence during the production of texts. Ceraso provides a heuristic for bringing users into planning and writing processes, giving them a voice at a point when they can bring some specificity to what is often only a very broad and general picture of the applications of technical communication.

Looking at connections between broad issues and applications and actions often requires us to stop and reexamine key concepts or terms we have been using without much thought. For Clay Spinuzzi, the concept of "context" represents just such a term: deceptively simple but hiding an enormously complicated and powerful set of assumptions about how we should look at the world around us. Rather than innocently assume we already know the answer to his question ("How Can Technical Communicators Study Work Contexts?"), Spinuzzi shows us the missteps we can take if we are not careful in defining not only what context means but how we can study context, use context to think strategically about our work, and communicate about context to each other. His heuristics provide several ways to study and represent the contexts for technical communication projects.

In "How Can Technical Communicators Evaluate the Usability of Artifacts?," Barbara Mirel considers the related issue of effectively connecting the details of contexts to documents that were developed in—and in response to—those contexts. As technical communicators create and work with texts in complicated situations, they need to understand how to assess the success of those texts, measuring and then revising them, and their own work practices, in light of careful thinking about how people actually use texts. The heuristics provided by Mirel can help you improve your individual working methods and your ability to respond to the needs of contexts.

As you might guess from the earlier chapters in this part (and book), technical communication increasingly involves not just single, concrete documents—a user manual, a product-testing lab report, a video tutorial—but often much larger projects involving tens, hundreds, or even thousands of individual texts of different types, worked on and deployed by nearly as many professionals and users. In "How Can Technical Communicators Manage Projects?," Stanley Dicks describes standard project management approaches (including time and personnel management), modifying them to fit the particular needs of technical communication endeavors. His heuristics can help you cope with the complicated management demands of projects ranging from short, information-dense, research-based reports to sprawling initiatives involving scores of resources and staff members.

J. BLAKE SCOTT

9 How Can Technical Communicators Work in an Ethical and Legal Manner?

SUMMARY

Because technical communicators help shape meaning as authors, they need to be familiar with the laws and ethical norms (including organizational and professional ones) that can affect their work and the various relationships (with employers, users, etc.) to which it attends. Because ethical deliberation is a *process*, technical communicators also need to know not only how to identify relevant principles and laws to consider but also how to position themselves in relations of power across communicative situations, engage with other stakeholders in an open search for the best course, and negotiate competing perspectives and obligations to arrive at a commitment to action. This chapter will take you through examples of and provide you with a heuristic for this process.

TWO TECHNICAL COMMUNICATION PROJECTS AT BIOINFO

The following fictional scenario presents several ethical and legal issues faced by a technical communicator working for BioInfo, a small e-learning company based in Baltimore.[1] This company conducts business with Vaccitech, a midsize biotech firm, and with MegaPharm, a large, multinational, research-based pharmaceutical company.

The management team at BioInfo is thrilled but also concerned after landing a big pharmaceutical company as a new client. A relatively new start-up company without much venture-capital backing, BioInfo had survived project by project, creating online courses about biomedicine for smaller biotechnology companies. Sometimes these courses were generic, predesigned courses, and sometimes they were customized to highlight the client's research and products for investors (existing or prospective), health-care professionals, and other audiences. Teams comprising scientists, technical communicators, programmers, and project managers developed the courses, which were typically modular (i.e., made up of separate but related segments that could be combined and arranged in various ways), designed for the web, and designed to serve both marketing and educational purposes. Trained in instructional design, among other things, Angela Williams is the head technical communicator on a BioInfo team. While most of the content for custom courses is provided and owned by the client, the content for generic courses (and generic versions of custom courses) is typically compiled by BioInfo from existing research.

Over the past year, BioInfo designed materials for Vaccitech, a private biotechnology company specializing in the development of vaccines and other medicines for infectious diseases. As it was working to secure FDA approval for its new oral (under the tongue) flu vaccine FluEase, the company asked BioInfo to produce new material about the product and the research behind it, to be placed on the investor portion of its website, alongside initial product sales forecasts. This material would also include more general information about the science and type of research on which the FluEase research drew. Vaccitech provided BioInfo's course development team with access to its FluEase research findings and a draft description of the general science, apparently compiled from several undocumented sources. Having created the content about the promise of FluEase and the research about it, Angela and the rest of the team's technical communicators found themselves in an ethical quandary when Vaccitech asked them to revise this content in a couple of ways. First, Vaccitech wanted them to remove references specifying that FluEase was found to be safe and effective only for people between five and forty-nine years of age (although trials were ongoing for children ages two to five). Because those outside of this age range comprise a large portion of flu victims and potential consumers for vaccines, the requested omission could create the impression of a larger market for FluEase. Second, Vaccitech wanted to add a section exaggerating the development stage of a new version of FluEase that would not require refrigeration for storage. Angela was concerned that these requested changes would mislead investors and other audiences as well as damage the e-learning company's reputation for providing scientifically accurate content. Unfortunately, this would not be the only communication-related ethical dilemma that Angela and BioInfo would face.

After Vaccitech secured FDA approval for FluEase, it attracted the interest of several big pharmaceutical companies and began a partnership with one—MegaPharm—for the production, distribution, and marketing of the new vaccine. For BioInfo, this meant developing expanded educational materials aimed at physicians, distributors, and others as well as new tutorials for the in-house training of the pharmaceutical company's sales force; both tasks involved a great deal of technical writing. Some of this content might later be adapted for marketing and instructional materials aimed at consumers as well. In fact, MegaPharm indicated that it might hire BioInfo to do further work for a direct-to-consumer marketing campaign and product labels and instructions if it was pleased with the first round of materials. Because gaining MegaPharm as a customer gave BioInfo a way into the lucrative pharmaceutical market, the stakes were high for this project. In addition, because the materials created for pharmaceutical training would be even more modular, BioInfo saw the opportunity to refine and secure property rights for its knowledge-management software in order to sell it to other e-learning companies.

MegaPharm provided BioInfo with most of the information to include in the educational material and training modules; unlike Vaccitech, however, it did not provide all of the scientific research findings to support this information. This led to a problem when MegaPharm wanted both sets of content to promote the increased efficacy of standard flu shots when used along with FluEase; beyond receiving no supporting evidence from MegaPharm, the scientists on Angela's project team had research from Vaccitech appearing to contradict this claim.

The BioInfo team faced another ethical dilemma when MegaPharm insisted on including images of patients who look younger than five and older than forty-nine years old, even though FluEase was only approved for ages five to forty-nine. And yet a third ethical challenge was posed by MegaPharm's directive that the physician- and distributor-directed material emphasize FluEase's ease of use and deemphasize its storage requirements. Angela and her team were even more unsure about how to negotiate the ethical issues involved in the MegaPharm project, partly because of their limited access to research and partly because of the high stakes of accommodating the client and users. Angela's team realized that pleasing this client and thereby positioning BioInfo for the larger pharmaceutical market was the ticket to BioInfo's financial stability.

This scenario, to which we will later return, raises a number of ethical and legal challenges faced by technical communicators of different stripes, challenges ranging from respecting intellectual property to protecting users from harm. By ethics, I mean deliberation about how to determine and act on what is good, right, just, beneficial, desirable, or commendable (see Dombrowski 2000, 7). Sometimes ethical and legal issues overlap, as with the Society for Technical Communication's (STC's) principles of providing truthful and accurate communication and respecting the confidentiality of a client's business-sensitive information.

In what follows, this chapter reviews some of what others have written about ethical and legal concerns for technical communicators before detailing an approach, adapted largely from James Porter (1998), that treats ethics as a situated process of negotiation for developing what the ancient Greeks called *phronesis*, or practical wisdom. Often, as the BioInfo scenario illustrates, this process requires stakeholders to weigh conflicting values that come out of interpersonal, organizational, professional, disciplinary, and larger sociocultural spheres of influence. Further, this process should involve, in the words of Porter (1998, 157), "some careful deliberation, some marshalling of options and alternatives, some care in determining the conditions of one's rhetorical setting, and some respect for the variances of human action." Although technical communicators need to know specific values, norms, conventions, and laws related to their work and the relationships to which it attends, the most important thing they can learn about this topic is *how* to engage in ethical negotiation, including how to position themselves in the power dynamics across situations (where power is understood as an ability to act effectively, influence outcomes, and achieve goals), how to discover the relevant principles and laws to consider, how to engage others in an open search for the best course, and how to arrive at a contingent commitment to action. This chapter offers a heuristic for arriving at ethical judgment and action, posing questions about the ideals of, consequences for, and your obligations to others. After discussing the process of ethical inquiry and presenting the heuristic, the chapter returns to the opening set of scenarios to illustrate the heuristic at work.

THE *PROCESS* OF ETHICAL INQUIRY AND DELIBERATION

Despite or perhaps because of the field's pragmatic emphasis on efficiency and effectiveness, ethics has become a significant concern for many researchers and practitioners of technical communication. Articles about ethics frequently appear in academic and trade journals, and the Association of Teachers of Technical Writing and STC have developed codes of ethics to guide their members. All major technical communication textbooks now have sections addressing ethical and legal issues. Several ethics-focused textbooks, such as those by Lori Allen and Dan Voss (1997), Paul Dombrowski (2000), and Mike Markel (2001), have been published for use in advanced technical communication courses, including courses dedicated solely to ethics.

Why consider ethics as *part* of technical communication, you might ask. Answering this question first entails defining technical communication and the role of the technical communicator. Jennifer Darryl Slack, David James Miller, and Jeffrey Doak (1993) argue for viewing the technical communicator as an active contributor to the meaning-making process as texts circulate across their contexts of production and use. They argue against relegating the technical communicator to the merely instrumental role of transmitter or sender of a fixed message determined by a technical expert such as an engineer, and they even argue that the technical communicator does more than translate an essentially static message to different audiences. Instead, the technical communicator is a full-fledged author who contributes to meaning making along with other producers and users, and who operates within and is constrained by channels of power, such as organizational pressures and cultural protocols. This view also attributes more agency to technical communication, which becomes less a transparent medium for transferring or even translating information than a means through which various actors negotiate and renegotiate meaning. When we position the technical communicator as an agent of meaning within specific power relations, we can also recognize, as Porter does, that technical communication necessarily entails ethical decisions about how to shape the parameters of meaning making and about possible effects; "whenever you write," Porter (1998, 50) explains, "you take a position and you establish a value vis-à-vis existing systems [of power]." In this way too, then, the study of technical communication should involve the study of ethics or ethical decision making.

In their discussions of ethics, technical communication researchers have drawn on various philosophical traditions, including those that emphasize duties and obligations (e.g., Kantian ethics), respect for difference and care of others (e.g., feminist ethics of care), impacts or consequences (e.g., utilitarianism), and the communicator's character (e.g., Aristotelian ethics). Although these and other approaches focus on different aspects of communicative acts, most emphasize the interpersonal or social relationships of those involved; according to Vincent Ruggiero (1997), respect for persons is a crucial value in most ethical systems, the core value behind numerous more specific obligations to others. Most legal concerns for technical communicators are also based on relationships and the obligations that accompany them. As Ruggiero explains, respect for persons is "not merely a theoretical construct but a practical standard for the treatment of others" (72).

This leads us to another common element of the field's various takes on ethics—that it is a situated, "practical art" (Markel 2001, 20; Bowdon 1999), a recognition illustrated by the widespread reliance on "real-world" scenarios and projects to teach it. Yet another common element in approaches to ethics is the recognition that ethical principles come from various spheres of influence, including personal beliefs (which themselves come from various sources), organizational goals and values, professional and disciplinary standards, and larger sociocultural value systems. Sometimes the spheres are in conflict, as in the potential conflict between the professional values of the BioInfo communicators and their organization's goal of financial stability. Taken together, these common takes on ethics endorse a definition something like the relational, situational, power-laden negotiation of principles derived from various value systems with the goal of arriving at the best course of action.

This definition doesn't quite capture legal concerns, which often overlap but can also diverge from ethical ones. Like ethics, the law has been formed out of various value systems and spheres of influence and is relational and at least somewhat context dependent. Most ethical codes, including that of STC, include legal mandates, such as meeting the terms of contracts. Further, some ethical principles, such as the technical communicator's duty to instruct and warn, are also legal ones. At the same time, the law is typically more codified and binding than ethical codes, though this doesn't mean ethical principles can't be binding and that the law is always defined or fixed. Intellectual property (IP) law regarding electronic texts, for example, is still very much in flux; emerging laws must negotiate the new forms of communication that new media make possible and the competing interests of respecting the property of communicators and enabling the dissemination of knowledge for the public good. Although technical communicators can seek help in clarifying their specific obligations under the law (indeed, some companies have staff specifically trained in legal issues), they still have a responsibility to know the basic types of relevant laws—including fiduciary duty (i.e., duty tied to legal financial relationships), the duty to instruct and warn, and respecting the intellectual property and privacy of others—and where to learn the details about them. Tyanna Herrington's *Legal Primer for the Digital Age* (2003) is a good resource for such details.

Although the definition of ethics above is a good starting point, its assumptions and enactment require further explanation. Through this elaboration, we can also compare an approach to developing phronesis to other approaches discussed by technical communication professionals. In addition to the recognition that technical communication is always embedded in and regulated by power relations, the approach advocated here assumes that technical communication is transsituational, that is, occurring across multiple, interlinked contexts. We can conceptualize these contexts along the life cycle of a text's production, distribution, interpretation and use, and continuing adaptation, and we can also imagine these contexts as overlapping zones of activity (see Scott 2004). Take, for example, an online software manual developed by a team of technical communicators. The text might initially be developed in an organizational context by a team of technical communicators working across company departments with programmers and marketing specialists, and then tested on possible users in a testing lab before being revised. It might then be distributed to and accessed by various types of users, from novice to expert, through specific web technologies, and then interpreted, used, and adapted by these different user groups in their specific situations or settings of use (e.g., at home, work). Meaning is made from the text *across* these contexts, from the authors' intentions to users' interpretations and adaptations. As they circulate across such contexts of use, technical texts are continually taken up and transformed, promoting certain values and enabling certain effects for those involved.

Along with moving and being transformed across multiple contexts, technical communication is, as Porter emphasizes, internetworked, and not only because it often moves across the Internet and World Wide Web. By internetworked, I mean created and performed across a network of communicative needs, interpersonal and social relations, and concrete settings or environments of use. Such an assumption expands the ethical "actors" of technical communication beyond the writers and their primary audiences to include various secondary audiences and other stakeholders impacted by the communication, larger discourse communities and social groups to which communicators belong, organizations or institutions that sponsor and regulate communication, and the technologies and media through which the communication is delivered. By discourse communities, I mean groups of people bound by a common interest who share and regulate specialized kinds of knowledge and ways of communicating (Anson and Forsberg 1990, 202). Considering these various actors and their sometimes conflicting value systems complicates approaches that focus on the technical communicator's personal sense of ethics or one common set of professional or social values.

Given that technical communication is constructed and reconstructed across multiple contexts by various actors, its ethical dimension is not fixed, absolute, or determined before and outside of these communicative contexts. Although Markel (2001) advocates an approach to ethics in which various stakeholders collaboratively deliberate about the most ethical course of action (127), he also suggests that this deliberation should be separate from and prior to determining what the communicative action will look like when implemented (21). Deciding the most ethical course of action, Markel argues, is more difficult than implementing it. In contrast, the approach to ethics offered in this chapter does not separate the ethical scene from the communicative one; indeed, ethical deliberation is a communicative process, and ethical issues are raised *through* the enactment of technical communication.

Rejecting a universal, fixed, and absolute system of ethics does not require abandoning a binding commitment to a specific ethical stance; the ethical approach advocated in this chapter is not completely relativistic and does not avoid judgment about what is best, right, just, or desirable. To explain this position, I have turned to the classical Greek concept of *nomos*, a term that can refer to norm, convention, custom, and law, all standards of conduct that are socially constructed, accepted, and regulated by specific groups or communities (Scott 1995). Although nomoi are not universal or fixed, the connection between norms and the law through this term suggests that nomoi can be binding, at least for a time. According to classics researcher Martin Ostwald (1969, 54), the concept of nomos "signifies an 'order' and implies that this order is, or ought to be, generally regarded as valid and binding by the members of the group[s] in which it prevails." In most cases, technical communicators will need to consider with others the relative importance of different types of nomoi—such as organizational goals, professional codes of conduct, business conventions, civic laws and regulations, and sociocultural traditions and belief systems—as they relate to specific sets of communication contexts. Some of these, including laws and professional codes, might be more codified than others, but even these must usually be adapted to the specifics of the situation at hand.

In his discussion of the term nomoi, Porter (1998, 30) defines it both as "the policies, conventions, and practices governing social relations" and "the processes of negotiating and constructing conventions." This notion of nomos as a *process* leads us to a similar conception of ethics as the process of arriving at and implementing the best course and, in a larger sense, the process of developing phronesis or practical wisdom. As Porter explains, phronesis can refer to the "art of considering divergent norms, principles, and conventions in light of particular circumstances that require action" (29). Beyond calling for a phronesis-building process, Porter (1993, 218) begins to discuss what this process entails: "Ethics is not a set of answers but a mode of questioning and manner of positioning. That questioning certainly involves principles—but it always involves mediating between competing principles and judging those principles in light of particular circumstances. Ethics is decision making—but it is decision making that involves question and critique."

Others, too, have outlined a process of ethical decision making. Ruggerio (1997), for example, outlines the steps of studying the situation, identifying relevant criteria, determining alternative courses of action, and deciding which one to take. Like Ruggerio's, most discussions do not lay out specific tactics for enacting such steps, however.

In what follows, I outline several tactics, grounded in nomoi, for ethical deliberation and decision making developed by Porter and others, illustrating them with the document-development process of user-centered design. User-centered design, sometimes called user-participatory design, is an approach to making a text more usable and better suited to its audience by inviting members of this audience to be active participants in its development, from the planning stage onward (see Johnson 1998, 32). Although it is more common for technical communicators to test a document's usability with the target audience after the document has been created, user-centered design requires a more "sustained dialogue between user and designer," one that places even more value on users' perspectives and knowledges (Salvo 2001, 289). Because of this more sustained relationship, user-centered design can illustrate some of the tactics of ethics-as-phronesis, which include (1) studying the specifics of situations across which technical texts are produced, distributed, and used; (2) mapping your position relative to others to whom you have responsibilities or obligations; (3) engaging in sustained dialogue with other stakeholders (including targeted users, when possible) about the best course of action, a dialogue that respects the unique perspectives, needs, and concerns of stakeholders; and (4) collaboratively arriving at a commitment to specific action.

The first tactic relates to Ruggerio's call to study the specifics of a situation, though it expands this to include the multiple situations across which a technical text circulates. One way to do this is to visually represent the multiple scenes of a text's production, reception, and use. Patricia Sullivan and James Porter (1997) propose drawing maps that show the relationships and flows of power among the actors, or stakeholders, in a particular communication situation. This first involves asking who will encounter, use, and be affected or impacted by the texts we create. Herrington (2003, 10), in her "axis of power" test, similarly calls on technical communicators to examine their positions in relations of power and the responsibilities such positions entail to others. Maps of communication relationships can also include other kinds of "actors" or influences on the communication, such as organizations, technologies, laws, and the texts themselves; indeed, doing so can help us remember that the various parts of communicative locales are more than just backgrounds for communication. Let's say, for example, that you are part of a team developing a technical proposal for a construction project. Your map would include the relationships among your team members, supervisors, and audiences assessing and then using the proposal. Your map would need to show your competing obligations to these various stakeholders, such as your duty to your employer and any legal obligations to the owners of the project derived from the terms of your proposal (e.g., cost, deadlines). Or let's say you were asked to write instructions for assembling and building an air cannon for a mechanical engineering lab on campus. In accounting for the contexts of production and use, you'd need to consider your specific legal responsibilities to ensure the safety of users and, if applicable, to respect the intellectual property of any existing instructions from which you planned to borrow. Although such maps can be a useful way for technical communicators to position ourselves, they should be recognized as the limited views of one stakeholder and as snapshots of larger, ongoing processes.

As Herrington suggests, the point of this positioning is to become better attuned to the obligations and responsibilities (including legal ones) inherent in relations of power as texts are taken up and transformed. Some technical communication scholars based such responsibilities on one's position relative to others: those with more power and expertise, in our case technical communicators with rhetorical power and expertise, also have greater responsibility to wield this power for the good of others (see Allen and Voss 1997, 9). Markel (2001, 160) puts it this way, pointing to the principle of due care: "the more powerful party is responsible both for providing special assistance and protection and for refraining from harmful acts that exploit the power difference." In a user-centered design process, being attentive to the power dynamics of relationships can help the technical communicators ensure that users are positioned to share their contextual knowledge in the text-development process and that safety and efficacy concerns around their contexts of use are fully addressed. Such attentiveness can also foreground any constraints on and risks from our responses to ethical dilemmas. In the opening scenario, such constraints include the technical communicator's position within the company, the company's contractual obligation to the client, and the company's need for financial resources.

Beyond becoming more aware of our position(s) in shifting relations of power, we can work toward phronesis by approaching ethics as an open process of mutual inquiry. By open, I mean without predetermined or fixed procedures for valuation. Instead, participants mutually constitute how to arrive at the best course as well as what that course will be. Porter (1998, 135) also advocates for such procedural openness, arguing that specific procedural moves must be developed and adjusted on the spot in light of the specific contexts, communities, and nomoi involved.

Like user-centered design, this approach to ethics entails engaging in a sustained dialogue with others about the best, most ethical course of action. Allen and Voss (1997) propose a process called "value analysis," in which those deliberating define the various stakeholders' interests and values, rank conflicting values in order of importance, and then base a collective decision on the most important values (see 20–21). Although they propose making this decision based on which path of action does the greatest good for as many stakeholders as possible (a utilitarian principle based on weighing the overall costs versus benefits), Allen and Voss seem to assume that this will be a relatively straightforward exercise and don't consider the possibility that some stakeholders will or should have a greater voice in making such a determination. Some stakeholders might have more at stake than others; poorly designed instructional materials, for example, might have the most negative impact on users needing them to perform an action safely and effectively. Like other scholars advocating a consensus-building model of ethical deliberation, Allen and Voss could further focus on difficult questions about who has or should have more power to shape decisions, about who wins or loses from a particular decision, and how technical communicators should weigh competing obligations.

Although the phronesis-building approach proposed by this chapter seeks a shared commitment to a particular course of action, it arrives at this end not by ignoring or squelching difference, but by acknowledging and embracing it, viewing it as a source of ethical invention. The goal is not to efficiently move beyond stakeholders' different perspectives, experiences, and values but to incorporate them into the communication's design, preferably through the direct participation of the stakeholders themselves. Underpinned by this ethical imperative, user-centered design assumes that incorporating the unique and different perspectives of users, preferably a cross section of prospective audience members, will improve the technical text.

Drawing on the feminist ethic of care developed by Nel Noddings and others, the approach advocated here moves beyond a respect for difference to a stance and practice of "caring" for and being responsive to others. As Porter (1998, 93) points out, an ethic of care requires one to "actively work for the welfare of others," to take "positive action" for and with others. It requires us to ask whose needs our work addresses and how our positioning as technical communicators calls us to learn about and be responsive to these needs. To some extent, this ethic is embedded in the technical communicator's role as user advocate and in the stance of user-centered design. As Robert Johnson (1998, 28) explains, the goal of user-centered design is to arm users with deeper knowledge about how technology works, so that they can perform tasks and solve problems in a sustainable way. He contrasts this goal with that of user-friendly design, which seeks to make tasks easy for users but can leave them unable to transfer "how-to" knowledge or troubleshoot when things go wrong.

Another difference between user-centered and user-friendly design is that the latter also involves the abstract consideration of users rather than their direct participation as codesigners and co–decision makers (Johnson 1998, 32). The concrete engagement of users throughout document development is the core tactic of user-centered design, and it's also a tactic associated with the stance of caring. Porter (1998, 154) explains that the feminist ethic of care was developed in response to more abstract ethical approaches based on rights, and Melody Bowdon (1999, 16) similarly emphasizes a feminist ethic of action "that values connection over abstract ethical principles and that grounds decision making in local, collaborative, and dialogic deliberation rather than universal rules or principles." A phronesis-oriented process of ethical deliberation, like user-centered design, strives for deliberation with, and not just on behalf of, other stakeholders. To the extent possible, the stakeholders of our texts should be invited to the table to deliberate about any ethical dilemmas connected to them. Beyond inviting others to engage in an open dialogue, we should take care in shaping how this dialogue actually functions. Michele Simmons (2007, 129) encourages us to take a close look at who participates in decision-forming discussions, and in what roles, as well as when and how frequently they participate. In the case of user-centered design, users are invited to provide input and share in decision making from the initial problem-defining stage onward. Rather than treating rivaling perspectives, opinions, and options as "noise" to be avoided or muffled, Michael Salvo (2001, 289) argues, user-centered design values them as useful material for the ongoing development and improvement of a technical text. This is why it is important for technical communicators conducting user-centered design tactics such as focus groups to facilitate open rather than leading or limited responses.

A phronesis-based approach to ethics involves arriving at a flexible judgment (i.e., one that can be adapted to changing circumstances) about the best course and commitment to enacting it. This can be the most difficult part of ethical deliberation, especially when differing perspectives can't be reconciled, but it is a crucial part of developing and enacting phronesis. Participants may, indeed, need to weigh various values, obligations, and consequences against one another, but the criteria for doing so should not be predetermined or fixed across contexts, except, perhaps, when a clear legal directive applies. My own justice-based preference is to give the most weight to the values and voices of those most directly affected by the ethical decision, especially when they are in positions of relatively lesser power. However you collectively arrive at a contingent judgment, you will likely, as Markel (2001) advises, need to consider what courses of action are actually possible given the power dynamics and other constraints involved. The result of this collective decision making might be the construction of new, more diversely informed ethical judgments and forms of technical communication (254).

A HEURISTIC OF GUIDING QUESTIONS

The following questions are offered as a set of invention tools, or heuristic, for moving through the process of ethical deliberation discussed above. Involving your relationships to others and the ethical obligations such relationships entail, these questions are not exhaustive and not meant to be followed in a lock-step manner. Depending on the ethical issue and the contexts out of which it arises, some may be more useful than others. Finally, some will need to be adapted to changing circumstances within and across specific contexts.

Embedded in some questions are specific nomoi about which you should be aware. Other technical communication scholars—particularly Porter, Dombrowski, Markel, and Herrington—discuss various types of ethical and legal principles in more detail and should be consulted in a fuller study of ethics. Some researchers have also developed categories of nomoi that might be useful in assessing the ethics of specific situations. Markel (2001, 128), for example, discusses a utility-justice-rights-care model that considers four types of ethical principles, and Ruggerio (1997, 73–74) categorizes principles as obligations, consequences, or ideals (i.e., notions of excellence). Most of the questions below focus on obligations that arise out of different types of relationships.

POSITIONING ACROSS SITUATIONS

- What are the contexts in which the technical communication will take place, in which the text will be created, distributed, encountered, used, and transformed? How might the text be taken up in unintended ways? Think through the text's possible life cycle.
- Who is involved in the communicative acts across these contexts? Who is likely to be affected or impacted by this communication? What do they have at stake, how do they stand to benefit or lose?
- What are the interpersonal and social relationships among these

various stakeholders? Who is linked to whom and how? How is power distributed across these relationships? How might these relationships and flows of power shift as communicative acts and contexts shift?

- What, if any, agreements do stakeholders have with one another, and what rights and obligations might arise from such agreements (adapted from Herrington 2003, 131)?
- How would you describe the discourse communities of the various stakeholders? What other spheres—including organizational, professional, and cultural—shape stakeholders' viewpoints and their relative position and power?
- What additional "actors"—including organizational or institutional structures, technologies, and communicative settings or environments—shape the communicator's relationships and constraints?

ETHICAL INQUIRY AND DELIBERATION

- How can you ensure the robust participation of other stakeholders? How could you engage them as partners in a mutually defined and reciprocally beneficial process of inquiry?
- How can you and other participants decide on how to proceed, and by what criteria? What roles, opportunities to provide input, and decision-making power will the various participants have?
- How can you strive to ensure that stakeholders' perspectives, knowledges, needs, values, and responses get voiced and incorporated into ethical problem solving?
- How can you encourage the expression and consideration of differing explanations of any ethical problems?
- How can you encourage the full and critical examination of different options for addressing the problem, how might these options be enacted, and what consequences might they have?
- How can you determine when values and positions are in conflict?

OBLIGATIONS TO AND CONSEQUENCES FOR USERS OR AUDIENCES

- What should you know about users or audiences, including both primary and secondary ones? What is the best way to get this information?
- How might you enlist users, and which specific users might you enlist, as codesigners of your communication?
- What obligation do you have to ensure the accessibility, clarity (i.e.,

understandability), accuracy, and completeness of your verbal and visual communication?

- What obligation do you have to accommodate users' needs and capabilities, including those of users with disabilities and users from different cultural backgrounds? When the needs of different users are in conflict, how will you decide which needs to accommodate? How can you ensure a fair distribution of benefits among users?
- What obligation do you have to ensure that no harm comes to users from your communication?
- If you are writing technical instructions, what are your legal obligations to describe the limits of use, provide clear and accurate instructions on all aspects of use, prominently warn about hazards that could arise from use, and tailor the communication to the audience's background and needs?[2]
- What ethical and legal obligations are tied to your role as a user advocate? How might you ensure that your communication empowers users to understand its subject matter and to mobilize this knowledge in a transferable, sustainable way?

OBLIGATIONS TO CLIENT

- What obligation do you have to offer quality services in an equitable manner?
- What obligation do you have to respect the privacy of client information and communication?
- What obligation do you have to abide by the terms of contracts with clients?
- What obligation do you have to protect your client from legal or financial harm?

OBLIGATIONS TO EMPLOYER

- What are your legal and ethical obligations to act for your employer's benefit and promote its interests?
- What obligation do you have to follow any code of conduct of your employer?
- What obligation do you have to abide by any contract with your employer, including a nondisclosure or confidentiality agreement (designed to prevent you from sharing specified materials or information with a third party)?
- What obligation do you have to respect and protect the confidenti-

ality of company information, including trade secrets and materials they are paying you to create?
- What obligation do you have to secure intellectual property protections for your employer, including copyright, trademark, and patent registration?[3]

PROFESSIONAL AND MORE GENERAL OBLIGATIONS AND IDEALS

- What obligations and ideals are tied to your professional status as a technical communicator?
- To follow the STC code of ethics, what obligations do you have to observe relevant laws and obligations, to communicate truthfully and accurately, to respect the confidentiality of others' business-sensitive information, and to fulfill contractual obligations in a timely manner?[4]
- What, if any, professional quality or regulatory standards should your text meet?
- What obligation do you have to safeguard public interests, such as human rights and environmental concerns?
- What obligation do you have to get permission to use copyright-protected words, images, and multimedia in your communication? To stay within the bounds of fair use of this material (which may not cover use for commercial gain)?
- What obligation do you have to get permission to use the code or design of web texts? To not mislead users (through framing, linking, or use of metatags) about the ownership, scope, and content of your web texts?

ARRIVING AT FLEXIBLE OR CONTINGENT JUDGMENT AND COMMITMENT TO ACTION

- How can you determine when values are in conflict? How will you and the other participants negotiate competing values about what is best, right, just, and/or desirable?
- What are the stakes of your communication for various stakeholders (audiences, employer, client), and how do you weigh these relative to one another?
- How will you arrive at a contingent judgment about the most ethical course of action?
- Which of the considered actions are most possible given the contextual constraints, such as your position in power relationships? Whose approval might you need to carry out the action?

- What can you learn from comparing your position to that of other stakeholders? About other aspects of these stakeholders' knowledge, interpretations, and values? From their forms of deliberation and their decision-making criteria?
- What can you learn about the negotiation of multiple and differing values and principles? About mutually developing a new set of shared nomoi and discourses?
- What can you learn about how to adjust ethical inquiry and decision making across related contexts?

ETHICAL DELIBERATION IN ACTION

In order to illustrate how you might apply some of these questions, let us now return to the opening scenario involving Angela and other technical communicators working for BioInfo. For both texts—the investor material for Vaccitech and the training/marketing material for MegaPharm—and their related contexts and ethical challenges, Angela works alongside other technical communicators and programmers, under the direction of a project manager and, ultimately, the company's management. Angela also works for the client, of course—in the first case a small company with which BioInfo has an established business relationship, and in the second with a larger, newer client whose relationship can help BioInfo make inroads into a profitable client base. Although we can think of Angela's client as her most immediate audience, her communication is primarily aimed at the users or audiences that her client is trying to reach and influence. Finally, Angela's communication will likely link her to others who are affected by how it is taken up and acted upon.

ETHICAL INQUIRY ABOUT VACCITECH PROJECT

One ethical challenge presented by the first module about FluEase and the science behind it is what to do with the general science material provided by the client. Angela will need to first determine whether the material—specifically its form of expression (whether verbal, visual, etc.)—is copyright protected or already in the public domain. If the former, she will need to get permission to use the material and give credit to the copyright holder; in addition to a legal obligation, honoring copyright is a professional obligation for technical communicators. In some situations, technical communicators can consider whether the legal doctrine of fair use enables them to disseminate a limited portion of copyrighted material for educational purposes (as long as it doesn't impede the profit-making goals of the copyright holder). In this case, fair use does not apply, given that the material would be developed, in part, for commercial gain and does not fall under any of the other exceptions. Although it might be tempting for Angela to limit her role to transmitter and accept the information given to her as is, such a stance could abdicate her responsibility as a cocreator of knowledge and meaning.

The second set of ethical challenges from the Vaccitech project—around the client's request that BioInfo omit information in order to suggest a wider market for FluEase and overstate the development of a newer version of the product—is slightly more complicated, as it involves additional stakeholders with competing interests and additional professional, organizational, and legal obligations tied to BioInfo's relationships with these stakeholders. BioInfo's primary audience, you will remember, consists of the investors who will use the information on Vaccitech's website to assess and make decisions about investing in the company. To these users and other website readers who encounter her text, Angela has an obligation to present fully accurate information and to protect them from harm, in this case possible financial harm from a misguided investment (see questions under "Obligations to and Consequences for Users or Audiences"). If Angela adopts an ethic of care and assumes the role of user advocate, she also has the ethical obligation to make needed information accessible to users (i.e., investors) and empower them to take productive action. Given the power and knowledge difference between Angela and her primary audience (considering that investors aren't able to shape the text and likely don't have the scientific expertise of Vaccitech or BioInfo), her ethical obligation to accommodate their needs is even stronger.

Other nomoi to consider revolve around Angela's roles as service provider and employer, as she has the obligation to protect the interests of both client and employer. The STC code of conduct urges technical communicators to offer quality services that meet contractual obligations to the client, and she might go further in determining how to protect her client's welfare; in this case, providing inaccurate or misleading information could cause investors to take legal action against Vaccitech and prompt FDA officials to further scrutinize Vaccitech's FluEase application. Angela must also consider the responsibility to promote the welfare of her employer, as she doesn't want her company's reputation to be damaged among its potential clients and investors because of a misleading and harmful text that she helped produce.

The forms of engagement and deliberation that Angela enters will be, to some extent, out of her control because of her access to the audience and role in the company. Although it might not be possible to enact a user-centered design approach in this case, Angela could attempt to get input from members of the investment community at some point in the text-development process, perhaps by asking them what they expect from this type of material and how they interpret the information she produced (see the prompts under "Ethical Inquiry and Deliberation"). To do this, she would likely need to secure the cooperation of Vaccitech, because she would be sharing their copyright-protected material before they publish it.

ETHICAL INQUIRY ABOUT MEGAPHARM PROJECT

The MegaPharm project—which calls for expanded educational/marketing and training materials—presents an even more complex web of stakeholders, relationships, contexts, and ethical/legal issues. Angela will need to consider and negotiate a wider range of sometimes conflicting nomoi around scientific conventions, professional ideals, legal obligations, and business obligations and goals. She will need to think through a longer trajectory of life cycles of her texts and their transformations, functions, and effects, particularly since the modules she creates could be adapted for or used in marketing and instructional texts targeting consumers (see "Positioning across Situations"). Angela will need to consider the higher stakes of her decisions for BioInfo, the client, and the various audiences or sets of users.

As in the Vaccitech project, one of the ethical challenges central to this one revolves around providing accurate and honest information, a principle that is a professional, cultural, and legal ideal and obligation. Again, Angela will need to deliberate about and determine (ideally with others) whether she has an obligation to users to verify the clients' claims about the product—in this case how its efficacy relates to that of the flu shot—and if the client refuses to provide verifying evidence, whether to resist including such claims (in this case, consider primarily the questions under "Obligations to and Consequences for Users or Audiences"). This first issue is perhaps a more straightforward one of accuracy, but the other two directives of MegaPharm blur the line of ethical communication a bit more.

The next issue—using possibly misleading images of patients apparently outside the approved age range for the project—could possibly be justified, given that the content would also verbally specify this approved age range and that the ages associated with the images are open to users' subjective interpretations, to some extent. Yet the accuracy and honesty of visual communication is a requirement of the profession and, in some cases, the law (see "Professional and More General Obligations and Ideals"). Like other pharmaceutical companies, Angela's client could face regulatory sanctions for using misleading images, especially if they appear in direct-to-consumer marketing. In addition, the courts have found makers of instructions liable for misleading or unclear graphics (Helyar 1992, 130–131). Technical communication researchers such as Nancy Allen (1996) have emphasized that the selection, emphasis, and design of visuals—from illustrations to graphs to tables—help shape readers' meaning making and therefore should be held to standards of accuracy and accessibility much like verbal communication. They have explained that questions of ethics apply not just to what information is included, but *how* it is included and presented. This is especially important given that not all users will focus on all or the same parts of a text to form their interpretations. Recognizing that she is first and foremost a user advocate should prompt Angela to move beyond the "functional utility" of her communication and to consider the various factors (including their knowledge and values) that shape users' interpretations of texts in specific contexts of use (Allen 1996, 99).

Yet another related accuracy question foregrounded by the MegaPharm project is what to do about the client's desire to emphasize some information about the project more than other, perhaps equally important, information in order to make FluEase more appealing to physicians and distributors, who will have to be concerned with the storage requirement of refrigeration. Although the module would include the storage requirement, this information would not be as accessible and memorable as the information about ease of use for patients.

The challenges posed by the MegaPharm project require Angela to first think through the life cycle of her texts ("Positioning across Situations") and then consider more specific obligations to and consequences for her primary audiences of physicians, distributors, and the client's sales force, consequences that, in turn, prompt consideration of additional users, namely consumers. As part of her ethical deliberation, she might weigh the stakes for the different audiences and her ability to take various actions (as asked in the section "Arriving at Flexible or Contingent Judgment and Commitment to Action").

Beyond possible financial impacts on these audiences, Angela's communication choices could have health and safety impacts: physicians and distributors could improperly store the medicine, physicians could recommend that patients receive an unnecessary combination of flu shots and FluEase, and physicians, later fueled by consumer demand, could administer FluEase to patients outside the approved age range. Thus Angela's obligation to protect others from harm may be even greater than in the Vaccitech project. This obligation to some of these users, especially consumers, might also be greater because of their relative lack of knowledge about and access to the full information about FluEase. Angela's team and client have more knowledge and communicative power than users, and thus she has a greater responsibility to protect such users.

As in the Vaccitech project, Angela may not be in a position to enact substantial user-design strategies and tactics; BioInfo's pharmaceutical client may not allow her to directly engage prospective users while the content is being developed. Angela will therefore need to negotiate the competing principles of respecting her contractual responsibility to her client and its property and creating the most accurate, accessible, responsive, and empowering texts possible. Perhaps she could persuade MegaPharm that conducting focus groups or testing with users would benefit everyone by leading to a better product. She might at least make suggestions about including the perspectives and concerns of other stakeholders and raise questions about the ethical and legal obligations and effects of not doing so.

Finally, in this project as in the last one, Angela should also engage others in her company and consider her obligation as an employee to work for her company's benefit. In this case, too, the MegaPharm project presents a higher-stake dilemma, given the future business opportunities to which it could lead. But the higher profile of this project and the texts Angela will create for it also likely poses a higher risk of damaging her company's reputation if these texts lead to harmful consequences or depart from legal and other commonly accepted nomoi.

TOWARD PHRONESIS

Fully appreciating the ethical dimension of technical communication requires first adopting a robust view of the technical communicator as a cocreator of knowledge rather than just a straightforward transmitter or translator of it. What we present and how we present it shapes how others make meaning; this implicates us, therefore, in others' interpretation and use of our texts across the various contexts in which they are encountered and taken up. We, as technical communicators, would do well to familiarize ourselves with specific nomoi—socially constructed, dynamic, and provisionally binding codes of action—relevant to the ethics and legality of our work. Such nomoi are tied, in many cases, to our relationships with and obligations to others, whether coworkers, employers, clients, various primary and secondary users or audiences, professional discourse communities, or other groups.

As this chapter defines it, ethics is more than a set of nomoi; it is a process, more specifically the relational, context-dependent, power-laden negotiation of, judgment about, and commitment to enact the best course. More important than knowing specific nomoi ahead of time is learning *how* to identify and negotiate the sometimes competing nomoi inherent in our relationships with others, how to identify and negotiate the perspectives, values, knowledges, and forms of deliberation that other stakeholders bring to an ethical challenge. To some extent, we can discern such nomoi and ways of negotiating them from texts about laws and social codes (e.g., the STC code of ethics), but we can also attempt to mutually discover them through the direct engagement of others. This chapter proposes that the *process* of ethics involves recognizing our roles and power in relation to others, open inquiry and deliberation with others (to the extent possible, given our positioning), and the mutual arrival at a contingent judgment and commitment to action. That is, what technical communicators most need to know about the ethics and legality of our work is how to develop and arrive at phronesis, or practical wisdom.

DISCUSSION QUESTIONS

1. Look at how technical communicators are defined in three different technical communication textbooks. How do these definitions relate to the notions of transmission, translation, and meaning-making authorship that Slack, Miller, and Doak (1993) discuss? What relationships and ethical obligations do these definitions emphasize?
2. Would you negotiate the ethics of the MegaPharm scenario differently if you were an executive decision maker instead of a technical communicator? Why or why not? If you were applying an ethic of care, how might your negotiation of competing obligations differ?
3. Can you recall a time when one or more of your personally held principles clashed with a value held by an organization or business in which you worked? What was at stake, and for whom? How did you negotiate this conflict and determine what to do?
4. Find a set of poorly designed and/or poorly written instructions. Explain how they are flawed and any possible consequences of such flaws for users. To what extent could these flaws be viewed as ethical or legal problems, and what is the technical communicator's role in attending to them?
5. Do you recall the conflict around the Florida Palm Beach County butterfly ballot in the 2000 U.S. presidential election? Read Bruce Tognazzini's (2001) analysis of the conflict and see an image of the ballot here: http://www.asktog.com/columns/042ButterflyBallot.html. Considering the technical communicator's role as user advocate, how might

the ballot's design be considered ethically problematic? What steps might the designers of the ballot have taken to avoid such problems?

6. In his article "Is This Ethical?," Sam Dragga (1996, 257) poses a scenario in which a technical communicator reporting a company's sales figures to investors designs a line graph in reverse chronological order to make a trend of decreasing sales look like one of increasing sales. The designer does, however, correctly label the years. Using the heuristic, identify at least two types of ethical problems raised by this scenario.
7. Identify ethical and legal principles or advice on the website of the professional association in your field (if necessary, ask one of your professors for help). How does this information compare to the STC code of ethics, and how could you account for any differences? What types of obligations, ideals, and consequences are emphasized, and what seems to be missing?
8. What types of ethical principles are embedded in the student code of conduct at your university? Should these be followed to the same degree in all student situations? Why or why not?

NOTES

1. The scenario discussed in this chapter is fictional. Any resemblance to actual companies, products, or situations is unintended and coincidental.

2. See Helyar 1992 and chapter 9 of Markel 2001 for more on the legal duty to instruct and warn.

3. See chapter 11 of Markel 2001 and chapter 5 of Allen and Voss 1997 for more about intellectual property laws and principles.

4. See the full STC code on the web at http://www.stc.org/pFiles/pdf/EthicalPrinciples.pdf.

WORKS CITED

Allen, Lori, and Dan Voss. 1997. *Ethics in Technical Communication: Shades of Grey*. New York: John Wiley and Sons.

Allen, Nancy. 1996. "Ethics and Visual Rhetoric: Seeing's Not Believing Anymore." *Technical Communication Quarterly* 5 (1): 87–105.

Anson, Chris M., and L. Lee Forsberg. 1990. "Moving beyond the Academic Community: Transitional Stages in Professional Writing." *Written Communication* 7 (2): 200–231.

Bowdon, Melody Anne. 1999. "An Ethic of Action: Specific Feminism, Service Learning, and Technical Communication." PhD diss., University of Arizona.

Dombrowski, Paul. 2000. *Ethics in Technical Communication*. Boston: Allyn and Bacon.

Dragga, Sam. 1996. "'Is This Ethical?' A Survey of Opinion on Principles and Practices of Document Design." *Technical Communication* 43 (3): 255–265.

Helyar, Pamela S. 1992. "Products Liability: Meeting Legal Standards for Adequate Instructions." *Journal of Technical Writing and Communication* 22 (2): 125–147.

Herrington, Tyanna K. 2003. *A Legal Primer for the Digital Age*. New York: Pearson.

Johnson, Robert R. 1998. *User-Centered Technology: A Rhetorical Theory for Computers and Other Mundane Artifacts*. Albany: SUNY Press.

Markel, Mike. 2001. *Ethics in Technical Communication: A Critique and Synthesis*. Westport, CT: Ablex.

Ostwald, Martin. 1969. *Nomos and the Beginnings of the Athenian Democracy*. Oxford: Clarendon Press.

Porter, James E. 1993. "Developing a Postmodern Ethics of Rhetoric and Composition." In *Defining the New Rhetorics*, edited by Theresa Enos and Stuart C. Brown, 207–226. Newbury Park, CA: Sage.

———. 1998. *Rhetorical Ethics and Internetworked Writing*. Greenwich, CT: Ablex.

Ruggerio, Vincent Ryan. 1997. *Thinking Critically about Ethical Issues*. 4th ed. Mountain View, CA: Mayfield.

Salvo, Michael J. 2001. "Ethics of Engagement: User-Centered Design and Rhetorical Methodology." *Technical Communication Quarterly* 10 (3): 273–290.

Scott, J. Blake. 1995. "Sophistic Ethics in the Technical Writing Classroom: Teaching *Nomos*, Deliberation, and Action." *Technical Communication Quarterly* 4 (2): 187–199.

———. 2004. "Tracking Rapid HIV Testing through the Cultural Circuit: Implications for Technical Communication." *Journal of Business and Technical Communication* 18 (2): 198–219.

Simmons, W. Michele. 2007. *Participation and Power: Civic Discourse in Environmental Policy Decisions*. Albany: SUNY Press.

Slack, Jennifer Darryl, David James Miller, and Jeffrey Doak. 1993. "The Technical Communicator as Author: Meaning, Power, Authority." *Journal of Business and Technical Communication* 7 (1): 12–36.

Sullivan, Patricia, and James E. Porter. 1997. *Opening Spaces: Writing Technologies and Critical Research Practices*. Greenwich, CT: Ablex.

Tognazzini, Bruce. 2001. "The Butterfly Ballot: Anatomy of a Disaster." AskTog: Interaction Design Solutions for the Real World. http://www.asktog.com/columns/042ButterflyBallot.html. Accessed July 2011.

ANTONIO CERASO

10 How Can Technical Communicators Plan for Users?

SUMMARY

Traditionally, a technical communicator may have understood planning as either an element of project management (creating time lines and assigning tasks), or as audience analysis that determined document requirements (planning for document content and form). In recent years, however, users of technologies and technical documents have increasingly become participants in development and documentation processes, whether through increased user and beta testing, or through online platforms and software programs that encourage and enable user involvement and feedback. This trend requires organizations to revise their understanding of planning. Rather than an activity that sets out the course of a limited project in advance, planning can be seen as a way to build in long-term opportunities for user involvement and organizational response. Moreover, because technical communicators have often been positioned as mediators between organizations and users, they are often asked to negotiate the tensions between planning in advance and responding to users. In addition to its traditional meanings for technical communicators, then, planning today can also mean anticipating, coordinating, and responding to user feedback. In this chapter, you will discover strategies for an expanded meaning of *planning for users*: both planning in the traditional sense *and* planning as a means of organizing ongoing responsiveness to users.

INTRODUCTION

Dave has been very busy lately: at night, he takes graduate courses for a master's degree in technical communication; during the day, he works full-time as the senior technical writer at PSInvent (pronounced P.S. Invent), a small software company in Pittsburgh that designs point-of-sale and inventory management applications, mostly for college and independent bookstores. Dave is always pleased when these two worlds inform each other—when his work experience helps him understand the theoretical arguments from his classes, or when techniques and theories from his classes can improve his work. But he also notices when these two worlds don't quite match up.

Specifically, many of Dave's course readings have addressed user-centered design (Johnson 1998), rhetoric and usability studies (Johnson, Salvo, and Zoetewey 2007), and the role of the technical communicator as a technology-user advocate (Salvo 2001). While he finds much of this research fascinating, it doesn't always translate smoothly at his job. PSInvent's two owners have a traditional view of technical writing, viewing software documentation largely as a support add-on. They don't see Dave's work as a way to involve users in product development and documentation. Their view is reflected in Dave's position itself. He began at PSInvent in customer support; it was only when numerous customers asked to speak to "somebody in the manual-writing department" that the position was created for him.

Apart from a traditional view of the technical communicator's role, PSInvent's owners have three economic objections to user involvement. First, PSInvent is a small company in a highly competitive field; the owners see both production costs and time to market as paramount concerns. They have argued that practices like formal usability testing—which may even require contracting outside specialists—would add costs and slow down product rollout. In this sense, they resemble the technology developers described by Robert R. Johnson (1997, 370), who "often dismiss, or at least downplay, the importance of usability because user testing and evaluation do not fit into the 'bottom-line' of a given project cycle." Second, the owners are guarded about new application features. They believe that opening the development process to users might allow their competitors to "scoop" their features, since PSInvent's customers might request new features from competitor vendors. For this reason, all usability testing is done in-house by a small beta-testing team. Finally, because the company is still somewhat small (about thirty people), all employees are very limited in the time they can devote to user involvement.

Dave can draw on his course work to make the case for increased user involvement, but ultimately he must plan and create documents that assist PSInvent's customers. Given the organizational constraints, he must usually do so without the benefit of direct user involvement. Of course, the users of PSInvent's software and documentation will eventually become involved when they purchase and use the products—they will encounter difficulties and provide feedback, either through support calls or directly through Dave. So Dave must also develop ways for adjusting his plans and documents in response to user needs.

This chapter aims to provide solutions for technical communicators like Dave who must negotiate the multiple meanings of planning in real organizations. While technical communicators remain responsible for traditional planning of documentation projects (such as project management and document planning), they are increasingly called on to plan for and coordinate user involvement. Through a two-part literature review and a set of heuristics, the chapter shows how these kinds of planning can work together. The first part of the literature review introduces three models of audience analysis that can help technical communicators plan the content and form of their documents, given various kinds of organizational constraints. In order to capture the sense of planning as identifying concrete project tasks, the chapter turns to research showing how each of these models imply concrete work activities that can be included in a project plan. The second part of the literature review turns to the theme of user involvement and responsiveness. It shows that, far from simply collecting information and data from and about users, responsiveness requires technical communicators to actively and responsibly coordinate the goals and needs of organizations with the goals and needs of users. The chapter then presents a set of heuristics, divided into "planning for audience" and "planning responsiveness." The first part (appearing in table 10.1) provides a set of questions to consider as you plan the content and form of your documents, and links these questions to activities you might have to perform to effectively deal with different organizational constraints. The second part (appearing in table 10.2) presents heuristic strategies that technical communicators can use to ensure that their project plans remain responsive to user feedback. Together, these heuristics function to coordinate initial project planning with the increasing need to involve users in technology development and documentation processes. The last part of the chapter returns to Dave's work at PSInvent. As it happens, Dave has three documentation projects on his desk that require him to both plan for and respond to PSInvent's users. The final section shows how Dave uses the heuristics as he plans for those three projects.

AUDIENCE, PLANNING, AND RESPONSIVENESS

In her 1997 book, *Dynamics in Document Design*, Karen Schriver sets out three models technical communicators use to understand audience: classification-driven, intuition-driven, and feedback-driven analysis (151–163). The classification-driven model may seem the most familiar to you, as it is often the way we are asked to analyze audiences when writing in academic settings; classification-driven audience analysis proceeds "by brainstorming about the audience and by cataloguing audience demographics (e.g., age, sex, income, educational level) or psychographics (e.g., values, lifestyle, attitudes, personality traits, work habits)" (155). In the intuition-driven model, communicators draw on their own experiences to construct an imagined or ideal reader, either as a "composite of human characteristics" or "people they have met before that could be like the reader" or even as an ideal reader to whom they would like to appeal. Finally, the feedback-driven audience analysis model involves *real* readers, bringing potential audience members "into the design process in order to draw on their ideas to guide invention" (161).

While Schriver highlights the value of the feedback-driven model over the other two, she explicitly argues that these models "needn't be viewed as being on a collision course"; rather, "they can be used alternately, depending on what the rhetorical situation calls for" (162). While the feedback-driven model provides advantages over the other models, we can see organizational constraints and time constraints as elements of the rhetorical situation that may require the classification-driven and intuition-driven models in practice.

Instead of viewing these models as separate categories, we can think of them as particular work processes that are enabled or constrained to varying degrees by organizational contexts and by project scope. In some situations, feedback from users would not be explicitly designed into project plans, but would happen once a product, document, or interface goes live. In such cases, technical communicators rely on feedback from previous projects, or on classification-driven and intuition-driven models. In other situations, extensive work with users (ranging from usability evaluation and testing to active co-development) might appear "early and often," leading technical communicators to use techniques for gathering, evaluating, and responding to feedback. Moreover, the constraints can vary from project to project, depending on its scope or importance, or across time within the same organization. For instance, a short software feature update for a familiar group of users might be written without much user feedback, while a manual for a complex product might require extensive usability testing. A company might become more open to user involvement, moreover, if it feels more confident about its market position. The combinations of these models—always based on concrete situations—require different competencies for technical communicators.

The different models also imply very different kinds of concrete work *activities*—that is, the hands-on stuff you actually do to develop an analysis. Technical communicators using a classification-driven model, for instance, might develop charts that connect the technical education level of an audience to possible document features (such as definitions or illustrations). Those using a feedback-driven model might design and conduct usability tests, and write memoranda or reports that explain and contextualize user responses for product developers. Each of these concrete activities can be included in planning documents that give technical communicators and managers a better sense of the actual work involved in a project. The next sections will flesh out the competencies and concrete work activities involved in planning for audience and responding to user feedback.

ORGANIZATIONAL CONSTRAINTS AND PLANNING FOR AUDIENCES

It's important to remember that the goal of planning for audiences is to determine what needs to be done with the "document" that is being produced—whether it's a website, a report, or a manual. Marjorie Rush Hovde (2000, 430) summarizes the kinds of information about users that Johnson suggests can be determined by getting users directly involved in documentation processes: "(1) their vocabulary and conceptual understandings, (2) their habits of using the technology, (3) their analogies for using the technology, (4) their ways of processing information about the document, (5) their typical uses of the documentation." Such user profiles would help technical communicators make decisions about the content and form of a document. For instance, if a technical communicator understands that a manual will generally be used in a cramped space, she may rethink the overall dimension or size of the printed documentation.

Hovde suggests, however, that technical communicators can and do build such profiles not only through direct interaction with users (such as usability testing), but also through what she calls "indirect" tactics, especially when organizational constraints make direct user interaction difficult or impossible. Hovde followed two technical communicators, one who worked for a small software company, and one who worked in a twelve-person technical writing department at a large financial services company. In both cases, various organizational constraints made direct user interaction difficult, so her subjects relied on indirect tactics for audience analysis. Hovde discovered the following tactics that could be combined to create a "rich" image of users: "talking with users during phone support calls" and consulting logs that recorded these calls; "interacting with users face to face"; "drawing on the writer's own experience"; "interacting with user contact people within the organization," like technology trainers and sales staff; "studying responses sent from users"; and finally, running simulated user tests with others in the organization (rather than with users themselves) (411).

Each of these tactics has strengths and weaknesses when measured against a set of criteria for "an ideal tactic for gathering rich data about audiences" (411). For example, phone support calls and logs could identify problems the users were having with documentation, but they could not show how people used the documentation and software in normal conditions. Internal testing of documentation, similarly, could reveal problems encountered by a general group but "did not necessarily reveal specific insights about external audience attitudes and practices" (425). While the various tactics these writers used all had inherent costs and benefits—both in terms of the information the technical communicator could garner from the activity and in terms of the organizational support for such activities—Hovde suggests that when "writers, working within an organizational culture, used these tactics in varying combinations, they were able to get a richer picture of users than they would have received by relying on only one tactic" (426). Like Schriver, Hovde ultimately recommends more *direct* interaction with users. Nevertheless, the tactics she discusses can aid technical communicators with initial planning if direct interaction is constrained.

The tactics Hovde describes are varieties of a feedback-driven model, even if they don't qualify as formal user testing or user involvement. By contrast, Katherine Miles's article "Reconceptualizing Analysis and Invention in a Post-Techne Classroom" (2000) adds depth to intuition-driven models of audience analysis. Miles sought to learn whether embodied experience would help technical communication students model the tasks that users of an immersive virtual reality (IVR) environment would perform. Miles asked two groups of students to write instruction manuals that would help users film in an IVR environment. Both groups had some direct contact with a user. However, one group of students received only information and a two-dimensional representation about the environment (the class group), while the other group visited and experienced the fully immersive environment directly (the tour group). Miles discovered that the tour group, which had an "embodied" sense of the IVR environment, better understood who used the IVR space and why they did, while also grasping the problems that users of the space might encounter. The tour group was also better able to translate from audience analysis to the invention of their document's content and form. By materially experiencing the space and time of the environment, the tour group gained a more useful model of users' tasks and task environment.

How might this research on technical communication students inform work in organizations where certain kinds of user interaction—such as a site visit to the user's workplace—is constrained? Consider two ways a technical communicator could develop a task model for the handheld scanning device that grocery store employees use to scan in new inventory and change prices on current stock. She could develop instructions for the device by speaking with its developers, or even by scanning in a few sample grocery items at her desk. She might also set up a simulated grocery shelf to get a more embodied feel of changing multiple prices at once. Or she could work through a few boxes of inventory to get a feel for how the device shapes and is shaped by the user's physical posture and environment. The latter version, which we could read Miles as suggesting, would provide her with a more complete task model that could streamline audience analysis and invention. Of course, her best option would be to go to actual client stores, observing how the employees in those spaces use the device, interviewing them about the ways they use the scanner, and perhaps even participating in some inventory activities to gain the material experience of performing their tasks. In particular organizational contexts, however, she might have to sacrifice the direct site experience for an in-office simulation.

Miles's article illustrates how one kind of work process and activity can derive from intuition-driven analysis methods: having a feel for users' embodied experiences can help technical communicators plan better documents. For a set of processes and activities that illustrate a classification-driven method, we turn to Kirk St. Amant (2005). As Schriver describes it, classification-driven analysis can be focused on how cultural differences—from differences in values and perceptions to differences in the way tasks are organized and conceived—affect the content or design of documents. St. Amant presents a method that enables professional communicators and designers to better "localize" websites for international audiences, that is, to modify sites for the values and expectations of users in different cultural contexts (St. Amant 2005). Like the other researchers we've reviewed, St. Amant imagines an ideal scenario that would involve interaction with both users and localization specialists, but recognizes that considerations such as "speed and cost" may require designers to conduct audience analysis on their own (73–74). St. Amant is primarily concerned with how the visual design elements of international websites can violate cultural expectations. In one striking example, St. Amant describes the kind of confusion that can be caused by a culturally specific icon, like a mailbox, to indicate a mail function: "[T]he perception of a mailbox being a metal or wooden box that sits on a post and that has a red flag on the side is, essentially, an American one. In other cultures, a mailbox might be a small door in a wall or even a cylindrical metal container that resembles an American fire hydrant. As a result, international users might come to a web portal and expect to find a mail function. The features used to depict a mailbox in the portal's access mail icon, however, render that depiction unrecognizable to users from different cultures" (76–77).

Such miscues could turn up in any number of design decisions. They become difficult to predict in advance, especially as the range of cultural contexts that documents enter into multiplies with globalization. St. Amant mentions, for instance, the classic problem of color selection: because colors signify differently in different cultural contexts, selecting colors for a web design can be a thorny problem indeed. Such audience concerns need not be thought of as simply matters of design, either. The McDonald's website for Jordan, for example, opens up with the following statement: "100% locally owned and operated." As you drill down into the website, you find a section called "Facts about McDonald's," which notes, among other statements, that "McDonald's is a solely commercial company that doesn't interfere in or support any political or religious acts in any country or region" and "[i]n Arab countries, McDonald's is totally owned and operated by Arab and Muslim businessmen" (McDonald's Arabia 2011). Clearly, such statements imply an audience worried about the political influence of an iconic *American* corporation. The McDonald's Arabia site, in response, aims to emphasize the independence of local franchisees from U.S. geopolitical interests. Moreover, localization extends to the level of word choice. The term business*men* rather than business*people* would probably be revised out, considered sexist language in other cultural contexts. Technical communicators cannot become experts on every possible cultural context, so, St. Amant asks, what other methods might they use to develop such localizations?

St. Amant offers prototype theory, borrowed from cognitive psychology, as a possible solution. Prototype theory is a complex theory of how humans categorize new information and images by comparing them to established information. Viewed from the perspective of work processes and activities, St. Amant offers a two-step analysis that might be generalized. To simplify considerably, the designer would identify a representative site and create a checklist of its features, looking especially for patterns that would indicate acceptable ranges of associations; then the designer would test this checklist against other sites in the same cultural context. Through this process, technical communicators and website designers could identify and confirm acceptable and effective elements for their own sites. St. Amant's study could be extended to other kinds of documents, particularly as technology and general cultural trends make visual elements common across document types. For example, since the 1990s, it has become normal to include screen shots (or screen captures) in software documentation (Van der Meij, Karreman, and Steehouder 2009, 271–272). In order to determine the quantity and kind of screen shots that would be acceptable in a given cultural context, a designer could conduct an analysis of existing documents within that context, seeking acceptable formatting, captioning, and image sizes. St. Amant's article thus shows us how determinations about audience can be derived through the collection and analysis of existing documents already deemed appropriate and effective by those users.

This section of the literature review has focused on initial planning in two senses. First, for the sense of planning as project management, the review emphasizes work processes and activities to show how audience analysis can be included in project plans. Second, for the sense of planning as invention, the review emphasizes how specific analytic activities help technical communicators determine a document's content and form. In sum, these activities sketch out what a technical communicator would do as he or she creates documents, even where organizational constraints limit access to real users. We will return to these activities in our heuristic section. But first, the literature review turns to the ways technical communicators become coordinators of planning and response.

PLANNING AND RESPONSIVENESS

Since the 1980s, it has become a commonplace in the technical communication research to argue that involving real users in projects is an ideal way of organizing production. Schriver (1999) seeks to move technical communicators toward feedback-driven models, while Hovde (2000, 426) argues that the images of the audience constructed by the technical communicators in her study "would have been richer had they had more direct interaction with external users," because they then "could have focused more on how users actually used software and documentation in their work contexts." Both Schriver and Hovde rely on the assumption now common not only in technical communication, but across development and design fields: developers cannot possibly anticipate all the needs, practices, and activities that users will find for technologies. Researchers have also recognized that the same technologies can have very different kinds of users, so that "[w]ell-designed systems must include enough flexibility to allow situational learning and support that vary according to individual users and user needs" (Smart and Whiting 2002, 162). Whether in workplaces or in their everyday practices, people will find ways to use and invent fixes for technologies based on individual and social needs (Spinuzzi 2003, 1–24).

Quite literally, developers and users work with and even *know* the same technologies in different ways. The differences between developers and users can thus be defined as both functional (in that these groups can deploy technologies for different tasks and purposes) and epistemological (where epistemology is defined as the study of how we know things). If we assume that users may have different functional and epistemological relationships to technology, it becomes clear why feedback-driven models have attained such prevalence. These models help us bridge functional and epistemological gaps by providing data and other information about users, thereby helping technical communicators and developers understand how people *use* and *know* technical systems.

Functional and epistemological differences, moreover, can be grounded in cultural difference. Huatong Sun's 2006 article, "The Triumph of Users: Achieving Cultural Usability Goals with User Localization," turns the more general question of use toward specific cultural relationships. Sun addresses a problem much like that discussed by St. Amant: how can technology developers more effectively respond to local uses of technology across cultural contexts? Arguing that both localization specialists and conventional usability practices can miss the often messy social and cultural factors that shape people's use and understanding of technology, Sun (2006, 461) develops a framework of "cultural usability," a blend of contemporary theoretical approaches that "brings social-cultural contexts into concrete user activities." Sun illustrates the framework by analyzing text-messaging practices in the United States and China. She finds that the "concrete use activities" of text messaging are determined by both deeply embedded social and cultural factors (such as the conventional character of friendship within a given cultural context), as well as by immediate user concerns (such as working around a manager who forbids personal phone calls). These studies suggest that technology developers should not focus on perfecting technologies for local use so much as they should "look for ways to initiate a communication channel and to build a support network to enhance user localization and help repair the possible breakdowns" (478). If a continuous stream of communication and interaction is ideal, what technical communicators do, in part, is build mechanisms that facilitate ongoing relationships.

If the difference between technology developers and users were only functional and epistemological, the drive toward user involvement might simply mean collecting as much data from users as possible. However, the research on user involvement suggests that data collection is not enough; in addition to functional and epistemological difference, user involvement also has ethical implications. For example, Michael Salvo argues in several articles that technical communicators can use rhetorical methods to shape an ethical relationship between technology developers and users. Shifting the ground from functional observation and data collection, Salvo (2001, 276) argues that active *collaboration* with users becomes an "ethical responsibility on the part of usability professionals . . . to maintain a dialogic relationship between technology producers and consumers." Rather than a one-way flow of information and data from users to developers, Salvo seeks a genuine two-way (or dialogic) communication channel through which developers and users actively participate in shaping technologies. Salvo expands the ethical and political stakes of user participation in a 2004 article in which he redefines "information architecture" as an alternative to a form of design that merely improves on fixed projects by pulling information from users: "Information architecture is a process of designing working models of the technocultural future that values participation, access, and input from users, seen as both stakeholders and citizens, for whom and by whom technocultural designs are created. Information architecture is a design process compatible with democratic processes and with an active, engaged user population; indeed, information ultimately requires the active participation of its users, who are agents in the definition, design, and maintenance of the technocultural system" (64).

We can understand Salvo's idea of active participation in contrast to other ways of organizing the relationship between technology developers and users. The first way would exclude users completely from the design and development process, leaving those steps of the process to technology experts. The second way does involve users, but treats them as a mere means to an end: rather than being asked to genuinely contribute to technology development processes, users are merely mined for data, while technology experts retain control of production. Salvo's goal in proposing a dialogic approach is to offer an alternative to both these kinds of relationships; he wants us to thinks of users neither as passive consumers of technology built by experts, nor as simple instruments from which to draw information about different uses, but as *active participants*. Salvo (2001, 275–276) seeks to find ethical strategies for engaging users as active participants, approaches that account for the purposes and needs of developers and multiple kinds of users. If technical communicators are now routinely involving users in development projects, Salvo seems to ask in both articles, how can they do more than simply collect user information for the purpose of creating new—but ultimately inaccessible—technologies and plans?

Together, Sun's and Salvo's research points to a somewhat troubling point: if user interaction enters into planning with a fixed endpoint in view—such as the short-term "improvement" of a technology or interface feature—it will not only miss the multiple ways people use and innovate with technology, but also reproduce power relations that reduce people's control over the technologies they use and technological systems more generally.

User involvement, therefore, must move beyond simply cycling various user-generated data into production. For both Sun and Salvo, technology developers and technical communicators must establish responsible frameworks for interpreting that data and constructing channels that allow users to become active participants in development processes.

An open and responsible channel of communication between developers and users does not, however, invalidate organizational interests or developer expertise. Rather, it requires the far more complex activity of *coordinating* initial plans with user involvement. In a retrospective overview of usability research, Johnson, Salvo, and Zoetewey (2007) argue that technical communicators are ideally positioned for such coordinating work. Usability testing, they suggest, cannot produce active user contributions to technology development when authentic user involvement is reduced to evidence for adjusting initial plans. Neither, however, should user involvement simply overwhelm or substitute for the plans and work of technology developers and organizations. Instead, user involvement "becomes effective when informing ongoing processes of design and development in a timely way" (325–326). Matching an "ongoing process" of design with "timely" interventions functions as a way of coordinating organizational plans and user involvement. Johnson, Salvo, and Zoetewey suggest that technical communicators play the important role of locating timely points of intervention in ongoing plans, thereby coordinating the expert knowledge of technology developers and with the contextual knowledge of technology users. If technological development is becoming increasingly "participatory"—if users are increasingly asked to be involved in technology development—technical communicators should be viewed as the active and knowledgeable coordinators of that participation.

The goal of such coordination would be responsiveness. We can define *responsiveness* as the capacity of organizations to adjust plans, technologies, and documentation based on feedback from and interaction with customers and users. As this section has sought to demonstrate, user feedback can take the form of data that reveal functional and epistemological differences between technology developers and users; organizations are responsive to the extent that they can adjust to such differences. However, responsiveness also has an ethical dimension. It requires that organizations view users as active participants in development and documentation projects, rather than as simple data sources. Finally, responsiveness is always a *relative* capacity. Organizations can adjust plans, technologies, and documentation only within existing constraints and goals. This definition may certainly seem abstract. The next section points out, however, that such abstraction is necessarily built on concrete processes and activities. We will drill down toward what technical communicators can actually do to cultivate responsiveness and build it into project planning.

HEURISTIC: PLANS, CHANGE, AND ORGANIZATIONS

The literature review emphasized specific work processes and activities in order to tie the general models to concrete steps you can take as you plan projects. This section gathers those various processes and activities together into a set of heuristics that can guide your work with developers and users. First, we'll formalize the steps you can take as you learn to plan for audiences and respond to users. We'll then return to Dave's story to illustrate how project planning, planning for audiences, and planning for responsiveness can work together.

PLANNING FOR AUDIENCES

Many technical communicators may find themselves in positions like Dave's. They know that involving users will make both the technology and the documentation more effective, but they are limited by organizational constraints. Still, as Hovde's research demonstrates, such constraints can serve as opportunities for innovation. Specific activities that appeared in the research on initial planning include, for example, reviewing user responses to previous documents (Hovde 2000), discussing user needs with contact people within an organization (Hovde 2000), setting up in-house simulations of user task environments (Miles 2000), and collecting document samples from particular cultural contexts (St. Amant 2005). Each of these activities can be included in project plans as they are drawn up—each given a time frame for completion, each considered for its effects on organizational resources, each assigned to people working on the project.

Where organizational relationships provide more contact with users, technical communicators can become a key point of contact for coordinating developer plans and user involvement. The research on user involvement points to further specific activities that can facilitate such a role. While data collection would seem insufficient, it remains necessary; to this end, technical communicators would include various methods for user testing in their planning—from interviewing and observing users to collecting and analyzing logs. Technical communicators will need to stay

Table 10.1. *Planning for audiences*

Audience considerations	Activities to include in project plans
Identify characteristics for multiple kinds of users: Users will have a wide variety of characteristics that may affect their patterns of using the technology and any technical documentation that goes along with it. Here are some of these characteristics listed in the research. • Demographic and psychographic (Schriver 1997) • Ways of thinking about and using technology • Cultural factors • Level of technical knowledge, including vocabulary • Experience with similar technologies • Previous and current problems encountered by users	*When direct user involvement is constrained* Collect, review, and analyze typical existing documents and genres (such as St. Amant's [2005] prototype analysis); meet with user contact personnel (in support and sales); review user feedback on previous documents; contact users informally; draw on personal experience. *When direct user involvement is encouraged* Conduct usability tests; interview and observe different kinds of users; develop or draw on research frameworks that help you identify patterns and differences between users (such as Sun's [2006] cultural usability framework); research, devise, and administer mechanisms for continuous feedback; review and analyze user-generated feedback platforms.

current on research methods, while developing strategies for explaining seemingly abstract theoretical issues to stakeholders. Technical communicators would also need to research, argue for, and perhaps build mechanisms that facilitate genuine two-way (that is, in Salvo's terms, dialogic) relationships between developers and users, such as online user forums or social media engines. Finally, coordinating developer plans and user participation often requires a range of documents (such as assessment reports and usability test reports), and could include direct involvement in the development of the technologies themselves. By introducing these responsive activities into initial project plans, technical communicators can build flexibility into the implementation of those plans.

Table 10.1 lists considerations about audience that will help you determine the content and form of your documents. Along with these considerations, table 10.1 sets out the concrete work activities you might perform as you develop a plan for audiences and respond to user concerns under various kinds of organizational possibilities and constraints.

PLANNING FOR RESPONSIVENESS

When technical communicators are the people in organizations who can productively coordinate organizational plans with user involvement, developing strategies for responsiveness becomes a core competency. Put

Table 10.1. *(continued)*

Audience considerations	Activities to include in project plans
Identify typical and atypical user task environments: Under what physical conditions do users perform the tasks? How do elements like space, time, light, and bodily position and characteristics affect task performance? (Miles 2000).	*When direct user involvement is constrained* Meet with user contact personnel; meet with subject-matter experts and developers; construct activity simulations and use cases; interview users informally. *Sample document types:* e-mails, memoranda, and meeting notes; simulated use scenarios; interview questions. *When direct user involvement is encouraged* Visit user work sites and/or observe users working in their typical environments; develop sound usability tests that evaluate; interview users about their task environments. *Sample document types:* evaluation, usability, and assessment reports; usability procedures and protocols; interview questions.
Identify typical document-use patterns: Why do users tend to draw on technical documents? When do they turn to documentation? What formats are preferred? Do users help create documentation (as in user forums, or through user-generated content), or provide evaluative feedback about them?	*When direct user involvement is constrained* Review user-support call logs; review access logs of online or electronic support documents; review user feedback on previous documents; interview users informally. *When direct user involvement is encouraged* Conduct usability tests; interview and observe different kinds of users; research, create, and administer mechanisms for users to evaluate and generate support content; observe and assess independent user-support platforms.

another way, technical communicators can be thought of as "responsiveness managers." Table 10.2 sets out a series of concrete activities you can practice that will build your competence as a responsiveness manager.

Tables 10.1 and 10.2 describe activities that cover the expanded meaning of planning we've been considering. The activities in table 10.1 span the range from very traditional kinds of audience analysis and project planning that happened (and still happen) in closed organizations, to more current methods of interacting with and involving users in development and documentation processes. Table 10.2 supplements these activities by

Table 10.2. *Developing feedback scenarios*

Task	Explanation
Catalogue points in a plan when feedback may occur.	When you create timelines for projects, identify the points where various stakeholders are likely to intervene. If you know *when* and at what stages of a project you might have to shift focus, change your plans, or respond to feedback, you will be better prepared when those points arrive.
Identify stakeholders who may provide feedback.	Along with the *when*, it's useful to anticipate the *who*. Different kinds of stakeholders will likely have very different reasons for and interests in providing feedback on projects. Have a sense beforehand of why stakeholders might be doing so, in order to anticipate the kinds of feedback you are likely to receive.
Classify effects of feedback.	Feedback can have very different effects, depending on where it comes from and what stage of the project it affects. Consider the economic effects, the effects on time to completion, and the effects on artifacts and documents that the feedback may require you to produce.
Identify documents that may be needed to address feedback.	You may need to schedule time to produce usability reports, or assessments of user responses. You may also need to develop proposals that argue for change based on feedback.
Construct response scenarios and develop a constraints sheet.	Not all feedback needs to be or even can be addressed. If you identify the points in a project when you may receive feedback and the kinds of feedback you are likely to receive, you can begin to develop scenarios for possible responses. For example, an organization could have severe cost constraints that would make an interface redesign impossible at late stages of a project, even if users experience significant difficulty with interface features. If you recognize these limitations in advance, you can address such feedback productively (showing how the *next project* can be improved). Or, perhaps even better, you can argue early for changing the way users are involved in the development process to avoid such results.

building flexibility and anticipation into the planning process. If the first provides a standard roadmap for a project, the second gives you a sense of the possible detours you may encounter. Used together, they can prevent a project from getting lost. The next section shows how Dave uses the concrete processes and activities in tables 10.1 and 10.2 to plan his projects, understand his audiences, and anticipate potential difficulties at PSInvent.

DAVE'S THREE PROJECTS AT PSINVENT

Dave sits at his desk at PSInvent and reviews his agenda for the morning: he has to plan his work for three projects. The first project is probably the easiest to plan. It's a one-page feature sheet that describes an update for a particularly troubling software function. Dave usually polishes these feature sheets off in a day and sends them out to the customers by PDF, so he's planning to finish that up this afternoon.

The second project is a little trickier. PSInvent's sales team just landed a major new customer—a small chain of Christian bookstores in the St. Louis area. Because of the bookstores' particular needs, however, the clients requested a customized version of PSInvent's software to close the deal. The sales contact, programmers, and owners spent the last two weeks working out the details of the new functions. Dave, unfortunately, didn't attend any of those meetings, but he did receive an e-mail from his manager on Friday, asking him to begin creating custom documentation to go along with the software changes. "This might take a while," Dave thinks, "but I probably should review the user manual anyway, since it will need to be revised this year."

The third project is the most involved: a major revision of PSInvent's user manual to accompany a new version of the software. Elements of the new software have already gone to the in-house beta-testing team, but it's not projected to be released for another five months. Still, Dave remembers how much work was involved in revising the manual for the last version update, and that one didn't include complete interface overhauls like this coming version. Plus, the feedback Dave has received on the current edition of the manual has given him some ideas for improving the new edition.

THE FEATURE SHEET

Dave has wanted to write this feature sheet for some time. It relates to a function of PSInvent's software that allows customers to search for and organize purchase orders using different search fields, like ISBN and edition numbers. There was only one problem with the function before this update: it didn't really work. Bookstore employees consistently received database error messages when they tried to run a search; Dave was very familiar with this phenomenon, because about a third of his calls with users since the release of the latest version involved irritated customers trying to get the function to work. At one point, Dave almost started answering his phone by asking "Purchase Order Search problem?" He heard it was even worse in the support department. The only thing they could tell the customers was, "We're working on it." But now the function is fixed, and all Dave has to do is write up a brief description of the modified interface, since the instructions haven't changed from the manual. Technically, all Dave has to do to create the feature sheet is open up the feature-sheet template he has ready in Adobe InDesign, perhaps take a screen shot of the new interface, and craft the text, some of which he's received in the form of quick specifications from the development team. He's written so many of these that it usually goes quickly. Still, with this particular function, he wants to do a little more investigation, since another round of mistakes and breakdowns could really alienate the customers. He observed the feature working perfectly when the in-house beta team really tested it, so he thinks it should be fine, but he just wants to make sure before the sheet goes out.

Dave knows that customers receive updates like this one at different times, since some customers have their software set to update automatically over the network, while some customers update the software manually, usually after they receive the feature sheets. Have the customers who already updated the Purchase Order Search experienced any problems? Can he head these off with developers, or figure out how to guide those users before everyone else updates, and more support calls start flooding in? Before writing the feature sheet, Dave needs to understand whether the users are experiencing current problems with the new feature. Since neither the organization nor the small scope of this project allow for formal usability testing, Dave will rely on other activities, like conferring with user-contact personnel, reviewing support logs, and contacting users informally. He opens up his project management software and starts to enter a few tasks for later in the morning. He wants to speak briefly with people in the support department to learn whether any customers have experienced problems with the updated function. Because Dave used to work in support and still considers himself a support specialist, such visits are fairly routine, and the support people always make time for him. That should be no problem. But to save time, he'll conduct a few searches of the support log database, looking for mentions of the Purchase Order Search feature since the automatic update went out. He marks down one last task: call Celia. While PSInvent doesn't do formal usability testing, especially not on feature updates, Dave does informal user testing with some of the customers with whom he's forged relationships, like Celia. Since he knows that she updates automatically, she should have the Purchase Order Search function available. Dave will ask if she has a few minutes to run some use cases. If Celia has any problems, Dave can identify them and write a delicate memo to the programming team.

Between the InDesign template, the specifications he received from developers, his observations of the beta tests, his discussions with support and review of the logs, and a few test runs with Celia, he should have all he needs to write the document.

THE CUSTOM DOCUMENTATION

Dave opens up a new project file in his project management software: the custom documentation for the Christian bookstore chain. Whereas the issues involved in the feature sheet were almost painfully familiar to him, on this one, Dave knows this much: he doesn't know much. He received the e-mail announcing PSInvent's first out-of-town sale (always a big deal for a smallish business) at about the same time that he received his manager's request for a customized electronic version of the manual. Clearly, the job won't require a total manual revision, but will focus on those features and operations that the client wants customized. But Dave doesn't know what those are. What custom features are they requesting, and why? "In the future," he tells himself, "I really have to ask to be included in these stakeholder meetings: coming in at the end like this is just inefficient."

He also doesn't know much about the customer at all. He doesn't know what previous systems they might have used for point-of-sale and inventory (A local competitor? Pencil and paper?), an important point for understanding their task expectations and user knowledge. He doesn't know the size or sales volume of the individual stores, information that could help him improve even the sections of the manual unrelated to the specific customizations. He doesn't know how many employees they have, or what their demographic characteristics or familiarity with technology might be. If he had to admit it, he'd say that he didn't even really know what it is that they sell. So, on both the product side and the user side, Dave needs to collect a lot of information, and somewhat quickly. Since it's unlikely that he'll be visiting the stores in St. Louis, he guesses that gathering the information will mean *a lot* of meetings here in the office.

What Dave needs to do, then, is conduct a thorough audience analysis that will help him determine the content and form of the document. He starts to consider the concrete activities that will help him identify the characteristics (or, in Hovde's terms, build an image) of the audience, and get a sense of the users' task environment. His first steps in project planning will be to set up meetings with the person at PSInvent who has had the most contact with the customer: Eileen in sales. He'll want to bring a written list of questions that will help him clarify who the users of the software and the manual will be, what their level of technical expertise might be, and in what environment they'll be using the software and the manual. He'll also need to set up a meeting with the programming team, and perhaps assign them to write some initial drafts of the relevant customized manual sections (he knows he'll have to revise these). He enters both of these requests into the project management software, then turns to the current version of the manual. He wants to do a quick document inspection to find out if there are any obvious audience-related issues that jump out.

Dave is drawing on personal experience here, trying to put himself in the place of the audience in order to identify possible problems. As he scans through the current manual for sections that might need to be customized, he notices the screen shots that accompany the instructions for processing returns. The screen shots help walk users through the returns process by displaying the required interfaces and screens, using a set of book titles as example returns. When the manual was last revised, Dave still had PSInvent's major customer base—college bookstores—in mind. As a result, the screen shots display book titles that Dave considered relevant for a college bookstore at the time of the revision. What he realizes at a glance is that all the titles are for books on evolutionary biology and theory. Would this work for Christian bookstores? He looks more closely at the instructions to find that several titles are even listed in the text as examples, with arrows pointing to them in the screen shots. PSInvent has been distributing this version of the manual to all its customers, and Dave has never received any negative feedback on the screen shots or instructions before. Indeed, he's not even all that convinced that users in general would *care* about example titles in a software manual. Moreover, because he doesn't know that much about the client's business at this point, he doesn't know whether the content would be an issue, since there is a wide diversity of beliefs about evolution among Christians. He feels a little uncomfortable, thinking that he might be jumping a bit quickly to stereotype the new clients. At the same time, he suspects that he will probably have to switch out those screen shots and revise these sections of the text in ways that are less likely to conflict with the users' values. At the very least, he wants to run this by Eileen in their meeting. He enters "Ask Eileen about evolution screen shots" under his entry for setting up a meeting, then marks out a few hours for later this week to pull, edit, and insert new screen shots for those sections, and to revise the text.

THE MANUAL REVISION

The manual revision, if it's anything like the last one, is going to be a long slog. In fact, Dave already senses that it might be even more complicated than the last revision, since he has seen wireframes of some of the new interfaces, and they are quite different than the last version. That will mean new instruction and description text, hundreds of new screen shots, many e-mails back and forth with the programming teams. It's not so much a revision as it is a new manual from the ground up. Still, none of the software is even going to beta testing for another month, so Dave feels like he's ahead of the game. In fact, he sees the new manual as an opportunity to build better relationships between users and developers, and perhaps even effect a little change in the organizational culture.

Because the new software version and manual project is in its very early stages, Dave can create feedback scenarios that can help PSInvent anticipate, respond to, and coordinate stakeholder feedback as the project is implemented. Last time, the company had a project plan, but it was not flexible enough to deal with contingencies. For example, the last printed manual went significantly over costs, largely because the outside vendor they used for print production charged them for changed pages—the revisions Dave completed on the manual after the printer had set the pages. Of course, the pages had to be changed because Dave was discovering use problems for some software features and instructions through informal interviews. Management finally decided to send the manual out with some flawed instructions, largely because they were tired of paying additional costs to the printers. If Dave could respond to feedback on the documentation earlier in the process, PSInvent could reduce or even eliminate such costs: they could intervene earlier in the programming stage, and certainly avoid sending out a flawed manual. To this end, Dave begins identifying the various internal stakeholders (management, sales, development) and external stakeholders (existing clients, new clients, vendors) and identifying the kind of feedback and responses they might provide. What he's really interested in is finding those "points of no return," the times in the project where even constructive feedback fails, since he would be unable to respond to such feedback because of cost or release dates (like the printer fees). If he knows when the manual will be sent to the printer, he can gather informal feedback from users earlier, and prevent the problems they had with the last manual. He starts to list such other points of no return in a notebook.

While Dave doesn't think the owners will go for full-on usability testing just yet (and he doesn't feel qualified at this point to run such tests anyway), he's been toying with the idea of gathering feedback through electronic means, or opening up better channels of communication with the customers. Currently, PSInvent doesn't have user forums, primarily because nobody has had time to create and run them. He thinks he might have an opening to at least research and propose some platforms; PSInvent's owners have already expressed concerns that the company's lack of user forums might negatively affect the perceptions of potential customers. Dave also wants to look more closely at the online help software he uses to author content, and its features for allowing user-generated content. The software was recently upgraded to include such a feature, but his manager wanted to hold off on paying for an upgrade for cost reasons. If Dave will be working on what is essentially a full rewrite of the manual anyway, he may be able to leverage that activity to open ways for users to participate in the future—in more formal ways. Besides, if internal stakeholders in management and the programming team can see how the users respond to and use the software and documentation, that might move PSInvent toward increased user involvement all the way through the process. "At the very least, we might save on support calls," Dave thinks. He starts taking notes on the kinds of documents he might need to produce for the big project: a proposal for user forums, a memo requesting the authoring software update, assessment and evaluation reports of user activity on those forums, memos to developers and programmers, perhaps arguing for programming changes they won't want to make. "If we're going to plan for a major manual revision," Dave thinks, "we may as well also plan for a more flexible and responsive organizational culture."

CONCLUSION

If a historian were to view our era from a hundred years in the future, she might conclude that the last decades of the twentieth century and perhaps the first few of the twenty-first century marked a radical shift in technological and cultural production. Whereas both broad fields had tended to be dominated by gatekeepers and production specialists and experts in closed organizations, they began to be shaped more and more by the participation and involvement of consumers and users. In technical communication, these developments have focused on the involvement of technology users in development and documentation processes. Indeed, arguments for involving users in technology production that may have seemed novel or controversial even a decade ago may now seem obvious—largely because they anticipated what might now be the dominant discourse of technology. Of course, such portraits of social and economic change always look better in broad strokes; they get much more complicated when we examine the details. Organizations still have to develop plans and hit benchmarks and show growth to investors and stakeholders, and they often continue to do so without much user involvement at all. Moreover, co-developing technology with users can be either chaotic or insincere (and sometimes both!) unless such processes are effectively and responsibly managed. As a number of the researchers covered in this chapter have argued (and as Dave's story illustrates), technical communicators are ideally positioned to manage such change, but they always do so within the constraints of the organizations with which or in which they work. We might, then, accept the broad-strokes view of user involvement, but we always work in the detailed view. As you build your professional repertoire as a technical communicator, you should be developing competence in balancing and coordinating organizational and user interests. Part of such competence involves developing more detailed views of your audience in order to create more effective technology and documents within organizational constraints. Part of it may also mean inventing ways to involve users in order to create more effective and responsive organizations.

DISCUSSION QUESTIONS

1. Consider a technical communication project you've completed, or one that remains due. Which of Schriver's audience model(s) (classification-driven, intuition-driven, feedback-driven) most closely fits what you have done to better understand your audience? What—if anything—constrained your use of those models? Using table 10.1, create a list of the concrete activities you might undertake to learn more about your audience. Next to each activity, write an estimate of how much time it would take and who else (if anybody) would need to be involved.
2. Think of a major improvement you would like to make to a workplace or organization to which you belong. (If you are working on a longer proposal or recommendation report for class, you can use that project to work on this question.) Create a brief plan for how that improvement would be implemented. Using table 10.2, try to anticipate the kinds of feedback you might encounter as you implement your plan. Think especially hard about the points of no return, or the events in the plan that would make it hard to respond to further feedback.
3. Review Dave's proposed activities for the custom documentation project. On a sheet of paper or in a computer file, create a table with three columns. In the first column, list the kinds of information about the

audience that Dave will need as he plans the documentation. In the second column, list the activities Dave could undertake to gather that information. In the third column, describe how that information might affect the content and form of the documentation Dave ultimately produces. You can use the information in the chapter to get started.

4. Describe in detail what you actually do to plan for class projects (i.e., do you keep schedules and tasks "in your head," do you use a paper planner, do you create time lines or remind yourself with sticky notes?). If the ability to plan projects is a competency, how would you rate yourself?
5. Consider Dave's three projects again. Given your understanding of the projects, what is the most serious problem that might emerge as he tries to complete them? How might he avoid that problem?
6. Locate and read Marjorie Rush Hovde's article "Tactics for Building Images of Audience in Organizational Contexts" in the *Journal of Business and Technical Communication*. (Planning note: give yourself plenty of time!) Then, imagine you are entering a meeting with PSInvent's owners. Create a list of arguments you can make as a technical communicator for instituting more user involvement at PSInvent. What arguments might the owners make for keeping things the way they are?
7. In discussing the concept of active participation, the literature review specifies three kinds of relationships that can be established between technology developers and users, and suggests that such relationships have an ethical dimension. In your own words, describe the three kinds of relationships mentioned in the text. Then answer these questions.

   - Do you agree that merely gathering data from users is ethically insufficient because it doesn't really include them actively in technology development? Why or why not?
   - Do you think technical communicators have an ethical responsibility to include users more fully in the process of technology production? Why or why not?

8. Go to the user forum for the Ubuntu operating system (http://www.ubuntuforum.org) and browse through several topics, particularly under the "Main Support Categories" heading. How does the user-generated technical support you found on Ubuntu Forums differ from what you might find in a formal user guide? How might an organization use forums like this to assist in developing new technologies? How might users exert influence on organizations through such fo-

rums? How might a technical communicator use such a forum to coordinate software developer and user needs?

WORKS CITED

Hovde, Marjorie Rush. 2000. "Tactics for Building Images of Audience in Organizational Contexts." *Journal of Business and Technical Communication* 14:395–444.

Johnson, Robert R. 1997. "Audience Involved: Toward a Participatory Model of Writing." *Computers & Composition* 14:361–376.

———. 1998. *User-Centered Technology: A Rhetorical Theory for Computers and Other Mundane Artifacts*. Albany: State University of New York Press.

Johnson, Robert R., Michael Salvo, and Meredith Zoetewey. 2007. "User-Centered Technology in Participatory Culture: Two Decades 'Beyond a Narrow Conception of Usability Studies.'" *IEEE Transactions on Professional Communication* 50:320–332.

McDonald's Arabia. 2011. "Absolute Truths about McDonald's." http://www.mcdonaldsarabia.com/. Accessed August 16, 2011.

Miles, Katherine. 2000. "Reconceptualizing Analysis and Invention in a Post-Techne Classroom: A Comparative Study of Technical Communication Students." *Technical Communication Quarterly* 19:47–68.

St. Amant, Kirk. 2005. "A Prototype Theory Approach to International Website Analysis and Design." *Technical Communication Quarterly* 14:73–91.

Salvo, Michael. 2001. "Ethics of Engagement: User-Centered Design and Rhetorical Methodology." *Technical Communication Quarterly* 10:273–290.

———. 2004. "Rhetorical Action in Professional Space: Information Architecture as Critical Practice." *Journal of Business and Technical Communication* 18:39–66.

Schriver, Karen A. 1997. *Dynamics in Document Design: Creating Texts for Readers*. New York: Wiley.

Smart, Karl, and Matthew Whiting. 2002. "Using Customer Data to Drive Documentation Design Decisions." *Journal of Business and Technical Communication* 16:115–169.

Spinuzzi, Clay. 2003. *Tracing Genres through Organizations: A Sociocultural Approach to Information Design*. Cambridge, MA: MIT Press.

Sun, Huatong. 2006. "The Triumph of Users: Achieving Cultural Usability Goals with User Localization." *Technical Communication Quarterly* 15:457–481.

Van der Meij, Hans, Joyce Karreman, and Michael Steehouder. 2009. "Three Decades of Research and Professional Practice on Printed Software Tutorials for Novices." *Technical Communication* 56:265–292.

CLAY SPINUZZI

11 How Can Technical Communicators Study Work Contexts?

SUMMARY

Context is one of those vague terms we use to explain why something—a text, a tool, an interface—works differently in different conditions, circumstances, practices, or activities. But what is context, and how do we study it? More specifically, how can we figure out how texts are deployed in and changed by ongoing activities, and how do we communicate these analyses to interested parties? In this chapter, I introduce three heuristics—communicative event models, genre ecology models, and sociotechnical graphs—for understanding and explaining context.

INTRODUCTION: CONTEXT IS NOT A PUMPKIN

I regularly send my students to investigate organizations with which they're associated, with the aim of discovering communication problems and developing solutions. These students have written reports about organizations such as brokerage firms, libraries, tutoring centers, restaurants, coops, and even comic book shops. At each organization, students find hidden problems and frictions (there are always a few), pinpoint them, and develop textual solutions for them. That is, by investigating the organization's context, these students can find and develop solutions that make their organizations work better.

Here's a scenario loosely based on those studies.

Ana is interning as a writer for a local nonprofit organization. Although she's a good writer and knows the organization pretty well, she soon finds that she has trouble understanding and executing her duties—and she's not alone. Singh and Connie, two other writers on writing team A, also have trouble as they do the things writers do at a nonprofit: edit brochures, develop collateral, assemble proposals, write annual reports, update the website, and take care of a seemingly endless set of other writing duties. Team A seems to be overwhelmed. But another team of writers, just down the hall, seems fine with their workload.

In writing team B, Ben and Sharlee always seem to know what they're doing. They never seem to ask for help. They're more productive than team A, putting out more documents in less time. They're always ahead of deadlines, while team A struggles to meet them. Maybe they're geniuses, Ana thinks. But after interacting with Ben and Sharlee a few times, Ana decides that they're definitely not geniuses.

Well, then, Ana reasons, perhaps it has something to do with the context of each team. But what?

What is context? Bruno Latour, a philosopher of science, once observed that whenever someone spoke about context, they would make a certain gesture with their arms, sort of a circle that started at the clavicle and ended at the solar plexus. Context, he concluded, must be about the size and shape of a pumpkin (quoted in Lave 1996, 22).

Of course, Latour was joking. But his point was that when we talk about context, it often just amounts—literally—to hand waving. We tend to think of context as whatever surrounds or circles the thing we want to study. Without a firm definition of context, a bounded understanding, there's not a lot of commonality connecting these different interpretations of context. And unfortunately that means that there's not a lot of guidance for studying or understanding context. After all, we can't study *everything*. At the same time, we can't just pick out arbitrary parts of the environment or activity to focus on—if we do, we end up missing some important influences, and we have a very hard time comparing them.

So in this chapter, I discuss how we can study *the relationships among people and their activities*. In particular, I focus on comparing how similar people, doing similar work, can be compared to understand certain kinds of similarities and differences in their work. For the purposes of this chapter, we can call the differences context.

Context has been a recurring issue for technical communicators because documents' success has a lot to do with how they are interpreted and used locally (see Cross 1993; Paradis 1991). When technical communicators emphasize audience analysis, designing for work conditions, or using appropriate visuals, they are focusing on this question of how to tailor texts to work optimally under local conditions. That is, texts have to work ecologically within a given activity; there's no such thing as a perfect document that works well under all conditions.

So the *problem* of context is not too hard for technical communicators to spot. But solutions are more difficult: without points of comparison, how do we tell what aspects of a text we must tailor to local conditions, and how? Technical communicators need points of comparison, and they need methods that will allow them to make such comparisons methodically—and cooperatively with the people who will actually have to use these documents.

To see how this works, we'll follow Ana as she tries to figure out why team B seems more productive than team A at their nonprofit. Ana is interested in understanding the similarities and differences among the writers in her organization. More than that, she's interested in improving how writers write in that organization—including her own writing. To do that, she'll examine differences in context: how the writers' backgrounds, habits, tools, environments, relationships, ways of doing things, and so forth have led to different performance. How can Ana understand what these writers contribute to their organization's work? How can she be sensitive to their contexts while examining the similarities in their work?

In this chapter, I first review the state of context studies in technical communication, showing how context has been defined and studied. Next I describe a set of techniques for studying and describing workplace contexts. I then show how technical communicators can choose a focus for their study; collect data through observations, interviews, artifacts, and electronic systems monitoring; and analyze those data through event models, ecology models, and sociotechnical graphs. I illustrate the techniques and concepts through an extended example. Finally, I discuss how to communicate the results of the study to participants and employers.

LITERATURE REVIEW: WHAT WE KNOW ABOUT CONTEXTS

If we're going to study context—if, for instance, we're going to understand the similarities and differences in how writers at Ana's nonprofit do what they do—we need to specify what we mean by context. After all, anything *could* be part of the context—room temperature, the color of the walls, magnetic fields, childhood trauma, shoe size, humidity, genetic proclivity, recent headlines, and so on. It's simply not possible—or useful—to catalog everything that might impact a particular incident. Where would you stop?

At the same time, we really do need to better understand context. Research in technical communication suggests that texts work radically differently in different contexts. For instance, texts that are learned in school tend to work differently from texts at work (Russell 1997b; Schryer and Spoel 2005). Texts that work in one work context might function quite differently in another (Gygi and Zachry 2010; Paradis 1991; Spafford et al. 2010; Swarts 2004, 2006, 2007). Texts work differently in different disciplines or activities, leading to confusion when these disciplines must work together (Paretti, McNair, and Holloway-Attaway 2007; Swarts 2007, 2009). Texts develop differently in different cultures (Sun 2006). And texts can also function very differently even in the same place and activity, depending on the other texts and resources on which people draw (Haas and Witte 2001; Sherlock 2009; Spinuzzi 2003). In a real sense, without accounting for context, we can't really understand how texts work or evaluate how well they're designed.

To understand how texts work in context, then, we must *bound* the case, as qualitative researchers say (Yin 2003; Creswell 2006). We must narrow the range of things we examine—the range of data we collect. In technical communication, we have tended to systematically analyze people's activities and goals and their tools for accomplishing these, especially the texts that they use. So, for this chapter, let's define context this way:

Context: The set of observable differences in actors' material relationships within two or more instances of the same activity,

where

Observable: Evident in actual data: seen through direct observations, reported in interviews, shown in texts and artifacts produced or used by people in the activity (see Doheny-Farina and Odell 1985; Henry 2000). (For instance, if you *see* someone writing notes or they *tell* you about writing notes, you have evidence. If you assume that they're writing notes, you don't have evidence.)

Differences: Identifiable systemic variations between patterns of data, such as different sequences, different texts and artifacts, different combinations of texts and artifacts, different reported motivations or outcomes. To paraphrase Gregory Bateson (1979, 64), information is a difference that makes a difference. (For instance, if you notice that some people always write notes and others almost never do, that's a difference.)

Actors: People who effect or affect an ongoing activity, including goals and processes (Gygi and Zachry 2010; Spinuzzi 2003, 2008; Wegner 2004). (For instance, you may find that writers, managers, and marketers work closely on a brochure. Even though only the writers are writing the brochure, *all* are actors who make sure the brochure gets done.)

Material relationships: The ways that actors establish concrete patterns of interaction with each other and with other texts and artifacts (Slattery 2007; Spinuzzi 2008, 2010; Swarts 2006). (For instance, when writers work on a document, they don't communicate telepathically—they leave traces in the form of drafts, notes, stacks of paper, and so forth. What do they mark up, place, move, and exchange?)

Instances of the same activity: Implementations of an ongoing, repeated set of actions with a specific objective (Paretti, McNair, and Holloway-Attaway 2007; Schryer and Spoel 2005; Spinuzzi 2011). (For instance, people in different organizations, cultures, and locations might say that they are "assembling proposals" or "writing an annual report" or "editing brochures.")

If we use this definition, we can get away from the hand waving that so often characterizes discussions of context, and instead focus on things that we can actually see, record, and systematically compare. Since we're technical communicators, we'll investigate context—these observable differences—in two senses: communication and mediation (see Spinuzzi, Hart-Davidson, and Zachry 2006). (We could probably examine nearly any text as communicative *and* as mediational; these are two different aspects of texts.)

COMMUNICATION

The first aspect of context is communication: communication processes, events, and technologies. What kinds of information do people hand off, to whom, in what sequence? What differences exist in how people enact the same sort of communication?

By *communication* I mean the ways that people exchange their information, thoughts, writing, and speech with each other. Think of the kinds of texts we send to and receive from each other: printouts, e-mail, instant messaging, announcements, sticky notes, and the list goes on. When we examine how people communicate, we must follow the information and find patterns.

Much of our work can be understood as chains of communicative events in which people exchange such texts in a regular series and combine them in relatively predictable ways (Hart-Davidson 2002, 2003). If you look at how accountants work (Devitt 1991) or how people structure team meetings (Yates and Orlikowski 2002), how patent offices work (Bazerman 1994), or how people at an Internet startup put together reports (Spinuzzi 2010), you'll find that they exchange a lot of texts with each other, usually in a predictable sequence. The chain of communication is a chain of custody of a particular piece of information.

For instance, when Ana looks at how writers work in her nonprofit, rather than just getting the experience of particular writers, she'll follow the texts that they receive and send. For instance, when writers at Ana's nonprofit have to edit a brochure, from where do they get it? Does it come with notes or instructions? Do they get it over e-mail, from a server, or via departmental mail? When they finish the edits, where and how do they send the edited piece? Most importantly, if Ana looks at these communication patterns, will she spot similarities and differences between the teams? In particular, will she find that team A and team B use different communication patterns?

MEDIATION

The second aspect of context that we'll study is *mediation*, which allows us to examine parts of people's activity that are neither serial, nor explicit, nor necessarily even interpersonal. Think of the sorts of texts we *don't* typically share: shopping lists (Russell 1997a; Witte 1992), checklists (Spinuzzi 2002), annotations (Spinuzzi 2008; Spinuzzi and Zachry 2000; Swarts 2004; Wolfe 2002), and computer interfaces (Johnson-Eilola 2005). The texts can guide and constrain others' activities in ways that don't represent communication between individuals: for instance, people often write checklists and stack documents for themselves, helping them to chunk, categorize, and structure their own work (Spinuzzi 2002). People bring many different texts to bear on a particular activity all at the same time, resulting in an ecology of resources that is qualitatively and quantitatively different from a single text (Freedman and Smart 1997; Nardi and O'Day 1999; Spinuzzi 2003; Zachry 2000).

In technical communication, we tend to talk about genres, types of texts that evolve in order to meet particular needs (see Miller 1984; Russell 1997a; Spinuzzi 2003). These genres can be quite informal (e.g., shopping lists, stacks) or relatively rigid (e.g., recommendation reports, reference documentation), but in either case they work because our participants have seen them before, recognize them, and know how to apply them to a particular activity. That is, these genres serve to mediate the work of the participants.

When Ana studies mediation, she'll look for texts that the writers use to regulate their own work but that they don't necessarily share with each other. How do they keep track of what they need to do that day or that week? Do they use checklists, calendars, stacks, or filing systems? Do they color-code documents on their computer screens? Do they try to use their e-mail inboxes as to-do lists, as so many people do? And again, what are the similarities and differences, particularly between the two teams?

Both the communicative aspect and the mediational aspect are valuable for examining how texts enable work in different contexts. They become even more useful when Ana coordinates them to support patterns, as we'll see below.

HEURISTICS: STUDYING CONTEXT THROUGH COMMUNICATION AND MEDIATION

So instead of just hand waving, we can use a systematic approach to delineating specific differences and similarities across different participants, organizations, and activities. Ana, fortunately, can use three heuristics to help her delineate these differences and similarities: communicative event models, genre ecology models, and sociotechnical graphs.

To use these heuristics, Ana will first need to investigate the context: choose her case, collect her data, then analyze the data. Let's discuss these steps first, then return to Ana's case to see how she implements them to study context at her nonprofit.

You can follow these steps very formally, as in formal qualitative research, or informally, by using these heuristics to organize your impressions from quick walk-throughs and discussions with your audience. Ana follows them informally.

CHOOSING A CASE

We start by determining what to study. If we're looking at context (as defined above, the set of observable differences in actors' material relationships within two or more instances of the same activity), the case has to meet these criteria.

- *The same activity*. Those in the activity—not just you—must define it as the "same." For instance, in Ana's case, four writers are working in the same organization and taking on very similar duties (editing brochures, assembling proposals, writing annual reports). If you study people who identify different activities—say, grant writing, software documentation, and marketing—the comparison is more likely to be about disciplinary differences than context. So you'll need to study people who identify their activity as the same, as Ana does.
- *Two or more instances*. Compare at least two instances for differences: two individuals, two teams, two departments, two companies or communities or locations. The term *context* doesn't make sense without comparisons. In Ana's case, she is studying two teams with two writers each, for a total of four writers.
- *Observable differences*. Collect the same material data across all

Model	Focus of analysis	Example
Communicative event models (CEMs)	Common sequence of events What texts and people are commonly involved in these	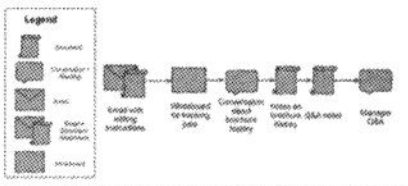
Genre ecology models (GEMs)	Common sets of resources used for regulating one's own work and others' work What texts are required across the activity and how they relate to each other	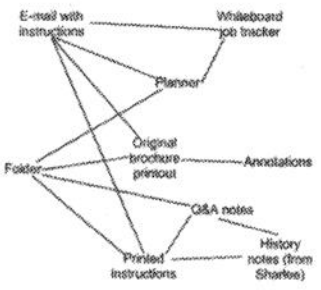 E-mail with instructions Whiteboard job tracker Planner Folder Original brochure printout Annotations Q&A notes History notes (from Sharlee) Printed instructions
Sociotechnical graphs (STGs)	How communication and mediation are coordinated What events and resources are commonly *associated* (AND) What events and resources can be *substituted* (OR) while still maintaining the activity	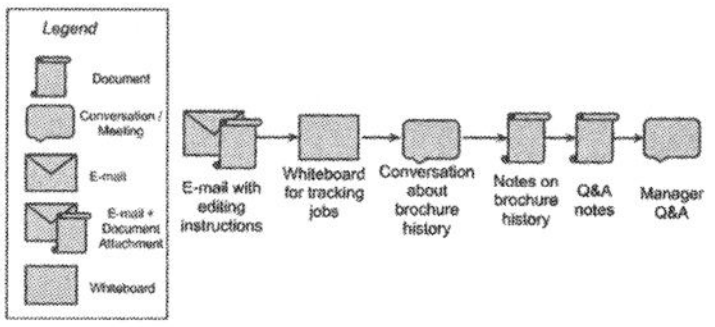 Legend Document Conversation / Meeting E-mail E-mail + Document Attachment Whiteboard E-mail with editing instructions Whiteboard for tracking jobs Conversation about brochure history Notes on brochure history Q&A notes Manager Q&A

Figure 11.1. The three heuristics for studying work context

instances. Don't mix and match, collecting just interviews at location A and just observations at location B, because you won't be able to compare the data directly for patterns. Don't rely solely on self-reporting, such as general interviews, because people often forget or misremember what they do and why they do it. Ana makes sure that she spends about the same amount of time with each writer as they work on similar texts, and she asks similar questions of each.

More pragmatically, as you choose a case, you'll need to make sure that you can

- *Obtain access* to the sites. Make sure that the participants and the person in authority at each site have given you written permission to conduct your study—don't just drop in and start talking to people! Ana makes sure she has permission from her manager and from team B's manager before she starts quizzing her coworkers.
- *Collect the same data* at each site. Establish an agreement, preferably a written agreement, in advance about what data you will collect and how you will collect it. In Ana's case, she puts together a written agreement that both managers can sign off on, and she provides a copy to each writer so that they know what she's doing. (If she doesn't, they might feel like she's trying to be their supervisor or simply trying to catch their mistakes.)

Now that you've established the case and the sites, it's time to figure out what data to collect.

COLLECTING THE DATA

To investigate context, we must collect a set of observable differences in actors' material relations at our sites. That means collecting measurable evidence that we can get from our research sites using systematic techniques. For our purposes, we'll focus on evidence related to our two dimensions of communication and mediation. Table 11.1 lists some formal data collection techniques and some studies in which they have been used. (You might use very informal versions of these if, like Ana, you're developing quick impressions rather than a formal study.)

When selecting data-collection techniques, you should

- *Select a combination of techniques*. Don't rely on just one technique, because each one has strengths and weaknesses. Instead, mix and match these so you can *triangulate* among them, comparing different kinds of data to give yourself a better idea of what is going on. For instance, Ana pairs observation and interviews, so that when she sees a writer doing something unusual, she can ask the writer about it later.
- *Select a technique that you can reasonably implement*. For instance, you might not be able to observe participants working in high-risk environments because you might distract them; you might not be able to get participants to agree to install data-logging software on their machines. Select techniques that will work with the constraints of the case, and make sure they're acceptable to participants before you begin the study. Ana knows that she can observe and interview writers during normal working hours, but she also knows that she shouldn't schedule interviews during the writers' weekly meetings. She also collects copies of the texts she sees the writers using, and she uses her phone to take photos of texts in context—such as the many sticky notes around Singh's monitor and the highlighted planner that Ben always keeps next to his keyboard.

ANALYZING THE DATA

Now that you've collected your data, you need to systematically analyze them to see what they mean. Below, I discuss the three heuristics for analyzing your data. But if you're after quick impressions, you can develop

Table 11.1. *Data-collection techniques*

Technique	Description	Sources	Advantages
Observation	The researcher visits the participants, observes them as they work, and takes notes.	Winsor 1996; Spinuzzi 2003, 2008; Doheny-Farina and Odell 1985; Swarts 2007; Slattery 2007	Direct observation; can see participants' work, including things that don't make it into their interviews.
Interviews	The researcher talks with participants about their work. Interviews are typically recorded and can be structured (a set list of questions), semistructured (a general list of questions, but the interviewer can also follow up on interesting issues), or unstructured (conducted during an observation).	Beyer and Holtzblatt 1998; Smart 2002; Winsor 1996	Direct interpretation; participants can explain why they do things and how these fit into their other activities.
Artifacts	The researcher collects artifacts that are used or generated by participants, such as drafts and notes.	Orlikowski and Yates 1993; Smart 2002; Winsor 1996	Structural features; in aggregate, can show development over time.
System monitoring	The participant consents to installing software on her/his computer that collects system events, such as when the participant opens files, views websites, or sends e-mail.	Hart-Davidson, Spinuzzi, and Zachry 2007	Like observation, but collects more data more accurately (although the range of data that can be collected is narrower).
Diaries	The participant periodically fills out forms describing what she/he is doing.	Hart-Davidson 2003; Sun 2006	Happens during the participant's natural work; maintains focus on specific questions; can happen without the researcher present.

Table 11.1. *(continued)*

Technique	Description	Sources	Advantages
Pictures	The researcher asks the participant to draw pictures or cartoons that describe her/his experiences.	Prior and Shipka 2003; Zuboff 1988	Helps people talk about experiences and feelings nonlinearly; can unearth tacit knowledge and associations.
Participatory design techniques	Participants and researchers collaborate in developing ways that describe solutions to participants' problems.	Muller and Kuhn 1993; Spinuzzi 2005	Unearths tacit knowledge; allows participants to be active rather than passive.

informal versions of these models to "eyeball" the data; you should still be able to sort out some differences in context.

Be warned: if you're studying context at different sites or groups, you may see differences in *individual* work as well as aggregate differences *between* different sites. For instance, Ana might find differences within team A as well as differences between teams A and B. That can make the analysis more complicated, but the tools below will help you map out these differences along the two aspects mentioned earlier: communication and mediation.

Let's look at the three heuristics first, then see how Ana uses them.

Analysis involves boiling down the data, so that we can make systematic comparisons without being overwhelmed by extraneous detail. We'll use three heuristics: communicative event models, genre ecology models, and sociotechnical graphs. These heuristics are basically diagrams that let you picture the data you've collected, so that you can systematically look for patterns of similarity and difference. These three heuristics let you visualize different aspects of the context—the differences that similar people encounter when engaging in similar activities.

Communicative Event Models: Analyzing Chains of Communicative Events

One way to represent the communicative "handoffs" is through communicative event models (CEMs). CEMs provide a simplified, easily comparable description of event sequences, a description that can help us detect patterns in people's work, compare patterns, and see sequential divergences. Any given action represents a choice that someone made in response to their context. In the CEM, these choices are essentially portrayed as strings of verbs and objects. If we were to apply CEMs to longer segments of work, we should be able to detect consistent patterns, identify larger units of interaction, and consistently explore places where sequences diverge across workers or conditions. But even in informal observations, we can sketch out common sequences that people follow.

In a CEM, you record communicative events (events in which actors exchange information by exchanging texts, speech, or other signs). Based on the kind of work, you define symbols for kinds of frequently occurring events (such as face-to-face meetings, e-mail, and phone calls). See figure 11.1. Once you put together enough strings of events, you'll start to detect patterns in how people communicate—in individual work, across people at the same site, and across different sites.

In Ana's case, she takes notes on how each writer communicates during her observation. How are they communicating with each other, and when? For instance, when Connie receives a brochure to edit, from whom does she receive it, what instructions does she get, and what information does she request in order to complete the brochure? How about Singh, Ben, and Sharlee? Do they follow similar patterns?

Genre Ecology Models: Analyzing Ecologies of Mediational Resources

Now let's pull back to see the whole set of texts and highlight some material actions that are not necessarily communicative. One way is to look at how people change their behavior and capabilities by using their tools. When you write a shopping list, or use a to-do list, or stack your paperwork in the order that you need to process it, you're using texts in this way—and chances are, you're using texts that you won't share with anyone else. They're not communicative, they're mediational.

Such texts are genres: relatively stable responses to recurrent situations. When we talk about a shopping list, for instance, we have a basic idea of what it might look like, based on what it's supposed to accomplish and on how similar texts have looked in the past. In your analysis, you'll look for text types that participants identify as types ("This is my to-do list"; "This is an inquiry e-mail") and text types that are repeatedly used at the same site to accomplish the same things. And in a genre ecology model (GEM), you map these texts, drawing lines betweens the ones that are used together.

In information-oriented work, such as the writers' work at Ana's nonprofit, genres mediate work in combinations; the mediation is *compound*. That is, people mediate their own work by using several texts at the same time. If you print out your e-mails so that you can write notes on them, or write to-do lists on your calendar, or put sticky notes on your computer monitor, or otherwise use texts in combination so that you can do things you wouldn't be able to do otherwise, then you know what I mean.

A GEM pictures this web of different genres (figure 11.1). It's a variation of a network diagram in which the nodes are texts connected with lines that show when people use texts together. Rather than a sequence of handoffs, the GEM depicts a set of mediational resources that people can pair, combine, and substitute for each other. Some of these resources are the same ones depicted in the CEM, but here they are seen in their mediational rather than sequential aspect.

The GEM answers two different questions.

- In a particular episode, it answers the question, "What genres are brought to bear during the episode but are not highlighted because they are not being used transactionally?"
- But in the aggregate—for instance, when comparing several different CEMs—it can also answer the question, "How did alternate genres get used to perform the same activity? Given *x* conditions, what genres are people likely to use to perform *y* type of activity?"

To put together a GEM, Ana looks for texts that the writers don't exchange but that they use to help them do their work. For instance, she notices that Singh writes down tasks on sticky notes and puts them around his monitor; when he finishes one, he takes it off and throws it away. Connie keeps her open tasks in her e-mail inbox. Ben and Sharlee write them down in their planners. All four writers tend to stack printouts on their desks in order of project importance.

Just as a purely sequential understanding of text use has its drawbacks, so does a purely mediational understanding. By itself, the GEM tends toward description without direction. That's not enough. We need to be able to coordinate the aspects of communication and mediation.

Sociotechnical Graphs: Analyzing the Coordination of Communication and Mediation

So far, we've mapped out our two dimensions of communication (via CEMs) and mediation (via GEMs). With the next model, the sociotechnical graph (STG), we'll coordinate these two dimensions to produce comparisons—differences and similarities—across sites. STGs have two dimensions: we can call them the AND and OR dimensions (cf. Latour, Mauguin, and Tiel 1992). These allow us to systematically compare different accounts such as diaries, interviews, and our own field observations.

- The AND dimension is that of *association*: Which elements must be associated to form a coherent claim? What sorts of communicative transactions occur in a particular work activity, and what resources support them?
- The OR dimension is that of *substitution*: Of the elements assembled, which can be substituted with others? How do alternate genres get used to perform the same activity? Given *x* conditions, what genres are people likely to use to perform *y* type of activity? (This is, of course, where the differences come in—the observable context.)

That is, STGs are basically matrices: tables that allow us to take the analyses from CEMs and GEMs and coordinate them, giving us a better understanding of the similarities and differences in how people conduct similar activities. They let us see what genres are associated and used by participants to accomplish a communicative event, and what genres can substitute for each other in some of these events. Think of an STG as a dashboard that lets you detect these associative (AND) and substitutional (OR) differences—differences in *context*. And as Ana will discover at the end of the chapter, these diagrams are good not just for analyzing your data, but also for presenting your analysis to others.

In Ana's case, she puts together an STG focused on how people edit brochures. Then she looks for differences between them—differences that she might not have seen otherwise. And she finds them, as we'll see in the next section.

EXTENDED EXAMPLE: WRITING AT ANA'S NONPROFIT

So now that we've established the data collection and heuristics, let's get back to Ana's case. The writers were all engaged in the same basic activity—writing documents for the nonprofit—but in two different teams down the hall from each other. It's the same organization, the two teams do the same types of writing, they have the same kind of equipment and workload. So why is one more productive than the other?

DATA COLLECTION

Ana decides to collect three kinds of data:

- *Observations*: Ana knows that the writers will be working on editing brochures during the upcoming week, so she arranges to visit each writer for about two hours. During each visit, she watches over the writer's shoulder, taking detailed notes about what the writers do, particularly when they receive, read, write, or send texts related to brochure editing. By observing them, Ana can tell *what* writers do.

- *Postobservational interviews*: After watching each writer, Ana briefly interviews the writer about what the writer did during the session. Ana has a list of questions to ask everyone, but she also makes sure to ask questions about things she noticed during the observation. For instance, she asks Singh about the sticky notes that he uses to track tasks, and she asks Sharlee when she began using a planner for organizing her day. Through the interviews, Ana now knows *why* the writers do what she saw them do.
- *Artifact collection*: Finally, Ana makes sure to get copies of all the texts that she can, from the brochures to the writers' notes to the instant messages they send each other. Some of these she can photocopy; some she fishes out of the recycling bin; some she photographs with her phone. Ana isn't sure she needs all these texts, but later she is glad she got them: as she goes over her observation notes and interviews, she can take a closer look at the texts. She can tell *how* the writers used them.

DATA ANALYSIS

To analyze the data, Ana uses the three heuristics I discussed earlier. CEMs allow her to examine the sequences that writers use to reach their goals. GEMs allow her to examine and track the many texts that the writers use to mediate and create conditions for successful work. Finally, STGs let her relate the other two heuristics, allowing her to systematically infer, test, and compare the activities of the two teams. To see how these work, let's take this incident that Ana observes while visiting Ben.

Ben receives an e-mail from his manager describing the changes he has to make to a brochure, along with its deadline. Immediately, Ben copies the date on his planner and finds time during the week to make the edits, then prints the instructions along with the original brochure. He puts these in a manila folder and labels it. He also writes the project name, due date, and his initials on a whiteboard on the wall; Ana notices that the whiteboard holds several such entries, some with Sharlee's initials. Next, Ben swivels around and asks Sharlee if she recognizes the brochure. She does—she edited it last year—so they discuss the changes she made and what she knows about the audience. Ben takes notes on a sticky note, then puts the sticky note on the instructions printout. He also writes down some questions he wants to ask his manager about the project. (Ben calls these "Q&A notes.")

After reading the rest of his e-mail, he takes the folder to his manager's office, and they discuss Ben's questions. Again, he writes his notes on a sticky note. Later, during the interview, he tells Ana that he likes to schedule everything right away and get everything in writing so he doesn't have to spend a lot of time looking for information later on. He also says that the whiteboard (which he calls the "project board") lets him and Sharlee see what projects they're working on so that they can plan together.

COMMUNICATIVE EVENT MODELS

First, Ana creates CEMs for each observation so she can see sequences in communicative events. In figure 11.2, Ana maps out the communicative events that were described in Ben's observation notes and confirmed through the interview and the texts he used. The figure depicts some of the communicative events in Ben's work, including receiving the e-mail, discussing the brochure with Sharlee, and discussing his questions with his manager. In fact, Ana notices that Ben follows this sequence nearly every time he receives an editing job. What's more, when Ana constructs a CEM for Sharlee's work, she sees that Sharlee follows almost exactly the same sequence. That surprises Ana, because Singh and Connie follow very different processes: different from Ben and Sharlee's, different from each other, and even different from one incident to the next.

By constructing CEMs for each participant, Ana is able to detect common events and patterns, but she is also able to observe material differences in actors' material relationships across the activities—in other words, she is able to detect contextual differences in how they communicated.

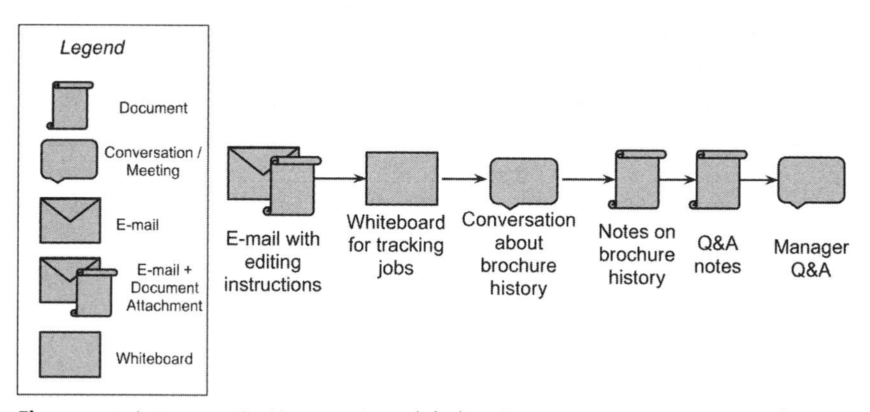

Figure 11.2. A communicative event model, showing communicative events from Ana's visit with Ben

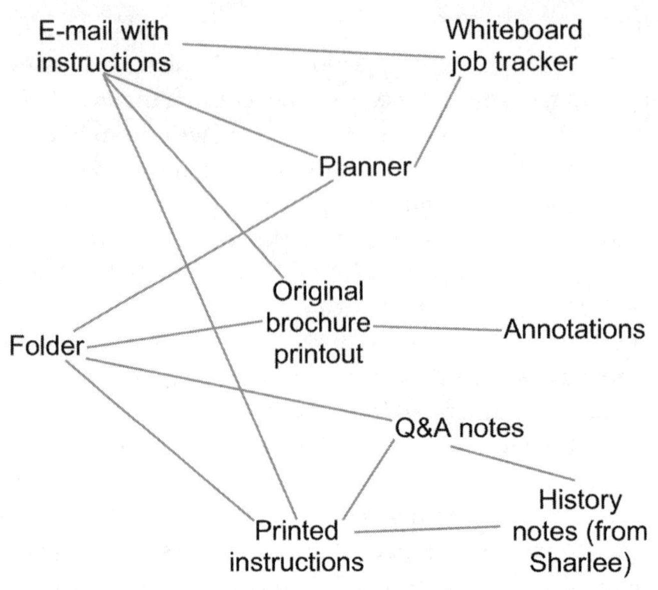

Figure 11.3. A genre ecology model, showing how Ben's genres are connected through mediation

GENRE ECOLOGY MODELS

Next, Ana uses GEMs to see how the writers are using and linking their genres. As figure 11.3 shows, Ben uses multiple genres when working on the brochure: the e-mailed instructions, the brochure, the folder, the planner, the notes he took when talking to Sharlee and his manager, the Q&A notes, the whiteboard, and others. Obviously, Ana thinks, he has a system for taking on these projects. Sharlee's GEM looked very similar, although Sharlee prefers to keep all her notes in a text file instead. By constructing GEMs for each writer, Ana is able to see differences in the resources the writers use when taking on the same task. Interestingly, Ana notices that the GEMs from team B look similar, while the GEMs from team A are much more idiosyncratic and involve more minor texts.

SOCIOTECHNICAL GRAPHS

Finally, Ana uses the CEMs and GEMs to construct an STG: a "dashboard" that lets her coordinate what she knows about communication and mediation among the writers. Table 11.2 shows Ana's STG. Notice that it shows the communicative resources (genres) Ana saw each writer using for each communicative event. The STG allows her to systematically spot differences as well as similarities in how activities are accomplished.

Table 11.2. *A sociotechnical graph, showing how writers defined communicative events and what genres help to mediate them*

	Receive	Schedule	Edit	Deliver
Ben	e-mail, brochure file	planner, whiteboard, printed brochure, printed instructions from e-mail, history notes, Q&A notes, folder, conversations	printed brochure, printed instructions, annotations on brochure, planner, conversation with manager	e-mail, attachment
Sharlee	e-mail, brochure file	planner, whiteboard, printed brochure, printed instructions from e-mail, history notes, Q&A notes, folder, conversations	printed brochure, printed instructions, annotations on brochure, planner, conversation with manager	e-mail, attachment
Singh	e-mail, brochure file	*sticky notes on monitor,* printed instructions, printed brochure, folder	*sticky notes on monitor,* printed brochure, printed instructions, *e-mails to manager with questions*	e-mail, attachment
Connie	e-mail, brochure file	*e-mail inbox,* printed brochure, stack	*e-mail inbox,* printed brochure, stack, *e-mails to manager with questions*	e-mail, attachment

Note: Italics indicate significant differences between individuals.

By coordinating the two views, the STG allows Ana to zero in on the material differences, the *context* (in the sense I've defined it) across the sites.

Looking at the STG, Ana notices that teams A and B use very similar steps and resources when they receive brochures and when they send out the edited versions. But in the middle, the scheduling and editing processes, things look very different. In team A, Singh and Connie have chosen their own processes for organizing their work: Singh writes reminders on sticky notes and sticks them to his monitor, while Connie keeps the e-mails in her inbox to remind her what she needs to work on. Both work on projects as they get to them, and both tend to e-mail their manager when questions occur. They don't have a way to share their status on their projects. In contrast, in team B, Ben and Sharlee have decided to use the same sets of texts, and they tend to coordinate more heavily with each other and with their manager. They also schedule times to work on each brochure in their planners, and they share status on their whiteboard. Finally, they tend to generate all their questions right away and have them answered right away.

So what do these models tell Ana? In terms of context, she concludes that team A works just as hard as team B—but team A doesn't have the processes or texts in place to keep itself organized. As Ana looks at the dozens of sticky notes on Singh's monitor and the dozens of e-mails in Connie's inbox, she thinks, no wonder team A has a hard time keeping on top of tasks—and she reconsiders her own system of stacking printouts on her desk.

CONCLUSION

Throughout this chapter, I've discussed a set of heuristics for investigating context. But now it's time to talk about communicating those findings. The beauty of these heuristics is that they're not hard to communicate to others. In a meeting with both teams and their managers, Ana puts these heuristics in a presentation and describes each.

"As you can see," she tells them at the end of the presentation, "even though we're doing similar work on similar brochures, we're doing it in different ways. And even though everyone is working hard, team B has figured out how to coordinate their work more closely, how to organize their work more tightly, and how to schedule it more effectively. In team A, we can easily pick up some of these ideas, and it will make us more productive and lead to fewer missed deadlines."

Singh and Connie are initially skeptical, but agree to try some of the changes. Although the transition is a little rough, within a week team A is working more smoothly and writing documents more effectively.

"Nice work," Ana's manager tells her. "Now if we can only improve on the model, maybe we can start outperforming team B."

Ana taps her STG thoughtfully. "Actually," she says, "I have a few ideas. . . ."

DISCUSSION QUESTIONS

1. *Context* has a specific definition in this chapter. How is it different from how you usually think about context?
2. Think of some task that you often perform—at work, while studying, in a campus organization, and so forth. Sketch out a CEM showing the different texts that you have to receive, produce, alter, and hand off during this task. Compare your sketch with that of others in your class. What differences do you see? What similarities?
3. Take the same task and sketch a GEM showing the different texts that

support it. What are some of the hidden "helper" genres that make this task successful? How does your diagram differ from others'?

4. Go to a place with a repeated activity, such as the checkout counter of a library, a coffee shop, a restaurant, or the front of a city bus. Take notes about what happens, especially in terms of what texts are used and who uses them. Using your notes, sketch out a CEM that represents the repeated activity. Summarize the "script" that people use when they perform this activity. What are the similarities and differences between people? Compare with someone else's CEM. What are the similarities and differences across contexts?
5. Using your notes from question 4, draw a GEM showing the texts people use in this activity. What are some of the more unusual texts that you saw? Compare with someone else's GEM. What are the similarities and differences across contexts?
6. Using the CEM and GEM from questions 4 and 5, create an STG. What is the minimum set of associated texts needed to perform this activity (AND)? What are some of the substitutions you and your classmates saw (OR)?
7. This chapter claims that an STG can help you compare contexts. What can you tell about the different contexts in table 11.2? What is the same? Do you think that a writer would have trouble moving from one team to another? Why or why not?
8. At the end, Ana doesn't have much trouble getting team A to change what they do. In real life, though, it's often much harder to get people to change what they do. What arguments might you have used to convince Singh and Connie to try the new system?

WORKS CITED

Bateson, Gregory. 1979. *Mind and nature: A necessary unity*. New York: E. P. Dutton.

Bazerman, Charles. 1994. Systems of genre and the enactment of social intentions. In *Genre and the new rhetoric*, ed. Peter Medway and Aviva Freedman, 79–99. Bristol, PA: Taylor and Francis.

Beyer, Hugh, and Karen Holtzblatt. 1998. *Contextual design: Defining customer-centered systems*. San Francisco: Morgan Kaufmann.

Creswell, John W. 2006. *Qualitative inquiry and research design: Choosing among five traditions*. Thousand Oaks, CA: Sage.

Cross, Geoffrey. 1993. The interrelation of genre, context, and process in the collaborative writing of two corporate documents. In *Writing in the workplace: New research perspectives*, ed. Rachel Spilka, 141–157. Carbondale: Southern Illinois University Press.

Devitt, Amy J. 1991. Intertextuality in tax accounting: Generic, referential, and functional. In *Textual dynamics of the professions: Historical and contemporary studies of writing in professional communities*, ed. Charles Bazerman and James G. Paradis, 336–357. Madison: University of Wisconsin Press.

Doheny-Farina, Stephen, and Lee Odell. 1985. Ethnographic research on writing: Assumptions and methodology. In *Writing in nonacademic settings*, ed. Lee Odell and Dixie Goswami, 503–535. New York: Guilford Press.

Freedman, Aviva, and Graham Smart. 1997. Navigating the current of economic policy: Written genres and the distribution of cognitive work at a financial institution. *Mind, Culture, and Activity* 4 (4): 238–255.

Gygi, Kathleen, and Mark Zachry. 2010. Productive tensions and the regulatory work of genres in the development of an engineering communication workshop in a transnational corporation. *Journal of Business and Technical Communication* 24 (3): 358–381.

Haas, Christina, and Steve Witte. 2001. Writing as embodied practice: The case of engineering standards. *Journal of Business and Technical Communication* 15 (4): 413–457.

Hart-Davidson, William. 2002. Modeling document-mediated interaction. In *ACM SIGDOC 2002 conference proceedings*, 60–71. New York: ACM.

———. 2003. Seeing the project: Mapping patterns of intra-team communication events. In *ACM SIGDOC 2003 conference proceedings*, 28–34. New York: ACM.

Hart-Davidson, William, Clay Spinuzzi, and Mark Zachry. 2007. Capturing and visualizing knowledge work: Results and implications of a pilot study of proposal writing activity. In *SIGDOC '07: Proceedings of the 25th annual international conference on design of communication*, ed. David G. Novick and Clay Spinuzzi, 113–119. New York: ACM. doi:http://doi.acm.org/10.1145/1297144.1297168.

Henry, Jim. 2000. *Writing workplace cultures: An archaeology of professional writing*. Carbondale: Southern Illinois University Press.

Johnson-Eilola, Johndan. 2005. *Datacloud: Toward a new theory of online work*. Cresskill, NJ: Hampton Press.

Latour, Bruno, Philippe Mauguin, and Genevi Eve Teil. 1992. A note on socio-technical graphs. *Social Studies of Science* 22:33–57.

Lave, Jean. 1996. The practice of learning. In *Understanding practice: Perspectives on activity and context*, ed. Seth Chaiklin, and Jean Lave, 3–34. Cambridge: Cambridge University Press.

Miller, Carolyn R. 1984. Genre as social action. *Quarterly Journal of Speech* 70 (2): 151–167.

Muller, Michael J., and Sarah Kuhn. 1993. Introduction to special issue on participatory design. *Communications of the ACM* 36 (4): 24–28.

Nardi, Bonnie A., and Vicki L. O'Day. 1999. *Information ecologies: Using technology with heart*. Cambridge, MA: MIT Press.

Orlikowski, Wanda, and JoAnne Yates. 1993. *From memo to dialogue: Enacting genres of communication in electronic media*. Cambridge, MA: MIT Sloan School of Management.

Paradis, James. 1991. Text and action: The operator's manual in context and in court. In *Textual dynamics of the professions: Historical and contemporary studies of writing in professional communities*, ed. Charles Bazerman and James Paradis, 256–278. Madison: University of Wisconsin Press.

Paretti, Marie C., Lisa D. McNair, and Lissa Holloway-Attaway. 2007. Teaching technical communication in an era of distributed work: A case study of collaboration between U.S. and Swedish students. *Technical Communication Quarterly* 16 (3): 327–352.

Prior, Paul, and Jody Shipka. 2003. Chronotopic lamination: Tracing the contours of literate activity. In *Writing selves/writing societies: Research from activity perspectives*, ed.

Charles Bazerman and David R. Russell, 180–238. Fort Collins, CO: WAC Clearinghouse and Mind, Culture, and Activity. http://wac.colostate.edu/books/selves_societies.
Russell, David R. 1997a. Rethinking genre in school and society: An activity theory analysis. *Written Communication* 14 (4): 504–554.
———. 1997b. Writing and genre in higher education and workplaces: A review of studies that use cultural-historical activity theory. *Mind, Culture, and Activity* 4 (4): 224–237.
Schryer, Catherine F., and Philippa Spoel. 2005. Genre theory, health-care discourse, and professional identity formation. *Journal of Business and Technical Communication* 19 (3): 249–278.
Sherlock, Lee. 2009. Genre, activity, and collaborative work and play in World of Warcraft: Places and problems of open systems in online gaming. *Journal of Business and Technical Communication* 23 (3): 263–293.
Slattery, Shaun. 2007. Undistributing work through writing: How technical writers manage texts in complex information environments. *Technical Communication Quarterly* 16 (3): 311–326.
Smart, Karl. 2002. Contextual inquiry as a method of information design. In *Content and complexity: The role of content in information design*, ed. Michael Albers and Beth Mazur, 205–232. Mahwah, NJ: Lawrence Erlbaum Associates.
Spafford, Marlee, Catherine Schryer, Lorelei Lingard, and Marcellina Mian. 2010. Accessibility and order: Crossing borders in child abuse forensic reports. *Technical Communication Quarterly* 19 (2): 118–143.
Spinuzzi, Clay. 2002. Modeling genre ecologies. In *Proceedings of the 20th annual international conference on computer documentation*, 200–207. New York: ACM. doi:http://doi.acm.org/10.1145/584955.584985.
———. 2003. *Tracing genres through organizations: A sociocultural approach to information design*. Cambridge, MA: MIT Press.
———. 2005. The methodology of participatory design. *Technical Communication* 52 (2): 163–174.
———. 2008. *Network: Theorizing knowledge work in telecommunications*. New York: Cambridge University Press.
———. 2010. Secret sauce and snake oil: Writing monthly reports in a highly contingent environment. *Written Communication* 27 (4): 363–409.
———. 2011. Losing by expanding: Corralling the runaway object. *Journal of Business and Technical Communication* 25 (4): 449–486.
Spinuzzi, Clay, William Hart-Davidson, and Mark Zachry. 2006. Chains and ecologies: Methodological notes toward a communicative-mediational model of technologically mediated writing. In *SIGDOC '06: Proceedings of the 24th annual international conference on design of communication*, 43–50. New York: ACM.
Spinuzzi, Clay, and Mark Zachry. 2000. Genre ecologies: An open-system approach to understanding and constructing documentation. *ACM Journal of Computer Documentation* 24:169–181. doi:http://doi.acm.org/10.1145/344599.344646.
Sun, Huatong. 2006. The triumph of users: Achieving cultural usability goals with user localization. *Technical Communication Quarterly* 15 (4): 457–481.
Swarts, Jason. 2004. Textual grounding: How people turn texts into tools. *Journal of Technical Writing and Communication* 34 (1–2): 67–89.
———. 2006. Coherent fragments: The problem of mobility and genred information. *Written Communication* 23 (2): 173–201.

———. 2007. Mobility and composition: The architecture of coherence in non-places. *Technical Communication Quarterly* 16 (3): 279–309.
———. 2009. Recycled writing: Assembling actor networks from reusable content. *Journal of Business and Technical Communication* 24 (2): 127–163.
Wegner, Diana. 2004. The collaborative construction of a management report in a municipal community of practice: Text and context, genre and learning. *Journal of Business and Technical Communication* 18 (4): 411–451. doi:10.1177/1050651904266926.
Winsor, Dorothy A. 1996. *Writing like an engineer: A rhetorical education*. Mahwah, NJ: Erlbaum.
Witte, Stephen P. 1992. Context, text, intertext: Toward a constructivist semiotic of writing. *Written Communication* 9:237–308.
Wolfe, Joanna. 2002. Annotation technologies: A software and research review. *Computers and Composition* 19:471–497.
Yates, JoAnne, and Wanda Orlikowski. 2002. Genre systems: Structuring interaction through communicative norms. *Journal of Business Communication* 39 (1): 13–35.
Yin, Robert K. 2003. *Case study research: Design and methods*. Third edition. Thousand Oaks, CA: Sage.
Zachry, Mark. 2000. The ecology of an online education site in professional communication. In *Proceedings of IEEE professional communication society international professional communication conference and Proceedings of the 18th annual ACM international conference on computer documentation*, 433–442. New York: IEEE Educational Activities Department.
Zuboff, Shoshana. 1988. *In the age of the smart machine: The future of work and power*. New York: Basic Books.

BARBARA MIREL

12 How Can Technical Communicators Evaluate the Usability of Artifacts?

SUMMARY
Presenting information to audiences does not guarantee that they will value it or understand and use it effectively or efficiently. Before delivering information to audiences, it is crucial to evaluate whether presentations are truly useful and usable, and, if not, to determine how to improve them so that they are. Once artifacts are in audiences' hands, it is also important to evaluate whether evolving use signals a corresponding need to add improvements. Evaluating artifacts is one of the main contributions that technical communicators make in a project. Technical communicators evaluate information products for usability and usefulness, and based on findings, they recommend and help guide improvements. To evaluate digital information, various academic disciplines train usability specialists, but technical communicators are distinctively skilled in assuring that such artifacts succeed rhetorically with audiences. To conduct evaluations effectively, technical communicators need to understand and master the rhetorical skills, usability methods, and complex evaluation choices necessary for strategic assessments and usability reports. This chapter will help you develop these talents.

INTRODUCTION

"If you build it they will come." The creative spirit behind this expression of innovation can become a technical communicator's worst nightmare if the innovation is built with little regard for audience needs and purposes. Without adequately taking audience into account, innovations are not likely to be useful or usable. The following scenario illustrates such a case.

Collins is a technical communicator in a small firm that produces biomedical venture software. She aims to create effective user manuals based on initial documentation that developers draft to accompany their products. She also evaluates the usability of newly developed web applications through user performance testing. When testing applications, she assesses the user manual and participants' uses of it as part of the tool evaluation.

The company in which Collins works develops life sciences applications for exploratory analysis. With them, scientists can explore huge volumes of diverse data and literature to uncover molecular-level relationships and processes that may influence mechanisms of a complex disease or other little understood physiological process. The applications let scientists query a database to retrieve information on genes, their functional traits, and their interactions. Retrieved information is displayed in networks of associated genes annotated by traits. The software provides other connected graphics for exploration as well, such as heatmaps (see figure 12.2). The visualizations are highly interactive and richly annotated with details such as functional attributes of genes, significance values for these attributes, biological pathways related to each gene, physical interactions between genes, and associated literature.

Collins learned today that she is to evaluate a new application and its developer-created manual. The developers of this application—BioConcept—believe the software and manual are both ready for user performance evaluation. The application aims to give scientists a quick means to search more than three million genes to uncover specific sets of potentially influential genes based on statistically significant functional traits. Collins is familiar with BioConcept from bimonthly meetings in her marketing group. Yet she has not been privy to the intricacies of design and development that went into creating this application. The developers' eagerness conveys that they expect BioConcept to effectively support users' needs and tasks even though it has not been subjected to either a user needs assessment or any iterative prototyping evaluation.

After an initial review of the software and manual, Collins sees that neither is ready for user performance testing. Neither meets common usability standards. If Collins tests these artifacts with users now, the tests will uncover only the problems already obvious to her. Collins decides on a different evaluation method known as heuristic evaluation. This assessment method involves applying a set of usability standards (also called heuristics) to interfaces and manuals. Usability heuristics have been established for a long time in the fields of technical communication and human-computer interactions. Collins has adapted these heuristics to the practices of exploratory analysis in the life sciences and validated them with usability colleagues and through pilot heuristic evaluations. She hopes that by running this usability inspection method, she can provide solid data to guide developers in improvements. After improvements, she hopes BioConcept and its manual will be ready for user performance testing.

At this point, Collins's situation highlights two choices for usability evaluations—inspections and user performance testing. As described later in this chapter, many more choices exist. A single chapter cannot explore usability in all its aspects for the various types of artifacts that technical communicators may assess. Consequently, the focus here is only on usability evaluation of software, web, and handheld applications. These industries are home to thousands of technical communicators like Collins, who may assess help systems, user documentation, tutorials, training materials, and user interfaces for websites, virtual realities, mobiles, games, tablets, software, and web applications. Additionally, in these evaluations, technical communicators implicitly have to assess aspects of a system's core functionality and architecture, because these aspects often contribute to usability and usefulness problems.

This chapter reviews usability research: it reveals the rhetorical nature of usability evaluation and provides findings from other studies on how to design and conduct your evaluations of digital artifacts. This review gives you details about evaluation methods such as usability inspection, field studies, formative and summative user performance testing, and mixed methods. It also covers success factors in writing usability reports. This background provides you with a good grounding for subsequently understanding three heuristic questions that should guide you in conducting evaluations:

1. What distinct skills do I have to apply to this usability situation?
2. What evaluation methods are best for the goals and circumstances?
3. What report choices will communicate convincingly and assure improvements?

To help you see how the questions apply to actual situations, we return to Collins's case at the end of the chapter.

LITERATURE REVIEW

In assessing artifacts, usability evaluators should assess both usefulness and usability. Usefulness is a value that users experience. It measures whether intended audiences find an artifact meaningful and valuable to their actual work flows and tasks as they want to do them. Usability, by contrast, is a property of the artifact. It measures and assesses whether the artifact's operations, displays, and content are easy to understand, use, access, learn, and navigate. Both usefulness and usability are critical but—importantly—they are not mutually inclusive. For example, if an application does not enable domain specialists to work through "their inquiries according to situational and professional demands, even the easiest-to-use application is not useful" (Mirel 2004, 33).

Usability has many disciplinary homes besides technical communications. For example, usability evaluation is part of the training in information science, human-computer interaction programs, industrial design, performance technology, and the learning sciences. Like technical communications, these disciplines all offer courses related to usability at the undergraduate, master's, and doctoral levels. Yet technical communicators are distinctive in bringing rhetorical skills and training to usability evaluations (Cooke and Mings 2005). Rhetorical skills make evaluators sensitive to what makes language and the presentation of information work well for a particular purpose, audience, context, and medium. Language and information presentations can be the printed words or digitally represented communications flowing from an interface visually, symbolically, textually, and tactilely.

The research literature emphasizes the benefits of rhetoric as a distinct skill for usability evaluations. For example, Johnson, Salvo, and Zoetewey (2007) argue that usability without a rhetorical component is an "applied science" that misses many nuances of how people construct meaning from information in nonformulaic communication (interaction) situations. As an applied science, usability functions to formalize people's "human" approaches to digital knowledge work, so that these approaches can be turned into specifications for software requirements, procedural steps, and use cases. What this view omits is that information flowing from an interface or help system is a language establishing a discourse between a user and a tool (and implicitly the tool or text developer). As Sullivan (1989) notes, how audiences dynamically access, browse, interpret, understand, and act on information that flows from user interfaces or help systems is mediated through language. The language communicates to users the affordances a tool offers—and does not offer—for their needs and purposes. Whether visual, verbal, or symbolic, the languagelike representations communicating to users at the interface have rhetorical dimensions. That is, the effect of a system or document design depends on whether it connects with users' actual purposes, contexts, domains, roles, reasoning, constraints, and conventions.

Consequently, assessing how interfaces or help systems are received, taken up, applied and valued by intended audiences requires rhetorical expertise. Anscheutz and Rosenbaum (2002) relate through several case histories how rhetorical skills have helped many technical communicators expand their workplace roles and responsibilities. In these cases, individuals take on lead roles in usability testing, user-centered system design, quality assurance, and user-interface design and development.

In addition to rhetorical skills, usability evaluators need to have a tool kit of methodologies from which to design and carry out evaluation projects for various situations. A large body of research, best practices, and guidelines exists that details diverse methodologies and protocols—both qualitative and quantitative, formative and summative (Rubin and Chisnell 2008; Krug 2005; Barnum 2002; Blakeslee and Fleisher 2007; Dumas and Redish 1993; Redish et al. 2002; Blandford et al. 2008; Sutcliffe et al. 2000; Cockton and Woolrych 2001; De Jong and Shellens 2000; Rosenbaum 2008; Ummelen 1997; Spyridakis et al. 2005; Kushiniruk and Patel 2004). Familiarity with this literature is important to guide the formulation of evaluation goals and methods and to assure that they fit a given multifactored usability situation. The implications of various situational factors for goals and methods also are discussed in the research literature (e.g., Barnum 2002; Dumas and Redish 1993). For example, the stage of artifact development is an important determinant of optimal methods for evaluation, as are the types and complexity of tasks afforded by the artifact and the diversity of its target users.

Ideally, usability evaluations are performed after many other user-centered activities for artifact design and development have already taken place. For example, context-based audience and needs assessments should occur and influence design before usability testing. Additionally, project management structures should be in place to assure that needs assessments and their influence on prototype designs are built into the front end of the development cycle from the start. These activities are addressed by other chapters in this book. If these activities do not occur before a usability evaluation, the evaluation is likely to uncover deep conceptual, architectural, and communication problems, at a point when it is harder to remedy them and can add significant costs and time to production. As Scotch, Parmanto, and Monaco (2007) show, if user performance testing demonstrates inadequate baseline usability in an artifact due to an absence of early assessments, users cannot conduct the full range of tasks required for evaluation. The evaluation, consequently, will not be able to uncover deeper problems that are likely to be detrimental to users' task flows in their actual analyses.

As part of their tool kit of methods, usability specialists can consider conducting various modes of usability inspection (Nielsen 1993; Hollingsed and Novick 2007)—cognitive walk-throughs, expert reviews, and heuristic evaluations. In cognitive walk-throughs, a usability specialist or domain expert performs predefined tasks that cover the types of goal-based interactions and reasoning that developers intend the application to support in actual use. Outcomes show actions afforded by the interface that are prone to error or other difficulties. In expert reviews, usability specialists, domain experts, or both intuitively evaluate whether artifacts reflect acceptable quality, based on their respective professional expertise. Heuristic evaluations, mentioned earlier in Collins's case, apply severity rankings to judgments about satisfying the specified standards (heuristics). Severity rankings might include these levels:

- Level 5: a catastrophic error causing irrevocable loss of data or damage. The problem could result in large-scale failures that prevent people from doing their work. Performance is so bad that the system cannot accomplish business goals.
- Level 4: a severe problem causing possible loss of data. A user has no work-around, and performance is so poor that the system is universally regarded as "pitiful."
- Level 3: a moderate problem causing wasted time but no permanent loss of data. There is a work-around to the problem. Internal inconsistencies result in increased learning or error rates. An important function or feature does not work as expected.
- Level 2: a minor but irritating problem. Generally, it slows users down slightly, involving poor appearance or perceptions, and mistakes that are recoverable.
- Level 1: a minimal error. The problem is rare or is tied only to minor cosmetic or consistency issues (Wilson and Coyne 2001).

Heuristic evaluation standards are often generic, and better outcomes occur when standards are adapted to the domain targeted by an application (Cockton et al. 2007; Mirel and Wright 2009; Cockton and Woolrych 2001). In terms of all inspection methods, research shows that no single method can achieve adequate usability (Nielsen 1994). Rather, these methods need to be combined with other evaluation approaches, such as user performance testing, interviews and surveys, and usage-log analysis.

Evaluations through user performance testing are either formative or summative. Formative testing occurs early and often, and it generates findings that can progressively improve an artifact during development. In formative testing, qualitative methods are often used because, at this early stage, too little is known about factors contributing to usefulness for a given class of tasks and work flows to have valid constructs available for quantitative analysis. Formative scenario-based testing and qualitative analysis help define such constructs and standards of excellence. These constructs and metrics often become the criteria by which an artifact is assessed in subsequent summative testing.

Summative testing occurs at the end of development and validates an application. Summative evaluations determine if an application and its documentation effectively and efficiently address audience, purpose, context, and tasks and if they meet standard usability criteria. This testing is typically quantitative. This chapter does not describe summative performance testing: the precision with which its quantitative methods need to be applied are covered in other sources, for example, Sauro and Lewis 2009; Evans, Wei, and Spyridakis 2004; Kirakowski 2005; Saraiya, North, and Duca 2005; Gray and Salzman 1998; and Hughes 1999.

Whether an evaluation is formative or summative, the evaluator needs to determine the appropriate unit of analysis for the user performance test. The unit of analysis specifies the focus. For example, the focus might be on individual or collaborative performance, or on user performance on predefined tasks or user-defined tasks. In evaluations of early yet fully functional prototypes, evaluators often run user performance tests in the field to identify contextually embedded demands for usefulness. Evaluators gather data as users perform their actual software-supported work in naturalistic work settings (Mirel 2004). In such field studies, the unit of analysis may be individual or collaborative performance of just certain tasks or performance of any on- and offline tasks having to do with a targeted problem. Field studies with the latter unit of analysis—any on- and offline tasks related to the problem—are especially effective for evaluating usefulness. This unit of analysis requires comprehensive data collection. It involves collecting observation and interview data. It also involves gathering and analyzing log data (automatically generated records of users' interactions with a program/website when they are not being observed); diaries; and other self-reports (Jarrett et al. 2009; Ivory and Hearst 2001; Spyridakis et al. 2005).

A usability evaluation method that often comes after field studies focuses on testing user performance on just certain specified features of a prototype (Snyder 2003). The unit of analysis is typically an individual's use of the specified features for predefined tasks. In this testing, evaluators often work hand in hand with developers to run quick cycles of prototype development, usability testing, prototype revision, and retesting. This approach is called rapid iterative testing and evaluation (RITE). It looks at program-defined, low-level actions, often actions that apply to higher-level tasks and work flows that users revealed during earlier field testing. Evaluators test just a handful of users (five or so), observing and gathering data on them as they interact with the prototype. Data may include time on tasks, errors, success rates, and recovery times from errors as well as satisfaction feedback (Medlock et al. 2005). This testing method is optimal when team members are experts in usability and when the targeted user tasks are well-structured. In these cases, usability experts are able to translate findings into redesigns and improvements quickly and to collaboratively set priorities and implementation choices with developers and other stakeholders (Rosenbaum 2008).

Because of organizational circumstances, as in Collins's case, usability specialists in an organization may not be able to conduct early field studies and iterative prototype testing. Organizations vary in their awareness of the importance of incorporating usability processes into the product-development cycle from inception on. In cases in which no early field tests or RITE testing has been performed, usability performance tests take on the burden of having to assess both usefulness and usability. To combine the evaluation of usefulness and usability, for example, predefined tasks and scenarios in the user performance test need to generate (1) findings on usability issues of concern (e.g., smoothness of flow of interactions with features and functionality, error management) and (2) findings on fitness to purpose (adequate task support for the actual reasoning and domain knowledge that users apply in their work). As cited previously, a good deal of research exists to guide informed decisions about test design and methodology for user performance testing in controlled environments. Research in the literature also addresses various issues specific to certain media (e.g., mobile, virtual reality, speech recognition), cross-cultural systems, degree of task complexity, domains (e.g., health, finance), demographic segments, and accessibility.

In field studies and controlled settings, user performance evaluations need to include appropriate sampling methods and sample sizes (Koerber and McMichael 2008). Strict criteria exist for quantitative studies, and it can be helpful to consult with statistical experts in designing these studies. For formative and largely qualitative evaluations, samples of users can be recruited through convenience sampling or purposive sampling methods. Convenience sampling, that is, selecting users based on availability alone, is not sufficient when users need to bring specific knowledge and experience levels to the tasks or scenario. A better method for this situation is purposive sampling, which involves selecting users based on set criteria for user traits. Typically, twenty users is a good sample size and acceptable for qualitative evaluation (Miles and Huberman 1994).

Conventional methods for running user performance evaluations include having at least two usability evaluators observe user performance sessions, upon consent from users. The evaluators ask users to think out loud as they conduct predefined tasks or scenarios. That is, they ask users to articulate thoughts such as intentions, choice points, and reasoning as they proceed through their work. The usability evaluators also train users briefly in think-aloud processes before testing (Boren and Ramey 2000). During the task performance sessions, software interactions and think-alouds often are video- and audiorecorded by screen-capture software. Evaluators also decide whether to set time limits on the performance of a task, based on whether such constraints are consonant with their evaluation goal. Evaluators observe each user session without intervening, and take notes to guide later analysis of the raw data. At the end of each user session, they often ask users to fill out standard satisfaction surveys (Brooke 1996; Kirakowski 1996; Kirakowski and Cierlik 1998). Satisfaction surveys—their reliability, metrics, significance of outcomes—have their own art and science, and usability evaluators should be familiar with the literature about them (Sauro and Kirklund 2005; Lewis 2006; Sauro, and Lewis 2009; Bangor, Kortum, and Miller 2008). Evaluators debrief after each session, sharing their perceptions and highlights.

To analyze the formative user performance and satisfaction data, evaluators typically follow standard qualitative methods. They characterize patterns, exceptions, and themes in users' performance behaviors, knowledge, and affective reactions (Krippendorff 2004; Barton 2002; Creswell and Clark 2006; Blakeslee and Fleisher 2007). To start, they holistically view the video and audio recordings of user sessions several times. Then, for each user session, they tag various uses of artifact functions and features and user behaviors for important traits and indicators of performance efficiencies and effectiveness. These may include, for example, types of information-seeking behaviors (e.g., access, monitor, search, browse, extract, chain, analyze; Makri, Blandford, and Cox 2008); demonstrated program-related problems, impasses, and errors; error-recovery instances; categories of interactivity (e.g., selection, filtering, navigating within a screen, navigating across screens, backtracking); elapsed time on certain tasks; and task boundaries. Screen-capture analysis systems and content analysis tools can expedite these analyses.

Evaluators also transcribe the think-alouds and interview responses, and analyze them to abstract patterns shared across user cases. Additionally, they extract exceptions and themes characterizing expressed and acquired knowledge, modes of reasoning, affective reactions, and patterns and exceptions during performance and post hoc interviews. Unlike qualitative methodologies that aim primarily to build grounded theory, these qualitative analyses may have theory-building outcomes, but they also give high priority to action. This action includes theoretically grounded rationales, recommendations, and user-oriented specifications for enacting design modifications and improvements.

As the research literature suggests, large amounts of data are gathered and analyzed to generate assessments and to recommend improvements. It is no small feat to turn findings from evaluations into communications and recommendations that prompt developers and other stakeholders to set the right priorities for improvements and to construct truly effective improvements. Evaluation specialists agree that writing high-quality reports "with recommendations that are taken seriously by the product team" is a core aspect of usability testing (Brady 2004, 67). Unfortunately, this aspect of an evaluator's role is underresearched. Few studies cite results about the best ways to compose effective formative evaluation reports and recommendations to direct evolving improvements and priorities for greater usefulness and usability.

Standards and guidelines for writing effective summative reports are better established and disseminated than standards and guidelines for formative evaluation reports (Industry Usability Reporting Project 2001). A well-delineated report format for summative testing, supplemented with examples, can be found at the National Institute of Standards and Technology (NIST) website (http://zing.ncsl.nist.gov/cifter/TheCD/Cif/Readme.html).

Despite a relative paucity of research into formative evaluation reporting, investigators have gained some important insights into effective strategies for structuring reports. For example, a qualitative analysis of formative reports and recommendations composed by seventeen teams of experienced usability professionals shows fifteen organizing patterns characterizing these reports' section structures. In general, the researchers argue that some or all of the elements in table 12.1 should be included in reports (Theofanos and Quesenbery 2005).

As an overarching composition principle, researchers highlight the need for rhetorical effectiveness—something that should be second nature to technical communicators (Theofanos and Quesenbery 2005). For

Table 12.1. *Elements to include in evaluation reports*

1. Title page	6. Overall test environment	11. Metrics
2. Executive summary	7. Participants	12. Quotes, and screen shots
3. Teaching about usability	8. Tasks and scenarios	13. Conclusions
4. Business and test goals	9. Results and recommendations	14. Next steps
5. Method and methodology	10. Recommendation details	15. Appendices

example, research stresses that writing choices should be shaped by the business context, its conventions, constraints, and priorities. Rhetorically, choices in reporting also should be shaped by the writer's relationships with the intended primary and secondary readers, their prior knowledge, their likely assumptions and misconceptions, and the questions that the evaluation tried to answer. Finally, rhetorical choices in reporting depend on the current phase of the development cycle, the type of artifact being evaluated, and the buy-in likely from the audiences.

Researchers also suggest content strategies. These strategies include presenting recommended improvements in the form of screen mock-ups with call-outs, including screen shots that depict the problem, and providing quotations from users as a means to bring the audience "in touch with users . . . and building awareness of user needs" (Theofanos and Quesenbery 2005, 34). Another means for eliciting a positive response from audiences composed of developers or managers is to include usability successes as well as problems. Researchers find that it is also important to categorize problems in the report by the user experience issues to which they relate, not by the program feature. For well-structured work, many such categorical schemes are available in the research literature (Fu, Salvendy, and Turley 2002). For complex work, Blandford et al. (2008) and Sutcliffe et al. (2000) provide a number of problem categories and descriptions that achieve this effect well with diverse readers—developers, managers, and users (detailed in a later section). In addition, problems and associated recommendations should have severity rankings and provide clearly written criteria for each rank.

For all these issues covered by the research literature—the necessary skills for evaluation, the application of the right methods at the right time, and the development of persuasive reports—a set of heuristic strategies can help you systematically bring them into your actual usability evaluations. The next section looks at these heuristics.

HEURISTIC

A good deal of work goes into preparing for, conducting, and reporting on usability evaluations. Organizing this work by three main heuristic questions will help direct your efforts.

WHAT DISTINCT VALUE AND SKILLS DO I HAVE TO APPLY TO THIS USABILITY SITUATION?

Your rhetorical expertise will be relevant to every usability situation. However, each usability situation calls for different combinations of other skills and knowledge and emphases. Evaluating artifacts effectively depends on the right combination. Among the skills you will need to variously combine with rhetorical expertise are

- knowledge and skills of universal usability (accessibility for different disabilities);
- knowledge of niche-based communication media, for example, visualizations;
- awareness of standard resources and setups for usability laboratories;
- awareness of software packages/technologies that can facilitate data collection and analysis (e.g., software for screen capture, usage-log analysis, content analysis);
- knowledge and skills in the subject matter and domain in which you work;
- a working knowledge of human cognition and reasoning, from novice through expert, including cognitive psychology, sociocognition, social interactionism and constructivism, distributed cognition, and actor network theory;
- an understanding of the technical logic and technological efforts required for development in areas related to providing various support to users; and
- a working knowledge of the relationship between user needs and goals, on the one hand, and design and development choices for user interfaces, manuals, and software design, on the other hand.

WHAT EVALUATION METHODS/TEST DESIGNS ARE BEST FOR THE GOALS AND CIRCUMSTANCES?

All the questions that follow direct you toward matching methods and evaluation designs to specific goals and conditions of a usability situation.

- *Why evaluate?* You should state and convincingly support why a usability evaluation is in order.
- *What is expected?* You should find out what developers think needs to be evaluated and fixed—and compare it to what you think needs to be fixed. Ideally, these will be in sync; practically, they often are not (Howard 2008). Find out, as well, the assumptions and other obstacles that may lead stakeholders to misconceive the meanings and purposes of usability as well as their perceptions about evaluations that are feasible and problems that can be uncovered through specific evaluation methods.
- *What constrains the evaluation?* You should be clear about the resources, time, effort, and expertise that are available for usability

evaluations, all of which constrain the choices of evaluation goals and methods. Also, identify how much and what is already known about usability issues relevant to this application and what evaluations have produced this knowledge. For example, what needs assessments have been completed to guide application development thus far, and who conducted them? What are the quality and substance of the user-oriented results of the needs assessments? What development processes are followed, and where do usability assessments figure into the development processes?

- *Who are the users?* It is important to make it clear to yourself and stakeholders who you intend to define as the targeted primary, secondary, and tertiary audiences of the application and manual. Be clear how the target audiences affect the scope of your evaluation. It is not uncommon for audience definitions and their ranked importance to be unclear. Therefore, explicitly negotiate with developers and other stakeholders to reach a consensus about target audiences before the evaluation. For example, biostatisticians may experience tasks with the program differently from laboratory scientists.
- *How will the artifacts help users?* You need to construct evaluation goals and methods attuned to the goals of each audience—their tasks and the support and enhancements the artifact should provide for them. Set your scope wide enough to include tasks for which users will want to control interactivity.

WHAT REPORT CHOICES WILL COMMUNICATE CONVINCINGLY AND ENSURE IMPROVEMENTS?

The research literature on report writing discussed earlier suggests important issues to consider. Making choices about high-level section headings, for example, provides a good first cut in determining the scope and content of the report and necessary connections with primary and secondary audiences. Good candidates for high-level headings are detailed in the literature review.

High-level section headings alone, however, will not give your report a user-experience perspective. That perspective depends as well on your choices of content, emphasis, and subheadings within many of the major headings (e.g., business and test goals, screen shots and video). Of all the major headings, arguably the most important one for orienting readers to a user point of view is the results-and-recommendation section. In this section, certain subheadings are better than others for evoking a user experience perspective on reported problems and recommended improvements. For example, traditional subheadings that focus on program operations, functions, and features conjure a system focus rather than a user focus for readers. User-centered names for subsections vary, with many examples provided by Sutcliffe et al. (2008) as well as Blandford et al. (2000), as follows:

- Missing functionality for conceptual reasoning and user tasks
- Inadequate or partial functions that fall short of user needs and expectations
- Viscous support, that is, too many actions, high costs for small moves, difficulty in specifying the action sequence pertaining to a domain-based task
- Visual attention not matched to user needs, that is, inability to detect what needs to be detected, defaults that do not draw the eye to selectively important items or relationships for a task
- Clarity of "What do I see, and what can I do with it?"
- Clarity of "What did that do?" feedback
- Clarity of "How do I get to what I want to do?"
- Clarity of "Where have I been, and what do I know?"
- Imprecision in seeing and/or doing, that is, difficulty carrying out actions or discriminating

The examples above highlight the context of user experiences that you can further narrate, diagram, or otherwise represent within each subsection.

Why should you pay such close attention to your subsection headings? In part, doing so will help avoid the common phenomenon of developers glossing over such overused subsection headings as "ease of use" and "ease of access" and looking straightaway only for implicated program features. These readers miss the coherence and completeness that users need to experience to interact with an artifact usably and usefully, and subsequent development efforts may repeat the same ill-conceived design choices in other artifacts.

To show these three heuristic questions in real situations of usability evaluation, we return to the case of Collins and the usability evaluation of BioConcept.

EXTENDED CASE: A RETURN TO EVALUATING BIOCONCEPT

In Collins's case, usability has not been addressed until the end of the development cycle. Had the development team incorporated ongoing usability assessments into the development cycle from the start, BioConcept most likely would have been ready for formative user performance tests, as its developers hoped. Then Collins could have followed the testing methods described in the literature review. So far, the lessons learned from Collins's case include the following:

- If project management does not include early needs assessments and iterative testing, the evaluation situation is likely to be less than ideal.
- Heuristic evaluations and other quick inspection methods are a start, but they are not sufficient in themselves.
- In software and manuals, giving access to huge volumes of information is not sufficient for users' analytical purposes. Presentations must accord with users' *flows* of analysis. The right content in the right verbal and visual forms needs to be presented with effective explanations; and the right level of user-controlled interactivity must be included and communicated. These are, in part, rhetorical choices.

As the case so far shows, Collins needed communication expertise in this situation—as addressed by the first heuristic question. She saw scientists' potential interactions with BioConcept as a human-computer *communication* act. Specifically, Collins was concerned that the initial software lacked—and therefore did not adequately communicate—the full range of content and interactivity relevant to scientists' analytical purposes. In critically reviewing the software and the manual, she did not simply make sure it "contained" rich information and innovative features, regardless of how well their design fit users' purposes and context. Rather, she took a perspective centered on users' holistic interactive experiences with the artifact. Based on research findings and her own experiences, she believed that when evaluations focus only on separate components of this experience instead of the whole, they end up assessing effects of just those formalizable aspects of users' experiences, for example, the low-level actions of selecting a data item or clicking an interface widget for an operational step. It is important for users to conduct these operations easily, but ease of use is moot if they cannot do the tasks and functions that meet their analytical objectives in the first place.

Collins initially chose heuristic evaluations to identify obvious problems that needed fixes. Following standard practices of having several raters, Collins brought in three colleagues with whom she had conducted heuristic evaluations on similar tools. Table 12.2 shows an excerpt from their heuristic evaluation of BioConcept.

To extend the case further, after the developers revised the application and manual based on results of the heuristic evaluation, the artifacts still lacked many of the capabilities and interface designs that would match

Table 12.2. *Sample problems found in the heuristic evaluation*

Heuristic (usability standard)	Severity rating (0 = no problem; 5 = major problem)	Comments about usability
Does it clearly show if there are no query results?	4	Shows a *0* but should have a sentence like "No results found for 'csf1r'" or something similar.
Is it easy to reformulate the query if necessary?	5	Big problem: if you go back to the search screen from the explorer, you lose your query and the search results.
Are the results transparent as to what results are being shown and how to interpret them?	4	Could use a header that indicates what we're looking at; e.g., "45 biomedical concepts found matching 'cancer.'"
Is there ability to undo, redo, or go back to previous results (e.g., history tracking)?	5	No history tracking.
Are the mechanisms for interactivity clear?	4	Could use more labeling or tool tips.
Can you access the necessary data to assure the validity of results (i.e., the sources of the results)?	4	Can't get to sources (e.g., can't click on MeSH term and get to MeSH).
Can results be saved?	5	No saving option.

users' purposes and practices. For example, scientists would expect to compare many conceptual groupings to uncover overlapping genes among groups. Yet BioConcept did not give them an easy and efficient way to make comparisons.

The manual also had problems (see figure 12.1). It did not tie program operations to analysis-based task objectives. Nor did it explain the tool in ways that would connect and resonate with scientists' exploratory intentions and standards of practice and validation. Despite revisions, explanations were scanty in the manual, and those that were included were copied verbatim from an article that the developers wrote for an audience of computer scientists, biostatisticians, and computational biologists, not

Gene-Set Relation Mapping – Heatmap View [Explanation]

The values used in creating the heatmap are defined by the counts of the enrichment concept pairs that a gene belongs to, and the genes and concepts are clustered using complete linkage hierarchical clustering with the euclidean distance measure. Color of columns ranges from black (gene belongs to no enriched concepts) to bright red (gene belongs to the most enriched concepts).

BioConcept provides a heatmap view of your gene set and its enriched concepts. To view relationships by using a heatmap view: [Procedure]

1. In the Concept Explorer window, select the concepts that you would like to view. You may click the **Select All** link in the chart area if you would like quickly to select all of the enriched concepts.
2. Click the **Draw Heatmap** button on the bottom of the Concept Explorer window.

Heatmap Viewer

Heatmap Characteristics	Meaning
X-axis	Your gene list
Y-axis	Enriched concepts
Color	Redness indicates number of enriched concepts that include a specific gene.
Clicking *Draw Network Graph* button	Displays the Network Graph view.
Clicking *Network View* button	Displays the Network Graph view. When in the Network Graph view, this button changes to Heatmap view, which allows you to switch between both views.
Clicking *Explorer* button	Brings up Concept Explorer screen (click the X in the upper right-hand corner of the Concept Explorer screen to return to the Network Graph View).
Clicking *Export Graph* button	Creates a PNG file image of the screen in a separate window.

Figure 12.1. Explanations and procedures in the developer-composed documentation. This passage is taken verbatim from an article that is not aimed at the primary user audience. Procedures lack task-driven goals and outcomes as framing devices.

life science researchers (figure 12.1). Similarly, screen shots did not help users interpret displays in relation to a sample analytical task (figure 12.2).

Collins recognized these problems without needing to put the application and manual in front of users. Some problems are likely to require new functionality and interface modifications and will not be quick fixes. Collins faces a decision on usability evaluation at this point. Should she conduct a user performance test or push for further improvements first? Factors affecting Collins's decision about the next steps to take are tied to the questions presented above, on evaluation methods and test design.

Collins knows that any decision is going to involve trade-offs. On one hand, she could conduct user performance testing and focus on only the narrow set of discrete features and operations that are ready with baseline usability and usefulness for scientists' tasks. This choice would keep the development cycle on schedule. But it would let some major problems slide, such as insufficient support for making comparisons. On the other hand, she could push for more improvements before conducting a user performance test. This choice would require investing more time and effort than budgeted in the development plan. She would need to make a persuasive case to management for this greater investment. In presenting her case, she could discuss the likelihood of mismatches between BioConcept and scientists' complex analytical needs for discovery, which would likely frustrate early adopters and make them skeptical about the quality of the firm's software. These early adopters are vital to the company for early user feedback after deployment, and they might abandon their loyalty to the software.

Again communication skills are a necessity, now for persuasion. Better outcomes in product design and development require a shared understanding across stakeholders of why certain evaluations are being conducted at a given point, based on application quality and progress.

Collins decides to pursue the second option. This option will push back the production schedule—but do so at an early enough point to make such adjustments without prohibitive costs or revenue losses. Collins's persuasive memo to management (copied to developers) is different from the usability test reports addressed by the third heuristic question. The persuasive memo, nonetheless, serves similar goals by providing empirical support for the usability judgments that Collins makes. For this support, Collins may informally put the current versions of BioConcept and its manual in front of a "friendly" early adopter of the firm's software. A trial user performance can identify successes, flaws, and limits to cite in the memo that go beyond Collins's subjective expertise.

Ideally, management will concur, the tool and manual will be improved,

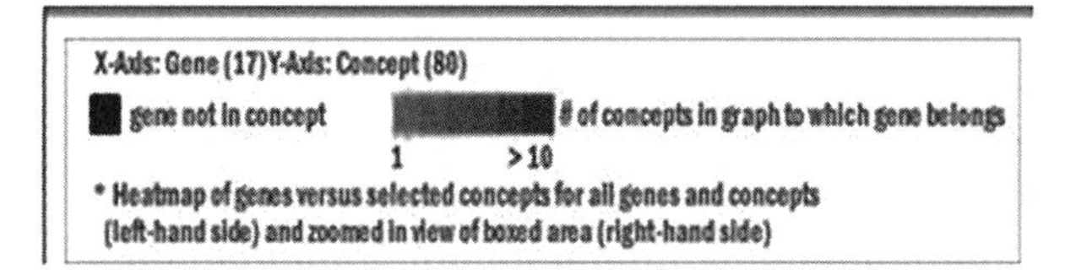

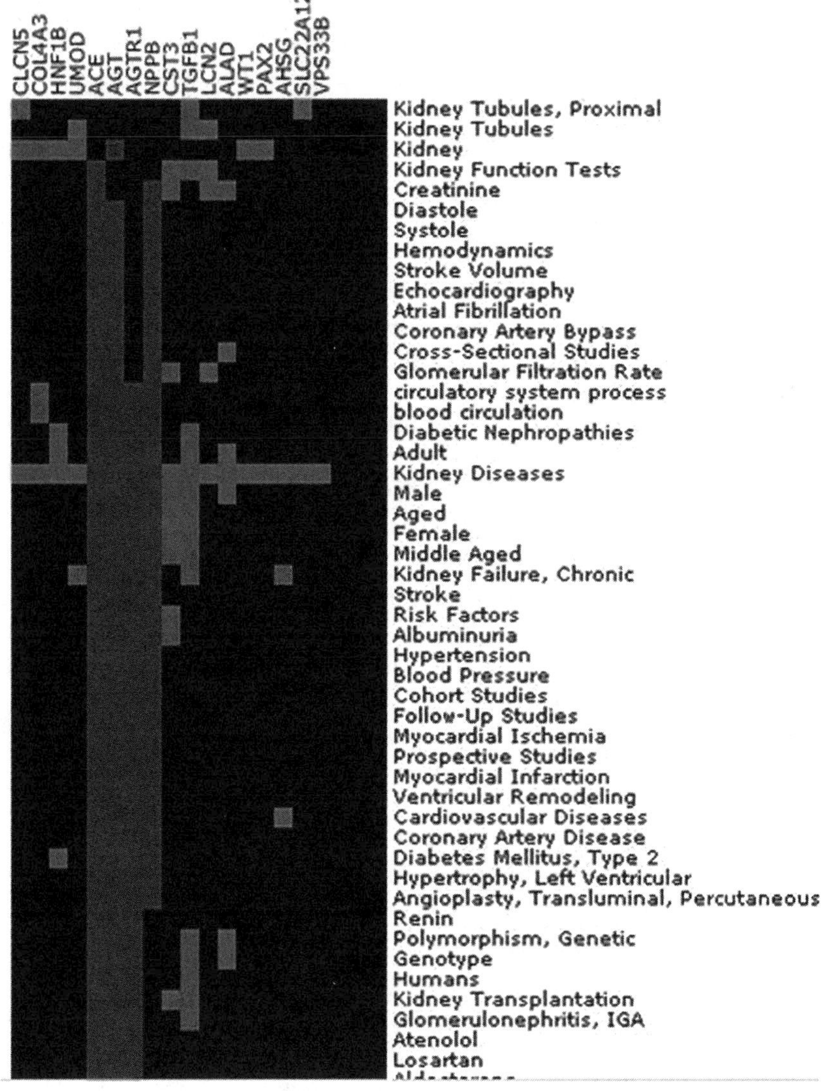

Figure 12.2. Heatmap view. The screen shot of the interactive visual presented in the documentation lacks labels and other explanations and cues to facilitate interpretations and applied meanings to an analysis objective.

and Collins will conduct the user performance evaluation. She will then report results in ways that effectively "talk" to stakeholders, addressing the issues implicit in the third heuristic.

At that time, she can present usability successes and decide on the structure of the problem sections. Because of her rhetorical orientation, she will organize sections by users' cognitive tasks, that is, by users' demonstrated task-based reasoning and behaviors. For the specific domain-based analysis conducted with BioConcept, the usability report subheadings might highlight users' comprehension of terms; their dependence on cues for staying oriented in analysis; their understanding of relevant molecular relationships based on specific workspace features, functions, and content; their validation of relationships; and the data manipulations they need and expect to do for specific types of task-based goals and reasoning. In using these higher-order cognitive tasks to structure the report, Collins will remind readers that choices about design modifications directly impact users' performance and cognition. She can subdivide these cognitive task categories into types of usability and usefulness problems and discuss tool features and functions implicated in each problem (Molich, Jeffries, and Dumas 2007). She will be guided by the categories synthesized from Sutcliffe et al. (2000) and Blandford et al. (2008) mentioned earlier in the heuristic related to writing reports.

Collins's report also includes severity rankings, her recommended priorities for fixes, and suggested designs for fixes. Application problems have behavioral consequences for users' practices and reasoning, and readers must be aware of these consequences. She aims for recommendations and suggestions with enough detail to guide the actual implementation of modifications. Importantly, she recommends fixes to problems that cause immediate usability shortcomings, but she also looks at the big picture. Her report will project the consequences of recommended fixes and enhancements. When stakeholders and Collins meet to review her report, they will work as a team to negotiate priorities in improvements. Her severity rankings and criteria will contribute to ultimate decisions about priorities for improvements.

Combined, these various structures and content will add impact to her report. But evaluation reports in themselves cannot carry the weight of the usability influence that Collins and others like her need to assert in their organizations. This influence is ongoing, interpersonal, and instructive. As Collins's case shows, her report on the user performance testing will be one part of a larger flow of communications about the quality and usability of BioConcept. From her earlier heuristic evaluation report through her persuasive report to management, she has been involved in negotiating with diverse stakeholders about improving BioConcept with convincing evidence. Successful collaborations and respect for usability roles in a product team are prerequisites for achieving the goals of usability evaluation reports.

CONCLUSION

This window into evaluation situations that you might encounter when assessing software, web applications, and/or documentation reveals that strategic choices run throughout your processes of setting evaluation goals, determining methods, and reporting. Strategically, it is important to actively assume the role of usability evaluator and justify your value and suitability for the role to the product team. Additionally, knowing methods and having a tool kit of skills are necessary but not sufficient. It is vital to use them advantageously. You need to devise goals for evaluation that accord with the situation at hand and select—and possibly mix—methods. Strategic approaches to reporting are also important, including your choices for framing the content, structure, and media of communications to stakeholders in ways that maximize the chances that top-priority problems and enhancements will be acted upon effectively. Throughout this evaluation work, you will be establishing your role and identity in the organization as an irreplaceable usability expert whose skills and knowledge in technical communications and related areas add value to the development and dissemination of products.

DISCUSSION QUESTIONS

1. What are some important criteria for judging usefulness? What are some important criteria for judging usability? Which criteria for usefulness and usability overlap with each other, and which are distinct from each other?
2. Reflect on this claim by answering the questions following the claim: "Knowing how to evaluate an artifact so that it achieves the purpose of having audiences go along and act effectively on it is a communication and rhetorical art." What issues related to usefulness and usability, respectively, are rightfully communication issues and why? What issues related to usefulness and usability are not communication issues and why?
3. Go to your university's library website, and use its catalog-search page to explore a topic that interests you. Evaluate the set of pages resulting from the search, using the extract from the heuristic evaluation instrument included in this chapter. Include comments and severity rankings. Compare your ratings and comments with three or four of

your classmates. On what do you agree and disagree? What criteria were you each using to determine "level of severity"? With these three or four classmates, try to agree on criteria for defining severity level.

4. Which of the usability methods mentioned in this chapter interest you most? What else would you like to find out about them? Write at least three questions.
5. For what aspects of the usability evaluation situations and approaches described in this chapter do you feel most prepared? For what aspects do you feel least prepared? Explain in detail your perception of your preparedness.
6. Examine an application or game that you like but that, at times, frustrates you. For the frustrating portion, write a real-world task that participants in a user performance test could perform. Craft the task so that it will enable evaluators to gather data on how easy and useful this "frustrating aspect" of the technology is (or is not) for this particular task. Write a brief rationale for your task, explaining what data it might generate that could provide evidence of the tool's shortcomings and strengths.
7. If you were writing a report for only the product team on the user performance testing that Collins ultimately will conduct, what elements and content would you include and why? If you were writing for only the director and marketing group, what elements and content would you include and why? If you were writing to everyone, what elements and content would you include and why?

WORKS CITED

Anscheutz, Lori, and Stephanie Rosenbaum. 2002. Expanding Roles for Technical Communicators. In *Reshaping Technical Communication*, ed. Barbara Mirel and Rachel Spilka, 149–164. Mahwah, NJ: Elsevier.

Bangor, Aaron, Philip Kortum, and James Miller. 2008. An Empirical Evaluation of the System Usability Scale. *International Journal of Human-Computer Interaction* 24:574–594.

Barnum, Carol. 2002. *Usability Testing and Research*. New York: Longman.

Barton, Ellen. 2002. Inductive Discourse Analysis: Discovering Rich Features. In *Discourse Studies in Composition*, ed. Ellen Barton and Gail Stygall, 19–41. Cresskill, NJ: Hampton Press.

Blakeslee, Ann, and Cathy Fleisher. 2007. *Becoming a Writing Researcher*. Mahwah, NJ: Erlbaum.

Blandford, Ann, Thomas R. G. Green, Dominic Furniss, and Stephann Makri. 2008. Evaluating System Utility and Conceptual Fit Using CASSM. *International Journal of Human-Computer Studies* 66:393–409.

Boren, Ted, and Judith Ramey. 2000. Thinking Aloud: Reconciling Theory and Practice. *IEEE Transactions on Professional Communication* 43:261–277.

Brady, Ann. 2004. Rhetorical Research: Toward a User-Centered Approach. *Rhetoric Review* 23:57–74.
Brooke, John. 1996. SUS: A Quick and Dirty Usability Scale. In *Usability Evaluation in Industry*, ed. Patrick W. Jordan, Bruce Thomas, Bernard A. Weerdmeester, and Ian L. McClelland, 89–194. London: Taylor and Francis.
Cockton, Gilbert, and Alan Woolrych. 2001. Understanding Inspection Methods: Lessons from an Assessment of Heuristic Evaluation. In *People and Computers XV*, ed. Ann Blandford and Jean Vanderdonckt, 171–192. Berlin: Springer.
Cockton, Gilbert, Alan Woolrych, and Darryn Lavery. 2007. Inspection-Based Evaluations. In *Human-Computer Interaction Handbook*, ed. Andrew Sears and Julie Jacko, 1172–1191. Boca Raton, FL: CRC Press.
Cooke, Lynne, and Sue Mings. 2005. Connecting Usability Education and Research with Industry Needs and Practices. *IEEE Transactions on Professional Communication* 48:296–312.
Creswell, John, and Vicki Plano Clark. 2006. *Designing and Conducting Mixed Methods Research*. Thousand Oaks, CA: Sage.
De Jong, Menno, and Peter Jan Shellens. 2000. Toward a Document Evaluation Methodology: What Does Research Tell Us about the Validity and Reliability of Evaluation Methods? *IEEE Transactions on Professional Communication* 43:242–260.
Dumas, Joseph S., and Janice Redish. 1993. *A Practical Guide to Usability Testing*. Norwood, NJ: Ablex.
Evans, Mary, Carolyn Wei, and Jan Spyridakis. 2004. Using Statistical Power Analysis to Tune-up a Research Experiment: A Case Study. In *Proceedings of IEEE International Professional Communication Conference*, 14–18. New York: IEEE Press.
Fu, Limin, Gavriel Salvendy, and Lon Turley. 2002. Effectiveness of User Testing and Heuristic Evaluation as a Function of Performance Classification. *Behaviour and Information Technology* 21:137–143.
Gray, Wayne D., and Marilyn Salzman. 1998. Damaged Merchandise? A Review of Experiments That Compare Usability Evaluation Methods. *Human-Computer Interaction* 13:203–261.
Hollingsed, Tasha, and David Novick. 2007. Usability Inspection Methods after 15 Years of Research and Practice. In *Proceedings of the 25th Annual ACM International Conference on Design of Communication (SIGDOC '07)*, 249–255. New York: ACM.
Howard, Tharon. 2008. Unexpected Complexity in a Traditional Usability Study. *Journal of Usability Studies* 3:189–205.
Hughes, Michael. 1999. Rigor in Usability Testing. *Technical Communication* 46:488–494.
Industry Usability Reporting Project. 2001. Common Industry Format for Usability Test Reports. ANSI/INCITS 354-2001. New York: American National Standards Institute.
Ivory, Melody, and Marti Hearst. 2001. The State of the Art in Automating Usability Evaluation of User Interfaces. *ACM Computing Surveys* 33:470–516.
Jarrett, Caroline, Whitney Quesenbery, Ian Roddis, Sarah Allen, and Viki Stirling. 2009. Using Measurements from Usability Testing, Search Log Analysis and Web Traffic Analysis to Inform Development of a Complex Web Site Used for Complex Tasks. *Human-Computer Interaction* 10:729–738.
Johnson, Robert R., Michael Salvo, and Meredith W. Zoetewey. 2007. User-Centered Technology in Participatory Culture: Two Decades "Beyond a Narrow Conception of Usability Testing." *IEEE Transactions on Professional Communication* 30:320–332.

Kirakowski, Jurek. 1996. The Software Usability Measurement Inventory, Background and Usage. In *Usability Evaluation in Industry*, ed. Patrick W. Jordan, Bruce Thomas, Bernard A. Weerdmeester, and Ian L. McClelland, 169–178. London: Taylor and Francis.

———. 2005. Summative Usability Testing: Measurement and Sample Size. In *Cost-Justifying Usability: An Update for the Information Age*, ed. Randolph G. Bias and Deborah Mayhew, 519–554. San Francisco: Morgan Kaufmann.

Kirakowski, Jurek, and Bozena Cierlik. 1998. Measuring the Usability of Web Sites. In *Human Factors and Ergonomics Society 42nd Annual Meeting*. Santa Monica, CA: HFES. http://www.ucc.ie/hfrg/questionnaires/wammi/research.html. Accessed February 12, 2010.

Koerber, Amy, and Lonie McMichael. 2008. Qualitative Sampling Methods: A Primer for Technical Communicators. *Journal of Business and Technical Communication* 22:454–473.

Krippendorff. Klaus. 2004. *Content Analysis: An Introduction to Its Methodology*. Thousand Oaks, CA: Sage.

Krug, Steve. 2005. *Don't Make Me Think*. Berkeley, CA: New Riders Press.

Kushniruk, Andre, and Vimla Patel. 2004. Cognitive and Usability Engineering Methods for the Evaluation of Clinical Information Systems. *Journal of Biomedical Informatics* 37:56–76.

Lewis, James R. 2006. Usability Testing. In *Handbook of Human Factors and Ergonomics*, 3rd edition, ed. Gavriel Salvendy, 1275–1316. New York: John Wiley.

Makri, Stephann, Ann Blandford, and Anna Louise Cox. 2008. Using Information Behaviors to Evaluate the Functionality and Usability of Electronic Resources: From Ellis's Model to Evaluation. *Journal of the American Society for Information Science and Technology* 59:2244–2267.

Medlock, Michael, Dennis Wixon, Mick McGee, and Dan Welsh. 2005. The Rapid Iterative Test and Evaluation Method. In *Cost-Justifying Usability: An Update for the Information Age*, ed. Randolph G. Bias and Deborah Mayhew, 489–517. San Francisco: Morgan Kaufmann.

Miles, Matthew, and A. Michael Huberman. 1994. *Qualitative Data Analysis*. Thousand Oaks, CA: Sage.

Mirel, Barbara. 2004. *Interaction Design for Complex Problem Solving: Developing Useful and Usable Software*. San Francisco: Elsevier.

Mirel, Barbara, and Zach Wright. 2009. Heuristic Evaluations of Bioinformatics Tools: A Development Case. In *Proceedings of the 13th International Conference on Human Computer Interaction (HCII-09) Lecture Notes in Computer Science 5610*, 329–338. Berlin: Springer.

Molich, Rolf, Robin Jeffries, and Joseph Dumas. 2007. Making Usability Recommendations Useful and Usable. *Journal of Usability Studies* 2:162–179.

Nielsen, Jakob. 1993. *Usability Engineering*. Boston: Academic Press.

———. 1994. Heuristic Evaluation. In *Usability Inspection Methods*, ed. Jakob Nielsen and Robert L. Mack, 25–62. New York: John Wiley.

Redish, Janice, Rudolphe G. Bias, Robert Bailey, Rolf Molich, Joseph Dumas, and Jared Spool. 2002. Usability in Practice: Formative Usability Evaluations—Evolution and Revolution. In *CHI '02 Extended Abstracts on Human Factors in Computing Systems*, 885–890. New York: ACM.

Rosenbaum, Stephanie. 2008. The Future of Usability Evaluation: Increasing Impact on Value. In *Maturing Usability*, ed. Effie Law, Ebba Hvannberg, and Gilbert Cockton, 344–378. London: Springer.

Rubin, Jeffrey, and Dana Chisnell. 2008. *Handbook of Usability Testing: How to Plan, Design, and Conduct Effective Tests*. 2nd edition. New York: Wiley.
Saraiya, Purvi, Chris North, and Karen Duca. 2005. An Insight-Based Methodology for Evaluating Bioinformatics Visualizations. *IEEE Transactions on Visualization and Computer Graphics* 11:443–456.
Sauro, Jeff, and Erica Kirklund. 2005. A Method to Standardize Usability Metrics into a Single Score. In *Proceedings of the SIGCHI Conference on Human Factors in Computing Systems*, 401–409. New York: ACM.
Sauro, Jeff, and James R. Lewis. 2009. Correlations among Prototypical Usability Metrics: Evidence for the Construct of Usability. In *Proceedings of the 27th International Conference on Human Factors in Computing*, 1609–1618. New York: ACM.
Scotch, Matthew, Bambang Parmanto, and Valerie Monaco. 2007. Usability Evaluation of the Spatial OLAP Visualization and Analysis Tool (SOVAT). *Journal of Usability Studies* 2:76–95.
Snyder, Carolyn. 2003. *Paper Prototyping: The Fast and Easy Way to Design and Refine User Interfaces*. San Francisco: Morgan Kaufmann.
Spyridakis, Jan, Carolyn Wei, Jennifer Barrick, Elisabeth Cuddihy, and Brandon Maust. 2005. Internet-Based Research: Providing a Foundation for Web-Design Guidelines. *IEEE Transactions on Professional Communication* 48:242–260.
Sutcliffe, Alistair, Michele Ryan, Ann Doubleday, and Mark Springett. 2000. Model Mismatch Analysis: Towards a Deeper Explanation of Users' Usability. *Behaviour and Information Technology* 19:43–55.
Sullivan, Patricia. 1989. Beyond a Narrow Conception of Usability Testing. *IEEE Transactions on Professional Communication* 32:256–264.
Theofanos, Mary, and Whitney Quesenbery. 2005. Towards the Design of Effective Formative Test Reports. *Journal of Usability Studies* 1:27–45.
Ummelen, Nicole. 1997. *Procedural and Declarative Information in Software Manuals: Effects of Information Use, Task Performance, and Knowledge*. Amsterdam: Rodopi.
Wilson, Chauncy, and Kara Coyne. 2001. Tracking Usability Issues: To Bug or Not to Bug? *Interactions* 8:15–19.

R. STANLEY DICKS

13 How Can Technical Communicators Manage Projects?

SUMMARY

Technical communicators need to know how to manage projects using both traditional, "waterfall" project management methods and newer, less time-consuming and more user-oriented methods/models such as agile development, iterative design, and extreme programming and documentation.

The most logical heuristic for project management is a flowchart that includes the universal characteristics of any communication project regardless of the method employed. That chart includes the project phases of planning, research and information gathering, composition/invention, reviewing and/or testing against quality criteria, revision, production, and dissemination. While technical communicators might not perform all phases on all projects, they need to be aware of which ones they are omitting and the possible repercussions of doing so. This chapter discusses the various methods/models and how each phase is envisioned and implemented within them.

Closely related to project management is the concept of time management, which is necessary for technical communicators to understand and use if they are going to succeed at managing their projects and meeting their deadlines. In some organizations, project and time management use combined technologies and are carried out simultaneously. Technical communicators should learn and use the technologies that others in the organization use to manage both their time and their projects.

INTRODUCTION

Ann Ross is a technical communicator who works for a financial services company that develops stock and securities management software and point-of-sale hardware/software systems. Point-of-sale systems involve cash registers, scanners, and similar equipment at checkout counters in stores, restaurants, and other businesses. She is currently working on three different projects: (1) developing online help for a web-based portfolio management program that will be the company's first web-delivered program, (2) a user's guide for a small point-of-sale system that is in the first draft stage and that will be delivered as hard copy with the product and as a PDF file online, and (3) a set of marketing materials for a large point-of-sale system aimed at department and chain stores, where dozens of machines will be configured in a network at each store.

Ann needs to have some systematic method of working on each of her three projects. In order to describe how she handles her multiple projects, this chapter reviews some of the more important literature about project management for technical communicators. It then looks at the methods used by communicators to manage projects such as Ann's, including planning, research and information gathering, composition/invention, reviewing and/or testing against quality criteria, revision, production, and dissemination. The chapter looks first at the more traditional project management methods and then at some more current, alternative methods intended to make projects more efficient and more likely to meet the needs of audiences. The chapter also examines the tools and technologies technical communicators use for product management, before concluding with a more detailed look at how Ann works through her three projects.

LITERATURE REVIEW

The most important book on technical communication project management is JoAnn Hackos's *Information Development*, published in 2007. *Information Development* offers a complete, overall explanation for how technical communicators can manage their projects to achieve consistently high-quality results. Critics might argue that the system that Hackos describes is so complex that hardly any organization would ever agree to follow it. My experience, however, is that there are organizations that do nearly everything that Hackos advocates, and that, more importantly, organizations that do very few of her steps usually produce inferior information products.

Only two other books have been completely devoted to technical communication management issues. *Publication Management: Essays for Professional Communicators*, edited by O. Jane Allen and Lynn H. Deming (1994), has a section on project management with chapters that in some cases are somewhat dated, but that nonetheless offer sound overall advice. My book, *Management Principles and Practices for Technical Communicators* (2004), does treat some aspects of project management, especially as related to estimations (127–137) and to personnel management on projects (23–110), but it is largely aimed at covering other management practices, such as financial management, overall personnel management, time management, training, and mentoring.

For an area that is as critically important to technical communication as project management, it is surprising that there are not more articles in the literature dedicated to the subject. The articles tend to be either descriptive, describing current project management practices, or prescriptive, offering methods to follow. One classic is Hackos's (1997) article explaining the process-maturity model, a method for assessing the maturity of an organization's project management and quality assurance methods for developing information products. Hackos borrows from similar maturity models in the software literature to construct a five-level system that ranks organizations as

- level 1, ad hoc;
- level 2, rudimentary;
- level 3, organized and repeatable;
- level 4, managed and sustainable; and
- level 5, optimizing.

This model has helped organizations analyze how mature their information process systems are and has helped technical communicators educate upper management about the need for more sophisticated project management systems for information development. Another important article is Carliner's (2004) survey and study of the management portfolios in large technical communication groups. Carliner provides detailed analyses of current project and people management in the field. While not addressing project management directly, Whiteside's (2003) article analyzes the skills that hiring managers consider most important for technical communicators, and project management ranks high on the list.

The Society for Technical Communication's annual conference proceedings contain a section on management each year, which usually includes some presentations on project management. Typical presentations might cover how to manage projects using a particular piece of project management software, how to manage relationships with other project team members, how to develop high-performance project teams, and similar topics. However, the two-page limitation for entries in the proceedings abbreviates the presentations to the point that they are rarely helpful for someone who did not attend the session.

Most of the other books and articles about technical communication project management treat one particular aspect of it, such as estimation, audience/task analysis, or agile development methods. For example, working remotely with virtual teams, including international teams, is treated in Duarte and Snyder's *Mastering Virtual Teams: Strategies, Tools, and Techniques that Succeed* (2006). Larbi and Springfield (2004) approach virtual teams from the point of view of the writers working on them.

ESTIMATION

Some technical communicators do not worry about estimation; they are simply assigned to a project that already has designated communication resources. This can be a problem if the designator was someone outside of technical communication who does not understand the time and resources necessary to develop high-quality documents. Other communicators spend considerable time and effort making estimates for projects. In the case of technical communication companies and consultants, good estimates are critical for staying in business. Hackos (1994, 153–195; 2007, 436–442) has extensive discussions of estimating, including methods for estimating various media such as paper, online, training, and video, as does Dicks (2004, 127–137), who offers a formula for doing estimates for both paper and electronic deliverables. Deliverables here refers to the information being delivered, which might be delivered physically (i.e., printed on paper) or electronically (i.e., in a help system or website or some other online method). David L. Smith's (1994) chapter is based strictly on performing estimates for paper pages, but it nonetheless has some solid estimation concepts. Judith J. K. Polson (1988) discusses the importance of preparing good estimates and provides a formula to help visualize the complex relationships among quality, functions, resources, and time on a project. Peter Zvalo (1999) discusses the differences between fixed cost (where a project is done for one total price) and hourly charges (where a technical communicator's work is billed by the number of hours worked times a cost per hour).

TASK ANALYSIS

In the research phase of their projects, technical communicators must find out who their audience will be and what tasks that audience will need to accomplish using the information being developed for the project. Hackos and Redish's *User and Task Analysis for Interface Design* (1998) provides full methodology for analyzing audiences and their tasks. The authors repeatedly stress the importance of working directly with end users of products under development to find out how they will use the product and what designs will optimize that use. In a similar vein, Houser (2001) argues that technical communicators should create and maintain databases of user characteristics and usability information. Such databases will help them plan their projects in the future while ensuring that the projects address user needs. Deborah Mayhew in *The Usability Engineering Lifecycle* (1999) provides extensive instruction on how to develop user profiles and to analyze the tasks that those users perform. Todd Warfel (2006) has developed a task-analysis grid that designers and communicators use to aid in analyzing audiences and their tasks.

ALTERNATIVE METHODS

The most traditional project management system is called the "waterfall" system because a diagram of its steps, with time allotted to one milestone and then to another and another and so on, resembles a sloping, downhill, left-to-right waterfall. Because the "waterfall" system can lead to lengthy development times and to products that do not meet the needs of customers, practitioners have developed project management methodologies that are aimed at reducing design and development time while assuring that products do indeed meet customer needs. Three of the most commonly practiced alternatives are user-centered design (sometimes called iterative design, because it is based on determining user needs, designing to those needs, testing the design against those needs, redesigning based on the test results, testing again, etc., until the user needs are met), extreme programming (wherein programmers or technical communicators work in pairs to develop materials that are determined by user "stories," testing what they develop with users, revising, testing again, etc., until the products meet the user requirements as expressed in the stories), and agile development (wherein project team members work in close proximity, meet daily, and concentrate their efforts on small parts of larger products, also testing them iteratively against user "stories"). Gould and Lewis's 1985 article issued a challenge to developers to make their designs more user oriented and their methods more involved with frequent testing, redesign, testing, and so forth, to better meet the needs of customers. Nielsen (1993), Rubin and Chisnell (2008), Dumas and Redish (1993), and Barnum (2001, 2011) have all provided books that treat usability testing and user-centered design. Extreme programming has been exhaustively described in a series of books in the XP Series published by Addison-Wesley, led by Beck and Andres's *Extreme Programming Explained: Embrace Change* (2004). Nuckols and Canna (2003) explain how they have adapted extreme programming to result in "extreme documentation." Agile development was started by a manifesto published on the web (Agile Alliance 2011a) and has been championed by the Agile Alliance (2011b) at http://www.agilealliance.com.

HEURISTIC FOR PROJECT MANAGEMENT

The "standard" project phases for technical communication work include planning, research and information gathering, composition/invention, reviewing and/or testing against quality criteria, revision, production, and dissemination (figure 13.1). Depending on the nature of one's work environment and the nature of a specific project, those phases may be performed by a single communicator or by a team. It is important for technical communicators to know how to perform each phase, even if their current work environment does not immediately require it.

It is important to conduct activities that add quality to information products while they are under development. Such activities include the planning at the beginning, determining the types of documents or structures of topics that will be used, meeting with customers or effectively gathering information about them, conducting expert review of drafts during development and performing some kind of validation assessment to ensure that the information products are accurate, doing some kind of usability assessment to determine that the documents work with their audience, copyediting the finished products, and, if necessary, performing

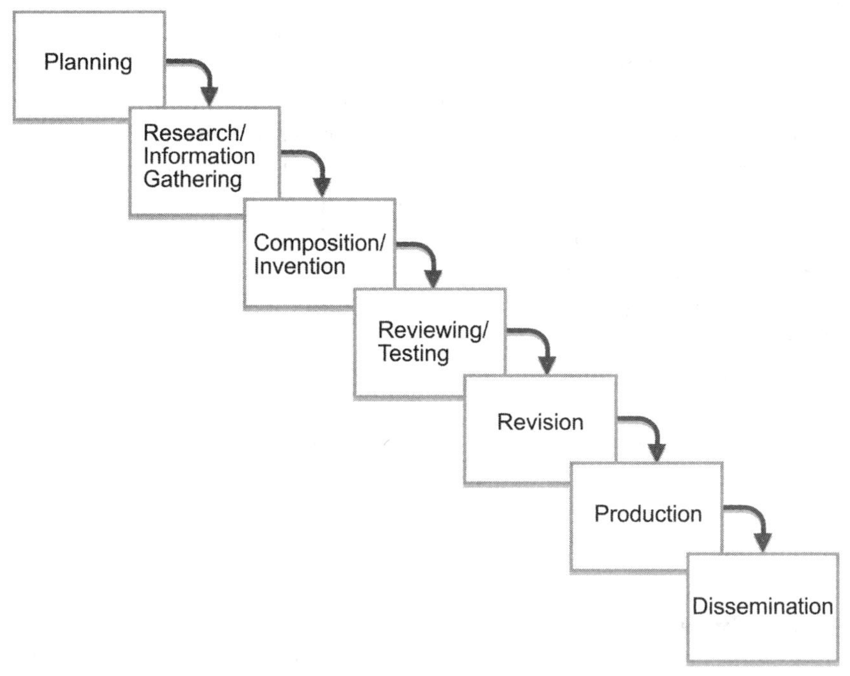

Figure 13.1. Heuristic: project management phases

production assessments to ensure high-quality printing or appropriate uploading of files to servers.

PLANNING

Hackos (2007, 315–330) recommends that 30% of a technical communication project should be spent on planning and design, before the first word is written or the first image is created. In mature organizations, technical communicators devote significant time to learning the organization's goals for a project, studying the users and the tasks they will be performing with the product or service under development, and designing information products that meet the goals both of the internal organization and the external audience(s). The technical communicators must also determine which media to use to develop and disseminate those information products, given the contexts in which the products will be used.

In this stage, the technical communicator examines any available documents that explain the product and what it will include. In hardware projects, such documents usually include engineering specifications and drawings. In software projects, they usually include a requirements document that describes how the program will work and that may contain proposed screens and interfaces. The communicator also analyzes the tasks that users will complete and endeavors to design an information product that will allow users to perform their desired tasks best with the product or service as it is described in the specifications and requirements.

Technical communicators can take one of two main approaches to planning and structuring information products: the document-based approach or the topic-based approach. The document approach assumes that traditional documents will be delivered, whether on paper or online. Such conventional documents might include installation guides, getting-started guides, user guides, instruction manuals, training manuals, procedures manuals, maintenance manuals, and so forth. Sometimes the delivery of manuals is dictated by conventions in the industry or within the particular organization, and technical communicators may have little say in what they will produce, even if it may not be the best solution. They attempt to structure their documents so that they match as well as possible the tasks that the user needs to do, although they frequently receive considerable pressure from engineers, programmers, and scientists to create documents that describe the system and each of its parts thoroughly.

Another approach to designing information products is to use topics. This approach requires the communicator to list every task and subtask that a user will need to perform, to envision the conceptual background information that users will need to know to be able to perform the tasks, and to obtain or create any reference materials (tables, charts, drawings) that users will need to perform the tasks. Each task, subtask, piece of conceptual information, and reference item becomes a topic. The communicator develops an entire list of the topics, and, just as with documents, imposes an organizing structure on them that will suit user needs for various purposes. One of the advantages of the topics approach is that the topics can be rearranged to create various types of documents. Once all of the topics are created and put into a database, different front ends can be designed to show the topics in different orders for use in different types of documents or in different media. For example, a training document typically takes users on a linear path through increasingly more complex information, and a technical communicator can create a manual that gets the topics in such an order. On the other hand, a reference document requires that topics be arranged so that access is easy even though they do not have any linear or hierarchical relationship to one another. Perhaps the technical communicator arranges the topics alphabetically, or according to which part of a system they relate to, or bunched with the larger tasks that they are part of. Such databases of topics make possible what is called single-sourcing, where topics are developed once and then used in multiple types of information products.

The planning stage typically concludes with the publication of an information plan or documentation plan that includes a discussion of the project's goals, the audience and their tasks and information needs, the proposed information products that will be developed, outlines or topic lists for those products, initial estimates of the time and resources that will be required, and milestone dates by which drafts will be sent for review and the final document(s) will be completed. Even if such plans are not required, technical communicators should always develop them and disseminate them to everyone associated with a project. In mature organizations, such planning is always done, and the plans must be approved and signed off by all of the appropriate internal groups.

RESEARCH AND INFORMATION GATHERING

In the second stage, technical communicators research users and their tasks and pull together as much information as possible to decide what types of information products to deliver and how the information within those products should be structured. The sources for the information will likely be varied. As mentioned in the previous section, there may be specification or requirements documents. There may be marketing-survey information or focus-group videos or transcripts. There may be information from customer support people about which aspects of the product or of similar products most often vex users. Technical communicators should try to interact with users directly to get information about how they will use the product or service to complete their tasks. This can be in the form of field visits, sitting in on training classes, accompanying installation or maintenance teams on site visits, sending out questionnaires, doing some telephone interviews with users, doing an online survey, and so forth. Other methods might include getting users to help participate in the design process by inviting them to join a blog or a wiki devoted to the information development effort.

It is important for technical communicators to seek information actively rather than to wait passively for it to be provided. Developers report that they wonder what value communicators add when the developers have to write up technical descriptions and the communicators simply "wordsmith" them or "pretty them up." Communicators should gather information aggressively and should not rely totally on subject-matter experts (or SMEs, as they are often called) to supply everything.

Once the technical communicators have gathered enough information, they can continue the planning process, ensure that their original plans were appropriate or modify them, fill out the structures they intend to use for their documents or topics lists, and prepare to develop their information materials. Technical communicators must continue to gather information throughout most projects, because of product design changes, unanticipated features being added, and all of the other factors that affect the process of creating something new. So the information gathering and planning stages do not end neatly on one day with the composition/invention stage beginning neatly the next morning.

COMPOSITION/INVENTION

Composition and invention are what most people think of as technical communication. Even many technical communicators believe they should bypass planning and information gathering and immediately begin drafting documents. They may also receive pressure from coworkers to start writing and stop doing so much planning and researching. Despite these pressures to begin developing information immediately, bypassing the important planning stages leads to information products that are not effective with their intended audiences. It also leads to high development costs when documents or topics have to be written and rewritten over and over because the optimum structure for the information must be discovered as it is being developed. More likely, the optimum structure will not be discovered, and inferior information will be delivered.

Once the planning and information gathering have been adequately completed, actual writing of the documents can begin. It is a good idea to develop a detailed schedule of deliverables for reviews, whether those are internal reviews with one's boss or the information team or larger reviews with the entire project team. Writing a 200-page document without having anyone look at any of its parts, or developing 200 topics without having anyone review any of them, is likely to lead to extensive rework. Further, the tendency to procrastinate and then have to work considerable overtime is mitigated by having numerous smaller deliverables due with short intervals between them.

Project management is often associated with record keeping, such as tracking hours worked, milestones met, money spent, and so forth. While some of that activity begins during the planning stages of a project, it begins in earnest during the composition stage, which is also the stage that many people associate with true project management. At this stage, technical communicators break the work down into smaller deliverables (which might include chapters, subchapters, or topics), establish due dates for all of the deliverables, and track their progress against those dates. This is also the stage where communicators are concerned with keeping records of the hours they spend on each information product and on the project overall. Mature organizations develop standard methods for reporting time spent and progress made. Many technical communication managers require a weekly report that briefly describes progress made, problems encountered, and the hours worked on each type of deliverable. It is important for an organization to track how many hours it took to do everything, so that they can do better estimates in the future when similar information products must be developed.

The "default" completion of the composition/invention stage for technical communication involves delivering a first draft for review to the development team and to other interested parties, which might include customer support, sales and marketing, and legal. It is important for technical communicators to define in their information plans what will and will not be included in the first draft of a document. In some organizations, it is expected that the draft will be complete, while in others it is permissible to have some gaps in the information. In some cases, it is expected that all graphics will be included, while in others it is permissible to include only rough hand sketches or simply labels. With publication software making formatting and layout "easy," many people expect the first draft to look like a finished document.

REVIEWING AND TESTING AGAINST QUALITY CRITERIA

Most technical communication work involves what is often called egoless writing. While there may be a technical communicator who is ostensibly in charge of the document, many other people contribute to, review, criticize, and affect the final form that the document takes. Technical communicators must learn not to own documents in the same way they own academic writing or personal writing. In fact, in most technical communication work, the communicator will not be identified at all, so the writing is, truly, egoless (Weiss 2002).

Reviews are generally carried out in one of two ways: individual reviews and tabletop reviews. Individual reviews, the more common of the two, involves the technical communicator sending to the reviewers a paper version of the document, or, if it is an electronic document, the files or a link to the files. Frequently, a transmittal sheet accompanies the document, reminding reviewers of the deadline date for returning the draft and pointing out any special concerns about missing information or incomplete sections that reviewers should focus on. Reviewers mark up the paper copy or add comments to the electronic files with corrections of factual errors, supplying missing information, copyediting, style suggestions, and anything else that they believe would improve the document.

The tabletop review involves all interested parties meeting and going through the document, page by page or screen by screen. This is most efficient when reviewers have previously received the draft and had time to do individual reviews prior to the meeting. Sometimes there is not enough time to do so, which means that the review team reads through the document during the tabletop session. The technical communicator takes notes on suggested changes to use when making the revisions.

The first draft of a document is often the first place that includes an overall picture of the product or service that is being developed or the communication that is being put together. For many of the reviewers, it will be their first overall view of the project's intended output. In some cases, serious disagreements can break out about the direction of the product, service, or communication. Technical communicators and the team members work together to compromise and to arrive at a final configuration that will work as well as possible from the varying perspectives of the team members.

REVISION

At the conclusion of the review process, the technical communicator is left with a pile of marked-up paper documents or one or more electronic files with reviewers' comments. The paper documents require that the communicator conduct what amounts to an individual tabletop review, going page by page through all of the documents at once—I have had paper documents reviewed by as many as twelve subject-matter experts. The communicator creates a master copy to which he or she marks up all of the suggested changes from reviewers, making special note of those places where reviewers' comments are unclear and require further elaboration or where two or more reviewers' comments conflict. Those incidents will require that the communicator get clarifications and resolve conflicts with the reviewers. In the case of electronic documents, some publishing systems support comments from multiple reviewers, so that there is only one electronic file, and each reviewer goes in and adds his or her comments. This has the advantage that the reviewers see each other's comments and can perhaps solve any conflicts during the review process. However, these systems tend to become unwieldy when more than a handful of reviewers use them.

The technical communicator takes the master paper document or electronic file and makes the suggested changes, adding any new information that has been developed while the document was out for review. The document then typically goes out for at least one more review cycle. While the default number of review cycles may be two, many types of documents, such as grant applications, proposals, sales materials, policy statements, and government filings, may require multiple reviews over a period of weeks and even months.

With the last review, technical communicators often perform some kind of checking or testing to help ensure the quality of the documents. Many organizations employ quality checklists against which an editor, peer reviewer (another technical communicator), and/or supervisor goes over the documents to verify that they meet organization standards for technical accuracy, organization, grammar and spelling, page layout, style, consistency, any industry standards that must be met, and any other criteria important to their audience and their industry. For hardware and software, some form of verification testing is usually performed, wherein the technical communicator, an editor, or perhaps a designated testing expert goes through the document and performs every task in order, to ensure that the document is accurate and contributes to the safe, effective use of the hardware or software.

Another form of testing, usability testing, is sometimes performed also. Here the document is tested with users or other people who are as similar as possible to the audience, to see if they can successfully and efficiently use the product to perform the tasks that users are most likely to want to perform, and to assess their attitudes toward the product. Usability testing is better done early in the development process when there is still time to make significant changes if major problems are uncovered, but because it is difficult to test with a less than complete product, the testing is often done around the same time that verification is done, sometimes in conjunction with the verification testing.

At some point, either because a deadline has approached and no further changes can be made, or when everyone on the review team agrees that the document is ready for release, the technical communicator creates the final copy and prepares it for production.

PRODUCTION

Production is the process of preparing the final information product for use by its intended audience. Production can be as simple as saving a file as a PDF and posting it to a web server, or as complex as having multiple sets of documents printed, copied to CDs or DVDs, and posted online in multiple formats. The process can take a few minutes or several months. Especially for complex projects requiring creation of several documents and several media, production must be treated like a project in itself, with milestone dates for tasks such as delivery of files to printers, shipment of printed materials to suppliers, linking of help files with associated programs, and mounting of files on servers.

Other tasks often done during the production stage include indexing and translation/localization, which should also be treated as projects within the larger project. While final indexing of a paper document cannot, obviously, be done until the document and page numbers are final, many of the tasks for creating an index can and should be done earlier, such as coming up with the terms to be used, inserting synonyms that users are likely to know with cross references to the terms used for the product or service, inserting index entries directly into the document file (if the publishing software being used allows), and deciding which entries for a term are the most important. Translation and localization refer to both translating documents into a language and making sure it will work with the versions of the language for which it is intended (for example, South American Spanish versus European Spanish). While translation cannot be completed until the document is complete, tasks associated with it can be done earlier, such as providing the translator with a glossary, sending drafts so the translator can estimate the time required and send a sample translation back for checking, and writing with a minimal vocabulary to make translation easier.

For paper documents, production requires printing. This can be as simple as making copies on a copy machine or printing multiple copies on a laser or inkjet printer. It can also involve working closely with a printer, specifying cover-paper weight, text-paper weight and type (clay, rag, cotton, and other materials are added to paper to give it different textures), binding type, and all of the other aspects of printed documents. Some large organizations have a separate production group that takes care of all of the production activities, while many smaller organizations must either hire outside firms or do the production themselves. With improvements in production technology, including inexpensive color laser printers, technical communicators can now perform production tasks that once were done on expensive printing presses, and many organizations have taken advantage of the technologies to drastically lower their production costs. They can achieve even greater cost savings by moving from printing to dissemination via various electronic media.

DISSEMINATION

In the dissemination stage, documents are delivered to their intended audiences. In some organizations, technical communicators are involved with ensuring that printed documents are delivered to the factory or warehouse from which they will be shipped. They may also be required to copy help, document, and website files to servers, so that customers can access them.

HOW TOOLS AND TECHNOLOGY ARE USED

Technical communicators generally use several tools for product management. The four most important technologies are spreadsheets, project management software, web-based collaborative and project-management systems, and time-management systems.

Project management involves tracking dates and budgets, and it is natural that spreadsheets are often used to do so. There are many spreadsheet templates available for project management, both delivered with the spreadsheet software and available on the web, ranging from free, simple templates to elaborate collections with sheets for nearly every conceivable project requirement. With projects that last a few months or less and that have budgets in the hundreds of thousands or less, spreadsheets can provide sufficient automation for project management.

With larger projects that last longer, dedicated project management software systems are often used. Microsoft Project is one of the most popular, but there are many alternatives, which range from free, fairly simple systems (see zoho.com, for example) to massive systems designed to manage large multiyear, multibillion-dollar projects such as shopping malls and power plants. Most technical communicators need only a spreadsheet, but if the larger organization uses another system, communicators should learn the system and use it for tracking their time and expenses, just as other developers do.

Increasingly, web-based solutions are appearing for collaborating on projects and keeping track of schedules, budgeting, exchanging documents, communicating, and managing all other aspects of a project (see eproject.com and basecamphq.com, for example). With more and more work teams distributed geographically rather than centrally located, and with broadband speeds increasing, a web-based management system works much more effectively than a system on a local network that cannot be accessed by those who are working at other locations. Web-based systems allow team members to be dispersed around a country and internationally. The systems provide messaging systems both for instant communication and for leaving messages and files for someone who is in another time zone and not working at the same hours.

Because most technical communication work is project-oriented, technical communicators work against a number of deadlines when drafts and final versions are due. A technical communicator who is working on several projects, not an uncommon scenario, may have dozens of dates to keep up with and deliverables to send out over the course of a few months. On top of the delivery dates, the technical communicator may have numerous other dates to track, including multiple meetings and appointments each week. Further, many communicators find it valuable to maintain lists of "to do" items, tasks that they need to perform in the short term and over the longer term. The short-term tasks are usually associated with the immediate projects they are working on, while the longer-term tasks are usually more conceptual and concern goals that will require protracted effort and cannot be completed in a few minutes or hours.

Keeping track of all of that data quickly becomes impossible without some kind of system, so most technical communicators use some type of time management system. There are several paper systems available for project management. Such systems can be maintained on paper, although for people with multiple projects and tasks, paper-only systems can become cumbersome. FranklinCovey (at franklincovey.com) and Day-Timer (at daytimer.com) make the two best-selling systems, which include notebooks from pocket size to full-sized 8½″ × 11″ three-ring binders, accompanied by tabs and forms that allow record keeping for almost any conceivable time and task. The Franklin Covey system is based on Steven Covey's famous book on time management, *The 7 Habits of Highly Successful People*. Another popular paper system (Allen 2008) is based on David Allen's *Getting Things Done*.

There are several types of more automated systems. Many e-mail programs include the ability to track dates and to add "to do" lists. Some, such as Microsoft Outlook, allow members of a development team to see and access each others' appointment calendars to make scheduling easier. There are also software add-ons for e-mail programs, such as Franklin Covey's PlanPlus, which runs with Outlook or independently in Windows. There is also a Getting Things Done Outlook add-in, based on Allen's book.

Some technical communicators prefer to use personal smartphones with appropriate software to keep track of their appointments and "to do" items, so they will always have the information with them. Others prefer to keep their project management systems on their laptops. Some use multiple systems, with a main system on their desktop computer and daily printed sheets to go in a pocket notebook.

Technical communicators should develop a time management system that allows them to meet all of their deliverable dates and to be aware of their meetings, appointments, and action items. If their larger organization uses a standard method, they should strongly consider using the same method.

ALTERNATIVE PROJECT MANAGEMENT MODELS

Performing all of the tasks outlined above in the traditional project management model can require considerable time and effort. This waterfall model is well understood and comfortable for many developers, but it is not the most efficient model, and following it can mean that organizations design products and services that are not appropriate for their audiences. Hence, several alternative models for managing projects have been developed. These models are often referred to as agile methods, even though one of them is also specifically labeled agile development. Practitioners now refer to any method that seeks to abandon the waterfall method and use more streamlined processes as "agile," whether they happen to align with the "official" agile methods espoused by the Agile Alliance or not. Three such models are user-centered design, agile development, and extreme programming.

User-Centered Design

In user-centered design, rather than developing the engineering specifications or program requirements used for waterfall methods, developers create user cases, which means that they put together typical strings of tasks that users would need to employ in their real-world work with the product. For example, a user wanting to employ a spreadsheet program to prepare an annual budget would have multiple tasks: filling in the columns for the months, listing the budget items in the left-hand columns, populating the fields with the actual budget amounts, inserting formulas at the ends of columns and rows to add up annual totals, saving the file, and printing the document. For each of those tasks, developers prepare a module, test it with users to make sure it works, redesign and test again if it does not work, and repeat the cycle until the task can be successfully completed by users with the required speed and efficiency. Due to its repeated development and testing cycles, this model is also often called iterative design.

Agile Development

In agile development, audience members and developers create "stories" that are scenarios of user tasks to be performed and features needed to perform those tasks. Such stories are similar to the user cases developed in user-centered design. The stories also describe how the processes will help the users achieve their desired results. The developers then design modules that lead to those desired results for each task, with constant customer interaction and testing to make sure that the design will meet customer needs. The modules are developed in intensive, short work cycles. At Lulu.com, for example, agile developers, including the technical communicators, work in two-week cycles, so that each module is begun and completed in two weeks; in rare cases, it might take longer (Jackson Fox, personal interview, February 19, 2009). Agile teams work together intensely, preferably in the same physical space, and are largely self-managed. Agile development is an iterative design method, and agile developers assume that each module they work on will undergo several rounds of development and testing. This may sound as if it would be difficult for technical communicators, because they have to try to explain processes that are being developed rapidly. However, agile development teams work together very tightly and often include the technical communicators right from the beginning. Because technical communicators work on each module straight through, they can contribute to the overall design of the product and better understand the philosophy behind the product and its operation.

Extreme Programming

Extreme programming is similar to agile development in that it is based on user stories. The stories feed into the design in the planning process and again in the testing process, when test scenarios based on the stories are used as the criteria for determining whether the product has achieved customer approval. Again, iterative design is used, with constant cycles of development and testing until each part of the product meets the customer test scenarios successfully. An interesting aspect of extreme programming is that it calls for developers to work together in pairs, which ensures that if someone leaves, there is still someone who understands what that person was working on. As in other agile development methods, extreme programming teams work closely together, and there is great transparency; all team members can see one another's work at any time. Nuckols and Canna (2003) describe a process they call extreme documentation, and they detail numerous advantages for technical communicators using such a system rather than the more traditional waterfall methods. On their extreme project, they sat with the programmers, which they reported as improving communication and respect among the team members. They also worked with the same customer stories and acceptance tests as the programmers, and developed and tested documentation modules alongside the programmers' development of the software modules. The result was less time pressure as the project neared completion and a greater sense that what they had developed was "complete and accurate" (9).

While most project work is still done using traditional waterfall methods, the newer, more agile methods are being employed more often to reduce schedule length and to get more usable products and services delivered more quickly. Technical communicators should know the steps for the traditional methods, but many will have to learn to manage products using alternative management models.

EXTENDED EXAMPLE

So how will Ann Ross manage her three projects: a web-based portfolio, the paper and online documents for the point-of-sale system, and the marketing materials? She will do so by working her way through the stages we have discussed here, going from planning through dissemination.

For the web-based portfolio, Ann is gathering information about the planned portfolio management program. She does so using several methods, including sitting in on programmers' planning meetings, reading over the original requirements document specifying how the system will work (including rough sketches of its user interface), looking at existing portfolio management programs on which the online one will be based, in part, and meeting with two customers who currently use her company's stand-alone portfolio management program delivered on a CD. When she has finished her research, Ann will prepare an information plan specifying the overall structure for the help system, including its main headings, a list of all the topics that will be included, an estimate of the time it will take to prepare them, and the milestone dates by which she will deliver the topics to programmers and test personnel for their review. Once the programmers, testers, project manager, customer support manager, and marketing representatives on the project team sign off on her plan, she can begin work developing the topics.

For the small point-of-sale system aimed at establishments with one or two cash registers, Ann has gone through the planning stage, where she sat in on marketing's focus group with small-store owners and listened to their responses about how they used their point-of-sale systems and which tasks were most important and were used most often; where she read over the engineering and software specifications for the system; and where she regularly attended meetings with hardware engineers and software programmers to learn of their plans for the system. She developed her own plan, which was approved by all, and has now nearly finished the first draft of the system. She has a firm set of dates for when the first draft will be delivered, and when she receives review copies back with corrections and comments, she incorporates all of those and makes further revisions based on the product development, delivers the second draft for review, receives "final" comments, prepares an index, and performs a final quality check before sending the final copy to the printer and saving it as a PDF file for online delivery. She also has dates for when she will send the final draft to translators in France and Germany, receive the first-draft translations, have the drafts reviewed by employees fluent in those languages, send the translations back for a second revision, and receive the final translations for printing and conversion to PDFs for online delivery.

For the marketing materials, Ann went through the planning stage by reading through marketing's description of the large department stores and chains that purchase multiple, networked point-of-sale systems and by attending joint marketing and engineering meetings when details of the system were being negotiated. After her plans for the materials were approved, she sent them through several review cycles with marketing, which, because of how much revenue such systems produce, wants to get the materials just right. She had originally planned for only two reviews, but the extra review cycles caused by marketing's lack of certainty have caused her to work overtime for the last three weeks to meet the final deadline date, which did not change. She has dates by which she will deliver the materials to the printer, go to the printer during the print run to do spot checks for quality, receive the final printed copies, do further spot checks to ensure quality, and deliver the materials to marketing for dissemination.

Even though the three projects involve three very different information artifacts—an online help system, a paper user's guide, and a set of

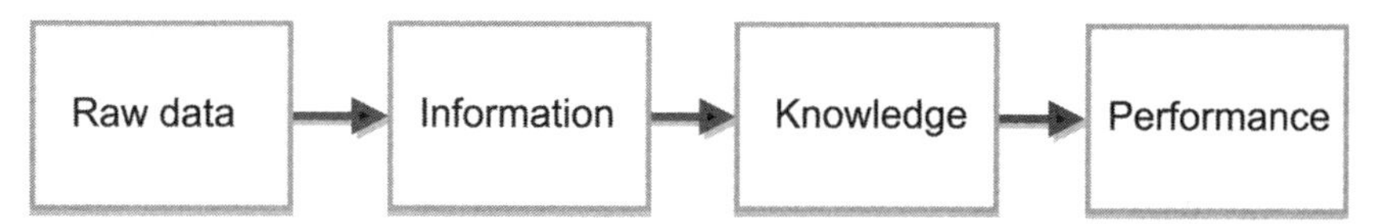

Figure 13.2. Continuum from data to performance

marketing brochures and flyers—Ann will follow the same basic set of steps for each of the projects. She plans and conducts very active research to gather raw data about the various systems. She imposes some sort of structure on that raw data so that it moves from being raw data to information. She adds further structure to that information and delivers it in various media so that it provides knowledge to its readers. In some cases, she has to provide that knowledge in ways that lead to performance of tasks by those readers. Hence, she goes through the process of gradually converting raw data into something that allows users to perform various tasks, following the conversion sequence in figure 13.2.

She has used a consistent process to achieve that transformation, including steps for planning and conceptualizing, imposing structure based on audience needs, developing drafts and sending them to other team members for review, revising and sending again, preparing the materials for production, managing the production and translation processes, and delivering the completed online and hard-copy media. To complete all three projects successfully, Ann has to manage her time effectively, switching from one project to another as information becomes available and as deadlines approach. Without a time management system that keeps a record of her "to do" list and the dates by which all of her deliverables are due, she would have no chance of getting everything done on time.

CONCLUSION

Estimates are that technical communicators spend only 20–30% of their time writing (Carliner, quoted in Hoffman 2008, 45–67). That means that they must spend much more time managing the other aspects of the projects on which they work. Those aspects include the steps that this chapter has followed: planning, research and information gathering, composition/invention, reviewing and/or testing against quality criteria, revision, production, and dissemination. Even if technical communicators are using alternative overall project management techniques such as agile development, they still go through all of the necessary steps. They do so perhaps in greatly compressed time periods and on much smaller modules, but to design and develop quality information artifacts, they still have to perform the basics. While excellent writing and editing abilities are necessary skills for technical communicators, they are not sufficient. To be successful and to create quality products, technical communicators must develop skills in project management and time management that allow them to complete their projects successfully.

DISCUSSION QUESTIONS

1. How do recent technological developments such as social networking and media affect the nature of technical communicators' work and the methods they use to manage their projects? Using such networking, would they still have to go through all of the steps listed in the heuristic? Or are there some steps that using social media would allow a technical communicator to eliminate?
2. Following up on question 1, how would the heuristic apply specifically to a blog, a wiki, or a forum, where development of information is not planned in any traditional way but rather is ongoing, iterative, and cumulative? Is it possible to replace traditional information types (documents, online help systems) with newer social media? If so, how would that affect the way technical communicators manage their projects?
3. Which of the major steps described in this chapter do you think will be hardest for you to learn? Are there any that you have not had to do in your courses and that you will have to learn when you get to the workplace? Where else might you go to learn more about them?
4. In the planning stage, how does a technical communicator read and understand engineering specifications or programming requirements without being an expert in the field? How does a technical communicator learn enough about the technology to be able to read and understand such documents? For the writing assignments you have done in your classes, how did you learn about the products you were developing information about?
5. For one of the course assignments in this class (or, if necessary, for another class), write a brief, annotated outline describing how you will do each of the main steps listed in the heuristic in this chapter. For each of the steps, include several sentences explaining how you will perform the step for your assignment. Bring it to class for further discussion.
6. Technical communication jobs are considered writing jobs, yet this chapter talks about all kinds of user analysis, planning, project tracking and testing, and other activities. Are there any jobs where technical communicators simply write and someone else does all of the non-writing work? See if you can find two technical communication job openings, one that stresses writing and editing skills, and a second

that stresses project management skills, and bring them to class for further discussion.

7. Why does a technical communicator need to know all of this? Isn't there project management software that handles it? Find three online project management sites and print the home page of each. Which one looks like it would be easiest to learn and use? Do the sites accommodate following the steps described in this chapter?
8. There is a lot of discussion in this chapter about reviewing and testing and revision. Don't technical communicators usually write a draft, have it reviewed once, make a few corrections, and then send it to production or straight for dissemination to the audience? Why would a project require multiple reviews and rewrites?
9. More and more organizations are publishing their information online. Shouldn't technical communicators be able to reduce the production stage to simply saving something as a PDF and posting it to a website? Are there some products for which online documents simply do not work?

WORKS CITED

Agile Alliance. 2011a. "Principles behind the Agile Manifesto." http://www.agilemanifesto.org. Accessed October 27, 2011.

———. 2011b. "What Is the Alliance?" http://www.agilealliance.org. Accessed October 27, 2011.

Allen, David. 2001. *Getting Things Done*. New York: Penguin.

———. 2008. "Getting Things Done and GTD." http://www.davidco.com. Accessed March 31, 2011.

Allen, O. Jane, and Lynn H. Deming. 1994. *Publications Management: Essays for Professional Communicators*. Amityville, NY: Baywood.

Barnum, Carol M. 2001. *Usability Testing and Research*. New York: Pearson / Longman.

———. 2011. *Usability Testing Essentials*. Burlington, MA: Elsevier.

Beck, Kent, and Cynthia Andres. 2004. *Extreme Programming Explained: Embrace Change*. 2nd edition. New York: Addison-Wesley Professional.

Carliner, Saul. 2004. "What Do We Manage? A Survey of the Management Portfolios of Large Technical Communication Groups." *Technical Communication* 51 (1): 45–67.

Covey, Stephen. 1989. *The 7 Habits of Highly Successful People*. New York: Free Press.

Day-Timer. 2008. "Day-Timer: Day Planners and Organizers Direct from the Manufacturer." http://www.daytimer.com. Last accessed March 31, 2011.

Dicks, R. Stanley. 2004. *Management Principles and Practices for Technical Communicators*. New York: Pearson / Longman.

Duarte, Deborah L., and Nancy T. Snyder. 2006. *Mastering Virtual Teams: Strategies, Tools, and Techniques That Succeed*. San Francisco: Jossey-Bass.

Dumas, Joseph S., and Janice C. Redish. 1993. *A Practical Guide to Usability Testing*. Norwood, NJ: Ablex.

FranklinCovey. 2011. "Planning and Organizing Products." http://www.franklincovey.com. Accessed March 26, 2011.

Gould, John D., and Clayton Lewis. 1985. "Designing for Usability: Key Principles and What Designers Think." *Communications of the ACM* 28:300–311.

Hackos, JoAnn T. 1994. *Managing Your Documentation Projects*. New York: Wiley.

———. 1997. "From Theory to Practice: Using the Information Process-Maturity Model as a Tool for Strategic Planning." *Technical Communication* 44:369–381.

———. 2007. *Information Development*. New York: Wiley.

Hackos, JoAnn T., and Janice C. Redish. 1998. *User and Task Analysis for Interface Design*. New York: Wiley.

Hoffman, Alan. 2008. "Technical Writing." http://www.monster.com. Accessed March 31, 2011.

Houser, Rob. 2001. "Why We Should Archive, Share, and Analyze Information about Users." *Technical Communication* 48:176–181.

Larbi, Nancy E., and Susan Springfield. 2004. "When No One's Home: Being a Writer on Remote Project Teams." *Technical Communication* 51:102–108.

Mayhew, Deborah. 1999. *The Usability Engineering Lifecycle: A Practitioner's Handbook for User Interface Design*. San Francisco: Morgan Kaufmann.

Nielsen, Jakob. 1993. *Usability Engineering*. Boston: Academic Press.

Nuckols, Carl E., and Jeff Canna. 2003. " Extreme Documentation." *Intercom* 50 (2): 6–9.

Polson, Judith J. K. 1988. "A Model for Management: Defending Yourself against Murphy." In *Proceedings of the 35th ITCC, STC Conference*, MPD-100-103. Washington, DC: Society for Technical Communication.

Rubin, Jeffrey, and Dana Chisnell. 2008. *Handbook of Usability Testing*. 2nd edition. New York: Wiley.

Smith, David L. 1994. "Estimating Costs for Documentation Projects." In *Publications Management: Essays for Professional Communicators*, edited by O. Jane Allen and Lynn H. Deming, 143–151. Amityville, NY: Baywood.

Warfel, Todd. 2006. "The Task Analysis Grid." http://zakiwarfel.com/archives/the-task-analysis-grid. Accessed May 26, 2011.

Weiss, Edward. 2002. "The Retreat from Usability: User Documentation in the Post-usability Era." *Journal of Computer Documentation* 19 (1): 3–18.

Whiteside, Aimee L. 2003. "The Skills That Technical Communicators Need: An Investigation of Technical Communication Graduates, Managers, and Curricula." *Journal of Technical Writing and Communication* 33:303–318.

Zvalo, Peter. 1999. "Pricing a Documentation Project Is Part Science, Part Art." http://www.writersblock.ca/summer1999/busword.htm. Accessed October 27, 2011.

PART 4 DEVELOPING FIELD KNOWLEDGE

Throughout your career, you will move back and forth between specific and broad issues facing people in the field, between contexts for work and applications and approaches for doing work. In this part, you will see ways to focus on advancing knowledge about technical communication by bringing in perspectives from related fields, translating (and often challenging and revising) knowledge and skills from these other fields to help you solve novel and emerging problems in your own work (figure P4.1). The questions addressed here provide you with a new set of perspectives and tools for rethinking some of the approaches and concepts we started out with, helping you to make them even more flexible and useful.

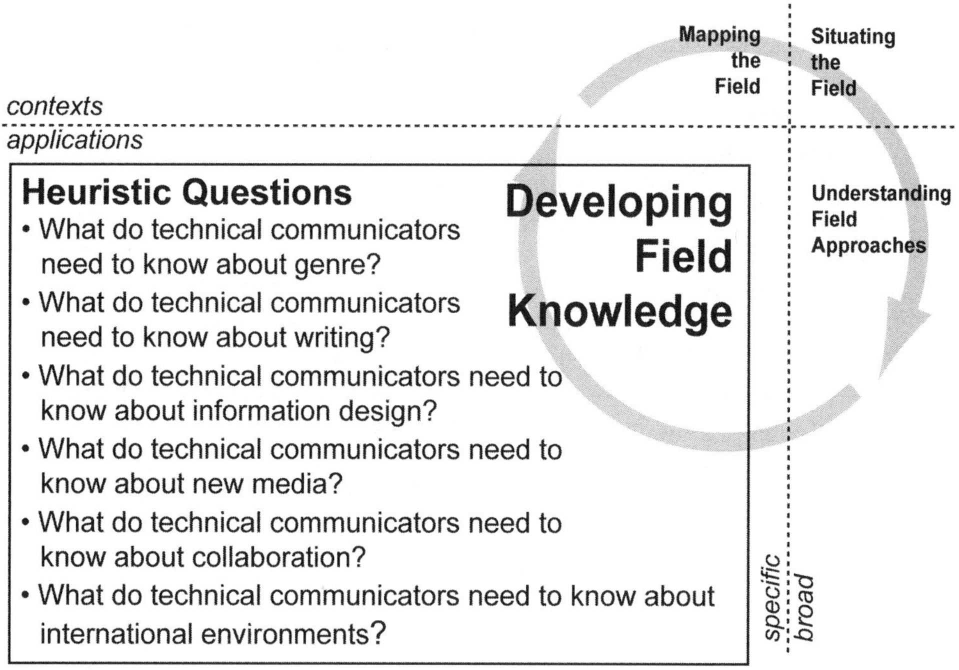

Figure P4.1. Heuristic questions for Developing Field Knowledge

In "What Do Technical Communicators Need to Know about Genre?," Brent Henze begins with a very basic but often confusing question: What genre (or type of document) should a technical communicator develop to address a problem? As the genre specialists that Henze draws on argue, this issue is more complicated than it seems, because document types and formats change from one context to another, and we often fail to realize it. Only the simplest problems and contexts allow you to take a type of document from one context and use it to easily solve a problem in another context. Successful technical communicators, Henze points out, need to examine genres in their original contexts, understand how they work there, and carefully revise those genres (often combining them with other genres) in ways that meet the needs of the new problem context.

In reexamining the issue of what format documents should take, you can also begin to see that technical communicators often do very different types of writing and communication from one project to the next or from one organizational context to another. In "What Do Technical Communicators Need to Know about Writing?," Ann Blakeslee and Gerald Savage draw on surveys of working technical communicators to illustrate the wide variety of types of writing that go on in the workplace. Blakeslee and Savage go beyond just describing the state of the field, however. They use this data to show ways you can answer these questions for yourself in new work situations: What type of writing do I need to learn to do *here* in order to be successful? How do I learn how to do it?

Although the term "writing" might have once worked to describe what technical communicators do, in many cases "writing" has expanded from lines of words in sentences and paragraphs to the much larger and more general issue of information design. In "What Do Technical Communicators Need to Know about Information Design?," Karen Schriver draws from empirical research on the design of both texts and graphics to construct heuristics for designing complex documents that are usable for people in specific situations. Her heuristics provide guidance on a necessarily broad range of issues, including the needs of specific kinds of users, the arrangement of visual elements on a page or screen, the selection of typography for different purposes, and the relationship of design elements to each other.

In "What Do Technical Communicators Need to Know about New Media?," Anne Wysocki continues the exploration of complex texts, examining how the added dimension of new media both enriches and challenges the work of technical communicators. Wysocki describes the dramatic changes going on in the media—as we move from static to interactive documents—that demand new approaches from technical communicators. The heuristics in this chapter not only deal with issues of learning new software and hardware, but also move on to more difficult but still crucial issues, such as the ways that new media invite different responses and responsibilities from designers and users.

As technical communication diversifies its strategies, structures, and media, projects in the field have begun to involve not only other technical communicators and managers but also team members in other fields who contribute their expertise. In "What Do Technical Communicators Need to Know about Collaboration?," Rebecca Burnett, Andrew Cooper, and Candice Welhausen discuss the challenges facing technical communicators who work in cross-disciplinary teams. In order to help you learn to navigate the multiple levels of complexity involved here, such as complicated projects and team members with varying (and sometimes contradictory) priorities and vocabularies, Burnett, Cooper, and Welhausen provide a coherent set of heuristics for working productively on team-based projects.

At the close of this part, Kirk St. Amant depicts an even broader world facing technical communicators: the international community, comprising both corporations that produce technical communication and people who use those products. In "What Do Technical Communicators Need to Know about International Environments?," St. Amant reminds us that international technical communication relies on many of the same principles of rhetorical sensitivity to audience and context; at the same time, answering those questions effectively in an international setting often requires the ability to see things in new ways. The heuristics provided in this chapter will help you develop new ways of seeing that are sensitive to the globalized world in which we all now live.

BRENT HENZE

14 What Do Technical Communicators Need to Know about Genre?

SUMMARY

Just as individual members of a language group depend upon every other member's familiarity with the language to facilitate group communication, writers and readers of documents depend upon shared textual conventions, applied in recognizably similar situations, to accomplish their goals. Those shared conventions of text and situation are "genres." Genres can simplify the technical communicator's work by constraining the range of possibilities in a given communication situation, and they can encourage innovation by helping technical communicators understand the goals of a text and envision a range of ways to achieve those goals.

An understanding of genre offers many benefits to the technical communicator. Genres can help technical communicators diagnose a document user's needs and produce documents that respond to those needs in situationally appropriate ways. Since genres arise as successful responses to recurring situations, careful study of a workplace's genres can also help technical communicators perceive how problems are solved and how work gets done in their workplace. Additionally, by understanding a workplace's existing genres, technical communicators can make more informed choices about the need to modify genres (or even to create new ones) in response to emerging workplace situations, technologies, and relationships.

INTRODUCTION

James (not his real name) is an economic development specialist for a county development commission. He helps county businesses with access to the county's private and governmental resources, provides research data for businesses and government agencies, and helps develop new programs to promote economic growth in the county.

James's boss asks him to survey county businesses regarding their resource needs and their satisfaction with the agency's services. His audience includes existing industries, community leaders, economic allies, and business-service providers. He has been trained as a technical communicator and has some familiarity with writing questionnaires for others, but he has never written a professional report for such a complex audience. He knows that he needs to address that audience, please his boss, and write something that will represent his organization well.

Faced with this new writing situation, how does he begin? In the rest of this chapter, I consider how technical communicators like James make sense of new writing situations, using an understanding of genres to guide their decision making and problem solving. I begin by summarizing several important developments in modern genre theory, focusing on the principles of social action, typification, choice, change, competence, correlation of genres, genre systems, and flexibility and originality. This theoretical understanding of genre offers both diagnostic and productive benefits that can help you diagnose communication more effectively by training you to think in terms of larger systems of communication, and of how communication operates on the social and cognitive levels.

Based upon these theoretical insights, I provide a set of heuristic questions that new technical communicators can use to diagnose their communication situations as "generic situations" and to develop an effective response. I'll return to the case of James to see how he can use this heuristic both to prepare an effective report and to improve his understanding of communication and decision making in his workplace.

THE RHETORICAL GENRE APPROACH: A LITERATURE REVIEW

The concept of genre has been around for thousands of years (Aristotle discussed genres in his *Rhetoric*). The most recognized usage of the term describes formal categories of literary or artistic texts: sonnets, problem plays, mystery novels, country love songs, and so forth. This "formal" approach to genre (that is, this approach that focuses on the form of a text) is useful for creating a taxonomy of text types, and even for understanding the relationship between different categories and subcategories of literary texts. (For instance, you could use it to explore the relationship between different kinds of novels: the detective novel, the political thriller, and the police procedural all might be considered as species of the genre of mystery novels.)

This formal approach can also be applied to nonliterary texts, of course. For example, in business writing, the genre of the business letter can be identified by a set of formal characteristics; it is generally a typed letter on company letterhead, using a friendly but formal style, and containing the following ordered elements: date, inside address, salutation, polite introduction, at least one paragraph of body text, conclusion, closing line, and signature. The genre of the business letter may also be broken down into several subgenres: good-news letters, bad-news letters, informational letters, cover letters, requests, and so forth. Each of these types of business letter may follow the generic form of the business letter, but also may possess additional identifying characteristics or patterns. For instance, a bad-news letter typically begins with a "buffer" paragraph, followed by a bad-news paragraph, one or more paragraphs of reasons, an optional paragraph with alternatives, and a polite closing.

The formal approach to genre is useful as a strategy for categorizing text types, and even as a source of templates to follow when producing texts. The approach tends to be prescriptive, however, and it oversimplifies the writing situation, making it seem as if a "cookie-cutter" approach to communication is adequate. More importantly, it does little to help us understand how writing is used to achieve specific purposes.

In the past three decades, researchers of rhetoric and technical communication have retooled genre as a rhetorical concept. In its rhetorical formulation, genre is seen not as a text type but as what Carolyn Miller (1984) has called a "social action"; or to put it another way, genres are one of the things technical communicators use to fulfill a specific type of purpose within a particular, recognizable, and recurring situation. What's more, the purpose and situation in question are both "socially situated"—there is a shared community understanding rather than merely an individual definition of the purposes and situations of genres. The community in question might be pretty big (college students) or relatively small (sales staff at the Celltech Company); the point is that members of the community all recognize certain types of communication situations and agree upon what sort of speech or writing would count as an appropriate response to that situation.

Following Miller's evocative description of genre as social action, the research and theory of genre has blossomed. I won't describe all of this work here, since much of the work seeks to develop keener analytical and critical tools, as well as keener teaching perspectives, rather than to develop new communication practices. For present purposes, I will focus on the major insights relating to the practice of communication in workplace settings, especially for novices—people who are just learning about their profession, their discipline, or their work setting. And there is much to consider here.

SOCIAL ACTION

First, let's consider what is meant by "social action" (and why it matters to view genre as a category of social activity rather than a category of textual form). All communication is social; we're always communicating with somebody when we write or speak. Even when we don't have a particular reader in mind, we're writing with at least a loose understanding that the thing we're writing will be read by someone, and that in order for it to be understood, we need to relate our intentions to the intentions of the reader.

But admitting that all communication is social isn't the same thing as admitting that genres are social. We need to take one more step. Whereas "communication" encompasses just about any form of symbolic expression between two or more people, genres are what Miller (1984, 24), among others, describes as typified rhetorical action. By "typified," genre researchers mean that the participants in a community recognize patterns in their communication forms and the situations in which communication is necessary.

For example, if you're in a classroom setting and your instructor asks you to introduce yourself to the class, you are in a typical rhetorical situation—one that is recognized by you and the others in the room as a common situation in this setting. Even before you say a word, you're already being acted upon by a genre—the genre of the "class introduction." How do you know what to say? You know because you've been in a similar situation before, either as a speaker or as a listener. Because the situation is familiar, you can also make some other assumptions about the situation. For instance, among all the many communication possibilities available to you, you can assume that you're being asked to produce a certain very specific type of response (a list of unembarrassing personal characteristics about yourself, beginning with your name, probably including your major, and lasting about ten seconds). In this situation, you recognize that people would be surprised if you passed around a copy of your résumé, or if you broke into song. Even if these responses communicated roughly the same information, they'd be surprising forms of response (to say the least) in this situation. That is, you recognize that your listeners would find your response unsuitable—as a communicative action, it falls outside the parameters of the genre.

But you also can count on something else. What is "typical" isn't only the form of your response, but also the situation itself. In planning your reply, you can count on the fact that your listeners will recognize the appropriateness of your response in this situation (assuming that you respond appropriately, that is!). Change a detail of the situation—from a technical communication class to a performing arts class, for example—and what counts as an appropriate response (a list, a song) also changes.

TYPIFICATION

What is it that makes any situation typical? There has been a lot of attention to this question. After all, strictly speaking, every situation is unique—you've never before been in that same room with the same instructor, the same group of students, on the same day, at the start of the same class, introducing yourself to those people for the first time. If every situation is strictly unique, then no response is ever truly a response to a "recurring situation," right?

Here's where the notion of typification comes in. According to Miller and others who have developed this concept, typification is not an individual process of identifying objectively identical situations, but rather a sociocognitive process in which situations come to be treated as examples of a type. Every situation is (in reality) different; what makes them the same is that we see them as the same—and, further, that our method of doing so is shared by others in our social group. After all, if you think that you're being asked to give your name and describe your "public self" to your instructor and fellow students, but nobody else in the room perceives the situation in those terms, then you're simply speaking out of turn, not participating in a shared understanding of the situation (that is to say, not employing the correct genre). But if the members of the group are all roughly familiar with how this situation works, then collectively the group will "typify" the situation and its responses as part of a recognized genre.

Let's look at a workplace example. Suppose that you're hired as a technical communicator for Celltech, a cell-phone manufacturer. As your first assignment, you're asked to write a "quick-start" sheet for unpacking the phone, installing the battery, hooking up the charger. You dutifully clear your desk, tinker around with the product until you're comfortable with how to perform the tasks, and then write up a list of the steps that you followed. You make some line drawings illustrating various steps, you arrange the text and illustrations on the page, you give the document a once-over, and then you submit the completed list to your boss: mission accomplished. The boss takes one look at the result and says, "That's not right—the procedure might work, but it's not a quick-start guide. You have too many steps, and it makes the product look too complicated. Plus, where did you get these pictures? That's why we have a graphics department."

What went wrong here isn't that your text failed; rather, the text was not recognized as an appropriate response within the specific rhetorical situation that you were responding to. After all, you weren't writing instructions in a vacuum; you were writing them in a specific context (Celltech), to be consumed by specific sets of readers (Celltech's customers, but, first, your supervisor and other members of the Celltech community).

How might you approach this situation differently? First, you'd inform yourself about how quick-start sheets are written at Celltech. You'd learn the genre, along with the variations that tend to meet the needs of this specific circumstance, however idiosyncratic they may be. For example, after reviewing several other instruction sets, you might discover that all of these documents use photos instead of line drawings, positioned in the same way on the page; they all use more or less the same verbal style—formal but friendly, with all instructional steps numbered and written as imperative statements. (You judge that these are firm genre conventions in this writing situation.) But there are variations in other areas. The length of the instructions sometimes varies considerably from one product to another. The text under each numbered instruction is a lengthy description in some cases, but shorter (or altogether absent) in others. Some products use two-color illustrations; others are printed in black and white. These variables, you judge, are less firm, indicating either that the technical communicator is free to play with these variables, or that there may be some underlying pattern corresponding to these differences—a pattern that you would need to figure out through further investigation.

In producing a quick-start guide for your company's product, you are not just writing a lone document. You are performing a well-defined and recurring social act that is recognizable to the other members of the work community that you've just joined. In a very real way, it's the whole community that is producing these instructions (by regulating what counts as the "right way to do it," among all the other possible ways), not you as an individual technical communicator. But by performing this social action, you're also identifying yourself as a participant in the group—someone who recognizes the recurrent situations that determine work in this setting, and who can make effective choices about how to respond to those situations in socially valid ways.

GENRE AND CHOICE

The next major insight of recent work on genre has to do with the issue of choice. The formal view treats genres almost as if they were templates: do *x*, *y*, and *z*, and you've written a business letter or an instruction set—call it the "Mad Libs" version of genre. Even the "genre as social action" view tends to suggest that writers engage in patterned, determinate acts; instead of the form predetermining what they write, it's the social situation—the social exigency—that determines the writers' actions.

Yet experience tells us that we generally have lots of options—what we write is not predetermined, and the work of a writer is the work of making effective choices. Many researchers of genre have sought to reconcile this contradiction by examining how genres actually operate in real-world workplace situations.

What they have found is that the power of genres is at once constraining and liberating. As Susan Katz puts it, "[genre conventions] constrain writers by limiting the form, style, language, and content that are appropriate in particular situations. Conversely, conventions enable by supplying templates, genres, and topics which can be useful to the writer at all stages of the writing process" (quoted in Artemeva 2006a, 23). Genre is constraining because the genre situation imposes certain limits upon the writer's approach (if I'm writing a business letter, I can't generally do it in iambic pentameter; if I'm writing a sit-com episode, I generally can't kill off the protagonist at the end of the episode). But genre is enabling because it gives me a place to start, both topically and structurally, allowing me to be creative within the scope of the genre's limits. If I'm writing a business letter, I don't need to decide how the letter should be formatted or what voice to use; I can focus on the best ways to present my argument. If I'm writing a quick-start guide, I can depend upon some of the conventions and expectations that the genre already provides. With those decisions already made, I can focus on details like the optimal order of steps and how much to write about each step, within the limits that the genre affords.

Moreover, if I'm convinced that the standard approach isn't the best approach for some reason, I can dream up unconventional approaches that may still qualify as examples of the genre in question. To adapt a metaphor from jazz music, the genre provides the theme and the structure, but allows the musician some room to play variations upon that theme, as long as those variations remain rooted in the theme and structure of the piece. As a writer, I may decide that some aspect of my rhetorical situation calls for a lighter tone, less formal style, more elaborate explanation, or a different type of illustration than is typical of the genre. If I vary every aspect of my approach, I risk losing the familiarity that readers depend upon when they read texts in this genre. But if I make limited strategic changes, and if I do so with specific purposes in mind, then I can carry my readers with me. The familiar features and content, and the familiarity of the communication situation itself, will help them interpret the unfamiliar bits properly.

GENRE AND CHANGE

This notion of choice helps us also to understand how genres evolve in response to new circumstances. As Schryer (1993, 200) notes, genres aren't absolutely "stable" or fixed. Rather, they are "stabilized-for-now," always subject to the pressures of changing situations, new writers and readers, and new content. Fortunately, it's pretty rare for every aspect of some communication situation in a given community to change all at once. But as conditions change, as technology provides communicators with new tools, and as new people enter and become experts in a community, so too do the community's genres change.

The above description of variations in genres suggests one mechanism for this change: individual variations that are responses to individual situations, if repeated in similar situations, can become just as much identified with the genre as were the supposedly "fixed" conventions that they supplanted. Berkenkotter and Huckin (1995, 6) observe that these changes are driven by situational differences as well as individuals' unique responses to the changing world, ensuring that genres are "always sites of contention between stability and change . . . [t]hey are inherently dynamic, constantly (if gradually) changing over time in response to the sociocognitive needs of individual users."

Researchers of genre have studied how genres change over time and discovered many instances of this process at work. For instance, Charles Bazerman (1988) and Dwight Atkinson (1999), looking at the genre of the scientific research article in different periods, discovered that several characteristics of this genre have changed, even though at no time did members of the communities in question ever declare the need for a change. The changes were incremental and situational, and they're only visible in retrospect: the discussion of methods becomes more prominent, the use of mathematical data increases, the voice shifts from first-person to third-person and from active to passive, and so on.

Why is this important? As a technical communicator, you need to be adept at recognizing just what parts of the genre can be budged and which are relatively fixed. Going back to the example of the business letter: some elements (such as the date, the salutation, and the courteous, professional style) are comparatively fixed (or, in Schryer's term, "stabilized-for-now"). If you wrote a business letter using slang or sentence fragments, your readers would doubt your competency as a member of the "social group" of professionals in the business environment—your text might be read as a failed attempt at a business letter. If you left out the salutation, readers might not even recognize your text as a business letter at all—in terms of its rhetorical situation, it would be unintelligible (and therefore it would fail to respond to the rhetorical situation).

Yet, other elements within the business letter genre are less fixed. For instance, most business letters span multiple paragraphs, but you could write a successful letter that consists of only one paragraph (or even a single sentence); it might be seen as unusual, but it wouldn't be rejected out of hand. Likewise, if you were writing to a business associate with whom you also have a personal relationship, you might opt for a more casual or more personal tone. Or if one of your intentions is to *encourage* a stronger personal connection with a business associate, you might strategically adopt elements of personal correspondence, stretching the business-letter conventions to achieve an unconventional purpose. Whereas questions about a recipient's family members or private life would generally be outside the scope of the business-letter genre (at least in current American business settings), the particular situation often calls for variations upon the usual theme. There are many conventions within a stable genre, but few absolute rules.

GENRE AND COMPETENCE

This flexibility matters a great deal when it comes to expressing yourself as a competent user of a genre, for, as Artemeva (2006a, 23) observes, "the better our command of genres, the more flexibility and freedom we can apply in using genres and the more fully we can express our creativity in them. Thus, even when acting recurrently in a recurrent situation, one can still express one's individuality when using a fully mastered genre." By knowing the boundaries (soft and hard), you can more comfortably stretch those boundaries when it's necessary to do so. Conversely, readers recognize that those writers who are the most versatile users of a genre are also the most competent. So not only does your increasing competence enable you to use genres more flexibly, but your fluid and flexible use of genres—your ability to strike the right notes with the genre—is a sign of your competence in that social situation.

As a newcomer to a work environment, you will naturally want to demonstrate competence. The effective use of that workplace's genres is one important way to do so. Competence is more than technical skill, content knowledge, or error prevention; it is also a matter of awareness, good timing, and judgment. Novice users of workplace genres tend to make two types of error. First, they make missteps that are the product of an incomplete understanding of how the genre operates in its specific context (say, not understanding the difference between a situation calling for a memo and one calling for a letter). Second, even when they perform a genre properly, they tend to stick with the safest (that is to say, the most stable) variations of the genre—they take fewer risks, because they're less comfortable deciding just how far they can play with the genre. To adapt a tennis metaphor, they don't aim for the lines as much. As their understanding of the genre and its situation grows, so too does their willingness to employ more of the genre's flexibility.

CORRELATION OF GENRES

So far the discussion has focused on genres as discrete entities. But, especially since the mid-1990s, genre researchers have become increasingly attentive to how genres work in combination, and how genres relate to the various roles played by members of a workplace or social group.

As early as the 1950s, long before the boom in English-language scholarship along these lines, Russian scholar Mikhail Bakhtin (1986) noted that every instance of communication is an utterance in response to previous utterances, and anticipating future utterances. Genres are not employed in isolation; all utterances (that is, all instances of speaking or writing) are in a sense *responses* to some prior utterance, and they instigate some sort of response, which itself instigates another response, and so on. The "rules" governing the appropriate types of response are genre rules; utterances of a genre tend to be made in particular, regularized patterns of relationship, and these patterns themselves reflect the relationships within the rhetorical community (e.g., workplace) in which they occur.

For example, if a friend gives you an insult, you have a limited number of conventionalized genres available to respond. You could respond with another insult, a joke, an angry retort, a defamation suit, or a punch in the nose (but not a purchase order, an anniversary toast, or a résumé). The punch in the nose could yield a return in kind, an apology, a curse, or a lawsuit. These conventionalized responses are not suitable merely because some "rule set" stipulates them. Rather, they're suitable because they *mean* something specific and conventional within the context and community in which they're used. To respond to an insult with a defamation suit might be interpreted negatively (as an overreaction, perhaps), but it would at least carry meaning. To respond with a résumé, on the other hand, would simply be unintelligible.

Thus genres are not merely standardized containers of message content. Genres establish, articulate, and reinforce socially recognized relationships, and these relationships play a role in how the content of messages is understood. Just as the shape, size, and design of a document often can prepare readers for the content within (helping a reader know whether to expect a novel, a shopping list, or a progress report), so too does genre cue readers' understanding of the social and structural relationships that workplace communication depends upon. Consider business correspondence. Business letters are often written in response to previous utterances (a prior letter, an advertisement, a phone call), and they provoke other typical generic responses. A client's letter of inquiry will prompt a letter of reply (either informational, good news, or bad news). A good-news letter may prompt a thank-you letter or an acceptance letter. The exchange of letters coincides with, and depends upon, the social relationship between (in this example) a company and one of its clients.

GENRE SYSTEMS

Beyond the basic notion that genres are interrelated (and that they reflect the conventionalized relationship patterns of rhetorical communities), there have been several specific models proposed for understanding the correlation of genres. Genre researchers have used various terms, including genre sets (Devitt 1991, 2004), genre systems (Bazerman 1994; Russell 1997; Yates and Orlikowski 2002), genre repertoires (Orlikowski and Yates 1994), and genre ecologies (Spinuzzi and Zachry 2000; Spinuzzi 2003), to describe these interrelationships (cf. also Spinuzzi 2004; Artemeva 2006a, 25–27). These concepts differ somewhat in their details, but they all endorse the notion that participants in a rhetorical community engage in recurrent, typified interactions using multiple, related genres. Consider these examples.

- Your boss sends out a memo calling for applications to head a new division of your company. This conventionalized genre—the call for applications—propagates a similarly conventionalized response—the letter of application.
- A government agency issues a request for bids on a new information technology services contract. Your IT services company responds to this request for bids by submitting a bid. (Leading up to that bid, other internal genres might also be used; for instance, memos, phone calls, strategic plans, recommendation reports, and other genres may be circulated, all resulting in the bid itself.)
- A customer writes a letter of complaint about a product manufactured by your company. The letter is delivered to your desk. First, you evaluate the complaint (perhaps by talking with a technical support specialist or reading reports of past complaints about this product). Based upon your evaluation, you reply to the customer, using one of several possible conventionalized responses: a letter of correction or apology; a refund; an explanation that may or may not assign responsibility to the customer, the shipper, or some other party for the failure; or perhaps a query, asking for more information about the complaint. Additionally, you may file internal

documentation reporting on the problem and its solution. If you discover that the problem is significant and affects other products or customers, you may need to compose additional messages as well, possibly even recommending a product recall or reporting the problem to an internal quality-control department or external agency, depending upon the scope of the problem.

As these examples demonstrate, certain genres propagate certain other genres in predictable patterns. These genres don't simply structure messages; they engage in work—going back to Miller's germinal insight, they are forms of social *action*, the way an organization gets things done. As a new member of the community, you are tasked with learning not only how to produce whatever individual genres are common in your position, but also when to produce them—in response to what events or signals.

Spinuzzi's concept of "genre ecologies" builds upon these other types of assemblage; in Spinuzzi's formulation, however, the interwoven genres in any workplace or activity network "are not simply performed or communicated, they represent the 'thinking out' of a community as it cyclically performs an activity" (Spinuzzi 2004, 114). Moreover, genres actually alter (Spinuzzi's word is "mediate") the nature of the activity itself; that is, the activity is fundamentally shaped by the nature of the genres that enact it, such that it is impossible to separate the social activity (say, fulfilling purchase orders) from the genres (the purchase order) that one uses to do that activity. The act of fulfilling an order is encoded in the "purchase order" form, and conversely, the form structures the act.

This notion of mediation elevates genres to the same level of importance (or agency) as the people who use them. Though people act through genres (for instance, by writing reports to influence a supervisor's decision making), genres also shape or "mediate" people and their activities by creating a system of authorized ways of doing things—ways that create certain possibilities for individual action, that foreclose other possibilities, and that even create and foreclose certain types of intentions. For example, within a genre ecology that includes various genres for lodging complaints or expressing dissatisfaction, certain activities associated with dissent are enabled and regulated—they're written into specific social relationships and communication modes. Within a genre ecology that does not offer customary genres for complaint, people either cannot participate in that activity, or, if they're sufficiently motivated, they are forced to employ existing genres in unconventional or illegitimate ways (such as by whistle-blowing)—ways that defy the social relationships of the group and that may cause disequilibrium in the genre ecology.

GENRE, FLEXIBILITY, AND ORIGINALITY

This account of genres propagating other genres and fitting together to reflect a social structure may make it seem as if technical communication is a purely mechanical process: standard letters come in; standard responses go out; standard forms are filed, calls made, and so forth. But there remains a great demand for invention, judgment, and creativity in how writers actually use genres. Each situation may call for certain distinct types of response, but each response is nevertheless unique.

Earlier we considered how genres invite writers to make choices within a set of genre constraints; as Katz suggested, these constraints can actually help writers move forward by limiting the number of choices that they need to make in any given situation. Schryer (2000, 2002) goes farther, arguing that working within a genre is an improvisation; communicators draw upon the conventions of genre as strategic resources that allow them to strike certain socially recognized notes (cf. Artemeva 2006a, 25). But from situation to situation, different decisions will be made, different judgments called for, different notes struck; ever-changing circumstances mean that our genre-based messages will change as well. In fact, genres are never truly stable; they are, to return to Schryer's (1993, 200) oft-quoted phrase, only "stabilized-for-now." They evolve in response to changing purposes, content, participants, and technologies. They can evolve only because those who use them are continually improvising—making small, situation-specific changes in their performance of the genres, drawing upon strategies that they have learned from other genres and settings, and, in some cases, flouting the conventions of the genre for strategic purposes.

Much of the genre research of the past decade has reported on various forms of genre change: writers' uses of novel rhetorical strategies and new, emergent, or hybrid genres in the face of new rhetorical problems (for instance, Artemeva 2006b; Bazerman 1988; Freedman 2006; Henze 2004; Yates 1989). In some cases, this genre change or hybridization can be a strategic effort to overcome the structural limitations of a rhetorical situation; Wendy Sharer (2003, 8) uses the term "genre work" to describe the "strategic blending of typified and innovative textual elements" intended to explicitly adapt the genre resources of privileged communities to a new set of purposes for which no effective genres exist.

New professionals (and people in new professional situations or identities) must be aware of the stability of the genres that they're expected to use in their work settings, but they must also be conscious of their flexibility—in other words, they should understand genres as resources to support strategic improvisation. Likewise, technical communicators need some understanding of how their workplace genres have evolved, and how they might evolve in the future. Not every writing task is the right moment to innovate, but often it's possible to alter the expected pattern of communication in order to solve problems.

HEURISTIC: UNDERSTANDING AND USING GENRES IN YOUR WORKPLACE

So how do you learn the genres of your workplace? How do you use genres once you've learned them? What can your knowledge of genres tell you about your workplace? As noted above, genres provide a kind of shorthand for professionals. If you're familiar with the basic set of genres typical of your workplace, including how and when they're used, you can rely upon those genres to simplify many everyday decisions. Knowing to use a particular genre doesn't eliminate all of the decisions that you'll need to make, of course, but it takes some variables out of the equation, giving you a starting point for engaging in the routine behavior of daily professional activities in your workplace.

The major theoretical perspectives on genre introduced above suggest a number of questions that we can bring to bear upon technical communication in the workplace. So let's turn now to another genre, the heuristic—a set of questions and tips that adapt the theoretical material above into a useful guide as you begin to use new workplace genres. I've grouped these heuristic questions into four categories: relationships, purposes, document characteristics, and learning. In each category, I offer questions designed to help you examine technical communication tasks and situations using the tools of genre. This heuristic can help not only when you are writing or speaking, but also when you're the recipient of workplace messages. As workplace-situated tools, genres can help you make sense of your work site just as they can help you participate in it.

CATEGORY 1: GENRE AND PURPOSE

The literature review discusses genre as one of the resources that technical communicators use to respond to rhetorical situations, and one of the key characteristics of any rhetorical situation is purpose—the communicator's purpose in creating a message and the recipient's purpose in reading or hearing (or, more broadly, doing something with) that message. So the first set of questions that you should ask about any situation involving technical communication genres is about purposes:

- What is the primary purpose of this genre? What result is it intended to achieve: a decision, an action or event, a reply, a

change in understanding or perspective? Remember that genres are not just document types—they are types of *action* intended to produce a result. What result typically ensues (and what is typically intended) when this genre is used?
- Does the genre have any secondary purposes? In addition to the direct outcomes intended for the work, consider what other outcomes, including relationship outcomes, your organization might be hoping for. For example, some messages have the direct purpose of informing but also the indirect purpose of establishing connections or promoting an organization's credibility.

CATEGORY 2: GENRE AND RELATIONSHIPS

Genres aren't simply interchangeable tools that anyone can use in any circumstance; rather, they reflect existing relationships between message creators and recipients. In some ways, genres even help to create those relationships, signaling how the participants in a dialogue should act with regard to one another. The second set of heuristic questions prompts you to consider the relational dimensions of genre. When you are faced with a communication situation, ask the following questions:

- What is my relationship to my audience or audiences, and what sorts of communication are typical in this relationship? Who are the players in this genre? Who normally writes it and who normally reads it?
- Do any others participate in some way (e.g., by contributing data or providing feedback)? What responsibilities are mine as the writer, and what responsibilities are shared with others (e.g., subject-matter experts, information providers, document designers, graphic artists, supervisors, or editors)?
- Where does the genre fit within the "genre system" or "genre ecology" of your workplace? Does it explicitly respond to other, initiating genres? Does it call for specific (either explicit or implicit) genres of response? Is there a range of possible appropriate responses?
- What else can I learn from the genre about the relationship that exists between these participants? Is the relationship hierarchical or nonhierarchical? Do the people involved have similar or different goals or concerns—are they collaborating? Disagreeing? Competing? Do they know each other or not? Is the relationship short-term or persistent?

CATEGORY 3: GENRE AND DOCUMENT CHARACTERISTICS

Even though contemporary genre theory considers rhetorical features other than form, the formal and stylistic characteristics of documents are nonetheless important clues to help us understand and use genres. Ask the following questions regarding the formal and stylistic characteristics of the genres you need to use:

- Are the documents in this genre durable or ephemeral? Are they filed, bound, shelved, or otherwise stored, or are they read and then discarded? Why? What does this fact suggest about the relational or rhetorical aspects of the genre? What does it suggest about how the document should be produced?
- Are documents in this genre read through in their entirety, or are they sampled piece by piece or section by section? Consider what the text structure reveals about how the text is meant to be read. Text structure often reflects reading behavior. For example, texts that have lots of clearly marked sections, heavy labeling, navigational tools (like indexes or tables of contents), or a "modular" feel are designed to support sampling/reference rather than cover-to-cover reading.
- If you have access to multiple examples of this genre, what variations (of structure, size, style, content, arrangement, or other elements) do you observe? Consider possible explanations for these variations. What seem to be the limits to these variations?
- What characteristics (of structure, size, style, content, or other elements) seem to be the most stable or "required" in this genre? Remember that genres regulate the choices available to you, but that writers have a great deal of flexibility in how they fulfill the expectations of the genre.

CATEGORY 4: LEARNING GENRES

Early in your career, you will spend much of your time learning to identify and use new genres, and all of the heuristic questions above will help you make the transition from novice to expert user of technical communication genres in your workplace. I'll end this section with a couple additional heuristic questions to help you in this learning process.

- Does your workplace provide written instructions or templates to help people write this genre, or do you have models from which you can work? Ask around; how did others in your workplace learn to use this genre? Collect copies of typical documents (electronic copies if possible), so that you can see your organization's genre conventions in action.

- Are there any "analogues"—similar, but more familiar genres that you can adapt to your purposes? What familiar genres does this one resemble, in form, content, medium, or purpose? Consider whether this genre is a version of something more familiar to you, or whether some of the same writing strategies could be applied in this genre. Even if this genre is distinct in some ways, your experience using other genres may help you make choices in this one. Be aware of the differences, but also exploit the similarities.

EXTENDED EXAMPLE: THE CASE OF JAMES, COUNTY DEVELOPMENT COMMISSION

Let's see how this heuristic works by returning to the real-world technical communication case introduced earlier. A communicator's decision-making process is rarely simple or linear, and not all of the heuristic questions will be equally relevant to every writing task. But as you'll see, the heuristic questions can help you make effective strategic choices about your writing tasks, based upon clues from your workplace and your prior experience with other genres.

As you may remember, James is a new technical communicator working on economic development for a county development commission. His boss asked him to conduct a survey of county businesses regarding their resource needs and their satisfaction with his agency's services, and then to share the results in a written report. James conducted the survey, then considered the data and reflected upon his technical communication task.

First of all, he considered the purposes of this report. The primary purpose stated in his boss's instruction was clear: to report information about the state of business in the county over the past year. The document therefore needed to represent the data that he collected as clearly as possible. The report was not necessarily intended to instigate specific actions, but its information might help businesses make more informed decisions in the future.

Although he had never written a report like this one, James knew that his boss had been collecting and reporting these data annually for several years. So, rather than start from scratch, James began by looking at samples from the previous four years. These samples, along with his boss's instructions, revealed to him that the report's purpose was more complicated than simply to report data. He was also informing stakeholders (primarily representatives of the local business community) about the agency's work during the past year on their behalf (another informational purpose).

Moreover, he was demonstrating the agency's stewardship of area business (an ethical purpose—that is, a purpose involving the agency's ethos, its credibility in the eyes of its constituents). All of these purposes were part of the report's primary purpose of informing business stakeholders about the local business environment.

Looking at the samples, James also recognized some secondary purposes of this assignment, mostly related to the multiple audiences of the annual report. In addition to its primary audience (owners of existing businesses), the report's audience also included area municipal leaders, prospective business owners (people considering a move to the region), and perhaps even interested citizens, each concerned about the state of area business for their own set of reasons. While the initial purpose—to report survey data—seemed pretty simple, James realized that the interests, motivations, and expertise of all of these different constituencies, along with his own organization's need to represent its efforts in the best possible light, made writing the report a challenging rhetorical task.

Next, James considered matters of relationship: his own relationship to the stakeholders in this rhetorical situation, but also the relationships between the annual report and other messages, resources, and knowledge that were part of the rhetorical ecosystem of the annual report.

As a representative of the development commission, James understood his relationship to business leaders as that of service: the agency provided a service to these leaders by making sense of the complex business climate, so that these leaders didn't all need to conduct their own research. The agency also used its data to promote the area's economic health through information and programs (another form of service, this time in relation to the citizenry as a whole, even people unlikely to read James's report). Conceivably, some readers of the report might have competing business interests, so James also knew that he had to remain as neutral as possible in presenting the survey results; it would harm his commission's ethos if the commission appeared to be favoring one business over another.

Indeed, James realized that even if he was the principal writer of the report, effectively the report's author was the agency itself, not him alone. His words had to represent the ethos of the agency. That ethos was not simply an effect of this single report; it arose from the full range of agency's products, including past reports, agency-led programs, and the reputations of organizational leaders. The report needed to be consonant with the agency's overall ethos, and the report could also depend upon that ethos to help fulfill its purposes.

Since the commission already had a good reputation as a useful resource for area business, James knew that he could rely upon that reputation and speak to his audience from a position of mutual respect. If, on the other hand, the commission had had a poor reputation among its constituents, James might have needed to approach the report somewhat differently, with the aim of repairing relationships or allaying doubts, while still presenting the data collected in his survey.

James also considered his task in light of the broader genre ecology of his workplace. He realized that this report sprang from prior messages (and other genres) and would instigate future messages. On the front end, the report responded to his boss's charge, but also, more broadly, to the range of business leaders' questions and concerns throughout the year, many of which guided decisions about what data needed to be generated and what questions asked. It drew upon existing data (collected by James himself and others).

On the back end, the report would play a part in the short- and long-term decisions of business leaders, politicians, and citizens of the area. After reading the report, some business leaders might decide to expand their businesses (or not) or to open new businesses. Legislators might decide to propose new laws or policies. The agency itself might use the report to help decide what new programs to offer in the coming year, and then to justify those choices to its constituencies. The report (if successful) would not simply be read; it would be the basis for actions, part of the large and complex puzzle of area economic decision making. Finally, the report was part of a series of such reports (one per year), and a part of a network of annual reports from a multitude of other agencies, companies, and departments, all of which characterized overlapping perspectives and interests. Even if each report could be read on its own, James realized that readers might also wish to look for trends in several reports together, and his report needed to support that kind of reading.

As a relatively junior member of the commission, James found that working on this project gave him a new insight into his organization's role in the region; it helped him see how the development agency's services affected other organizations, and how his agency's own status depended upon the work of others. These insights would help him in future projects, and in building strong relationships with his fellow employees and constituents.

These investigations of his rhetorical purposes, and how they fit into broader communication patterns and relationships, helped James plan his approach to the report. But they didn't completely determine all the details of form, content, and style that the report should have. He still had plenty of decisions to make.

In each previous report, he recognized several common features. All the previous reports fit onto two sides of one standard 8½″ × 11″ sheet in portrait (vertical) orientation. The text was black, and the headings and graphical elements used the same two spot colors—blue and green—that appeared in his office's logo. They also all used the office's preferred typeface (Palatino).

Organizationally, the reports were all modular, with several headings. In the most recent report, the first page used three text columns while the second page displayed several pie charts that represented survey data; previous years' reports were visually simpler, using single columns and fewer, less polished visuals. When he asked his boss about the differences, she said she'd wanted to make the report more "professional looking," but she didn't have the document design expertise—one of the reasons that she was asking James to do this year's report.

So James assumed that there was some flexibility in several document design elements (such as column layout and arrangement of graphics and data displays) as long as the results looked "professional." But other aspects of the document design were more stable, including the choice of colors (the organization's documents all used the standard blue and green palette), the document length and page size (his boss felt that busy clients wouldn't want to read a report of more than one double-sided sheet), and well-labeled, modular structure (different constituencies would be interested in different data, and few people would read the report top to bottom).

In light of his analysis of the commission's role, James realized that this report's function was a bit more complex than simply to report data. He considered the stated purposes of the survey and the report: to determine "the issues and problems that stunt industrial growth" in the county, and then to demonstrate the agency's commitment to respond. But the mission of his commission and its relationships with county businesses, government, and service providers indicated that the report also contributed to a *dialogue* with business leaders, service providers, and other stakeholders. It reported information, but in doing so, it also established goodwill, reinforced the agency's commitment to solving business problems, and demonstrated its joint partnership with the community and area businesses. James realized that maintaining this dialogue with stakeholders was just as important—in this report and in all of the agency's dealings—as the data-reporting function itself.

The report's formal features (its size, structure, feel, ease of reading, clarity, data representations, and consonance with prior reports) all worked toward these goals. The report's audience were extraordinarily busy people;

they would probably spend only a few minutes with this document, so the design and content had to be high-impact. Since he was presenting data trends that were fairly dramatic, James decided to foreground the data by displaying them in high-impact pie charts accompanying short blocks of explanatory text. Though pie charts are not very good at displaying detailed data, they're an excellent choice for depicting simple but dramatic trends.

He positioned the pie charts close to their corresponding text sections so readers wouldn't need to flip back and forth. Applying some principles that he had learned as a technical communication student, he also got rid of the long sections of narrative and descriptive text that had been used in previous years' reports, believing that a more concise presentation with data displays would do the job better.

Despite the need to be concise, James recognized that simply presenting the data might not sufficiently convey his agency's commitment and its partnership. So he added a brief, but somewhat more relationship-oriented, "'purpose-setting' introduction [in an] attempt to achieve credibility." However, James chose not to provide a lengthy analysis or conclusion because, in his words, "I wanted the results to speak for themselves. Since I had three primary audiences that this survey was for, I wanted them each to come to their own conclusions. Business owners needed to focus on what they could do better, service providers could focus on what they could do to help companies better, and community leaders and economic allies could see that our office was actually working (by the mere fact that we surveyed industries and distributed the results)."

Although he knew from looking at other data reports that an analytical conclusion was fairly conventional for such reports, he judged that in this case—with multiple audiences motivated by somewhat different and perhaps competing interests—it was more useful to flout that convention in favor of an overt data-reporting approach. He knew that there would be other occasions to discuss with each constituency what the data meant for them. James used the same professional, courteous writing style that he had developed in his regular correspondence with business leaders in other facets of his work—a style that reflected the mutual respect shared between the commission and its constituents, but also the commission's sense of responsibility, its role as a trusted steward of area business.

James's experience composing his first annual report of county business data was a success because he took the time to analyze his task in terms of the rhetorical situation and genre conventions that the task posed. He also understood that this report was only one of many genres that his office used in combination to achieve its mission of supporting business growth in the county. In turn, the learning (and strategic decision making) involved in producing this report gave James a fuller understanding of the mission, methods, and strategic goals of his agency—insights that he could apply in future writing tasks.

CONCLUSION

Technical communication genres are patterned, situation-specific responses to the recurring communication needs found in technical communication workplaces. As a technical communicator, you will spend much of your time reading and writing texts that conform to the genres of your workplace; understanding these genres is therefore important.

Technical communication genres might at first appear to be simple patterns into which the communicator inserts content. But an understanding of the social and rhetorical characteristics of genre can help technical communicators diagnose and respond to communication needs creatively, efficiently, and effectively. Genres bring together a writer's and reader's shared understandings of communication purposes and social relationships; communication is "typified" when the participants in communication share an understanding of the communication situation. By employing genre conventions, communicators reinforce this shared understanding.

Genres also offer technical communicators many choices and opportunities to vary their approach to any communication situation. By constraining some characteristics of a text, genres make it possible for communicators to vary other text features, allowing them to respond to the particular needs of their communication situation. In fact, as a technical communicator's genre expertise increases, his or her ability to adapt genres for strategic purposes also increases.

Technical communication genres usually work together in groups or genre systems in the workplace, helping the participants in a professional community to engage in a range of ongoing communication types. An expert technical communicator not only knows how to use specific genres, but also understands how these genres fit together in a sustained conversation. In fact, examining the genre system in a workplace setting is a good way to learn about the community itself, since genres (and genre systems) are one of the key tools used by communities to accomplish work. As the activities, goals, and challenges of a community change, the genres in that community's genre system also evolve, enabling the community to adapt to new circumstances. As a new technical communicator, one of your most important tasks will be to learn your new community's genres, and to use your growing knowledge of those genres to understand the values, priorities, challenges, and social relationships of that community.

DISCUSSION QUESTIONS

1. The first group of heuristic questions presented above discusses the connections between genre and purpose. Find a technical communication document for which you are a member of the intended audience (e.g., a product manual, a business letter, or a newsletter). Using the heuristic questions presented in this chapter, identify the document's purposes (including any secondary purposes). What outcomes should follow from this document? Which specific characteristics of the document's form, style, organization, or content point to these purposes or intended outcomes?
2. Consider the document described in question 1 from the standpoint of relationships. Using the second set of heuristic questions presented in this chapter, consider the relationship between the document's players—its writer(s), its reader(s), and any other stakeholders involved in the communication. What kinds of relationships exist between these participants? How do you know? What features of the document signal these relationships?
3. Are there any unexpected or anomalous characteristics in the document described above? If so, how would you explain them?
4. Find at least two other examples of the technical communication genre selected in question 1—examples from either the same company or a different company. Identify the formal, rhetorical, and stylistic elements that these documents all share. What tells you that these documents are all examples of the same genre?
5. Using the same set of documents that you used in question 4, identify any formal, rhetorical, and stylistic differences among these documents. In what ways does each document stretch, modify, or challenge the conventions of its genre? How would you explain these modifications? Do you see these modifications as examples of the flexibility of the genre, or do they fundamentally conflict with the genre as you understand it?
6. How transferable are genres? To what extent are genres specific to a particular workplace or professional community (e.g., Celltech, an Atlanta-based cell-phone manufacturer) and to what extent are they transferable to other, similar workplaces (e.g., technology manufacturers in general)? Imagine that you are moving from one technical communication job to another in a different company. What genre knowledge and experience do you imagine you'll be able to transfer, and what new knowledge or experience will you need to obtain in your new position?
7. This chapter argues that our responses to communication situations

tend to be generic (that is, patterned according to genres), and it describes the advantages of learning and working with the genres of the workplace. The chapter does not discuss any disadvantages of genre. In what circumstances (if any) might genres be a disadvantage?

8. As a technical communicator, how would you respond if faced with a situation in which no obvious genre exists, or the genre implied in a situation does not seem to support your rhetorical purposes?
9. The summary for this chapter notes that genres are "shared conventions of text and situation." But it also notes that communicators have a great deal of flexibility in how they compose documents in most genres. Just how far can a genre be stretched—how much can the writer flout genre expectations—before it no longer fits? Are there any textual characteristics that are so essential to the genre that they can't be altered?
10. As a student, you have read and written documents in several different academic (and perhaps other) genres. When you enter the workplace, you will likely have to learn a new set of genres. Which academic genres do you believe will most easily transfer to your intended workplace? Assuming that the two genre sets do not match up perfectly, how will your familiarity with academic genres serve you in your workplace? What can your expertise with one set of genres teach you about other sets?

ACKNOWLEDGMENTS

Many thanks to current and former students Ted Byrnes, Angela Connor, Joseph Dawson, Julie Martin, Ashley O'Neil, Doug Solomon, and Roxanne Tankard for generously sharing their experiences with genre in the workplace.

WORKS CITED

Artemeva, Natasha. 2006a. "Approaches to Learning Genres: A Bibliographical Essay." In *Rhetorical Genre Studies and Beyond*, edited by Natasha Artemeva and Aviva Freedman, 9–99. Victoria, BC: Inkshed Publications.

———. 2006b. "A Time to Speak, a Time to Act: A Rhetorical Genre Analysis of a Novice Engineer's Calculated Risk Taking." In *Rhetorical Genre Studies and Beyond*, edited by Natasha Artemeva and Aviva Freedman, 189–240. Victoria, BC: Inkshed Publications.

Atkinson, Dwight. 1999. *Scientific Discourse in Sociohistorical Context: "The Philosophical Transactions of the Royal Society of London," 1675–1975*. Mahwah, NJ: Lawrence Erlbaum Associates.

Bakhtin, Mikhail. 1986. "The Problem of Speech Genres." In *Speech Genres and Other Late Essays*, translated by Vern W. McGee, edited by Caryl Emerson and Michael Holquist, 60–102. Austin: University of Texas Press.

Bazerman, Charles. 1988. *Shaping Written Knowledge: The Genre and Activity of the Experimental Article in Science*. Madison: University of Wisconsin Press.
———. 1994. "Systems of Genre and the Enactment of Social Intentions." In *Genre and the New Rhetoric*, edited by Aviva Freedman and Peter Medway, 79–99. London: Taylor and Francis.
Berkenkotter, Carol, and Thomas N. Huckin. 1995. *Genre Knowledge in Disciplinary Communication: Cognition/Culture/Power*. Hillsdale, NJ: Lawrence Erlbaum Associates.
Devitt, Amy J. 1991. "Intertextuality in Tax Accounting: Generic, Referential, and Functional." In *Textual Dynamics of the Professions: Historical and Contemporary Studies of Writing in Professional Communities*, edited by Charles Bazerman and James G. Paradis, 336–357. Madison: University of Wisconsin Press.
———. 2004. *Writing Genres*. Carbondale: Southern Illinois University Press.
Freedman, Aviva. 2006. "Interaction between Theory and Research: RGS and a Study of Students and Professionals Working 'in Computers.'" In *Genre and the New Rhetoric*, edited by Aviva Freedman and Peter Medway, 101–120. London: Taylor and Francis.
Henze, Brent R. 2004. "Emergent Genres in Young Disciplines: The Case of Ethnological Science." *Technical Communication Quarterly* 13:393–421.
Miller, Carolyn R. 1984. "Genre as Social Action." In *Genre and the New Rhetoric*, edited by Aviva Freedman and Peter Medway, 23–42. London: Taylor and Francis. Originally published in *Quarterly Journal of Speech* 70:151–167.
Orlikowski, Wanda J., and JoAnne Yates. 1994. "Genre Repertoire: The Structuring of Communicative Practices in Organizations." *Administrative Science Quarterly* 39:541–574.
Russell, David R. 1997. "Rethinking Genre in School and Society: An Activity Theory Analysis." *Written Communication* 14:504–554.
Schryer, Catherine F. 1993. "Records as Genre." *Written Communication* 10:200–234.
———. 2000. "Walking a Fine Line: Writing Negative Letters in an Insurance Company." *Journal of Business and Technical Communication* 14:445–497.
———. 2002. "Genre and Power: A Chronotopic Analysis." In *The Rhetoric and Ideology of Genre*, edited by Richard M. Coe, Lorelei Lingard, and Tatiana Teslenko, 73–102. Cresskill, NJ: Hampton Press.
Sharer, Wendy B. 2003. "Genre Work: Expertise and Advocacy in the Early Bulletins of the U.S. Women's Bureau." *Rhetoric Society Quarterly* 33:5–32.
Spinuzzi, Clay. 2003. *Tracing Genres through Organizations: A Sociocultural Approach to Information Design*. Cambridge, MA: MIT Press.
———. 2004. "Four Ways to Investigate Assemblages of Texts: Genre Sets, Systems, Repertoires, and Ecologies." In *SIGDOC '04: Proceedings of the 22nd Annual International Conference on Design of Communication*, 110–116. New York: ACM.
Spinuzzi, Clay, and Mark Zachry. 2000. "Genre Ecologies: An Open-System Approach to Understanding and Constructing Documentation." *Journal of Computer Documentation* 24:169–181.
Yates, JoAnne. 1989. *Control through Communication*. Baltimore: Johns Hopkins University Press.
Yates, JoAnne, and Wanda J. Orlikowski. 2002. "Genre Systems: Structuring Interaction through Communicative Norms." *Journal of Business Communication* 39:13–35.

ANN M. BLAKESLEE & GERALD J. SAVAGE

15 What Do Technical Communicators Need to Know about Writing?

SUMMARY

Responses of twenty-four technical communicators to questions about the writing they do for their jobs revealed a heuristic that new technical communicators can use to determine ways to write effectively in the various roles and contexts in which they find themselves. This heuristic consists of questions that encompass the amount and quality of writing technical communicators do, the nature of that writing, the genres technical communicators produce and the rhetorical strategies they use to produce them, their writing approaches and processes, the knowledge and skills they need, and the personal traits and qualities they should have. Our data suggest the range of answers that writers might give to these questions and how those answers often depend on factors such as the workplace, the nature of one's job, the industry, the project, and even one's personal work preferences and styles. In addition to sharing the variety of responses to these questions that our data revealed, we also present an extended example to illustrate how newcomers to the field can use these questions to determine the writing knowledge and skills they may need to be successful.

INTRODUCTION

Siena just started as a technical writer in a department with twenty technical communicators. Her department is in a division of a large, multinational corporation. The division creates specialized business software, and her department produces all the instructional and reference documents for that software. As a new writer, she is assigned to a team with three other writers to document one piece of the software. The software her team is documenting is targeted at a well-defined user group. Her teammates have all been at the company for at least three years. Her supervisor, Allie, has been with the company for thirteen years.

Siena is about to be assigned her first writing task. In anticipation of her meeting with her supervisor, she jots down several questions. She has some sense, from conversations with the other team members, of what she may be asked to do and what might be entailed in doing it. She is still learning the organization, however, and trying to determine its expectations—and what happens if and when those expectations are not met. She is a little anxious because she knows there is a lot she does not know yet or has not done. For example, she has never created a complete and fully usable set of online help topics. She wonders how much assistance she might get from her supervisor and teammates; how much assistance she might need; and what, precisely, she will need to know to be successful. She just earned her bachelor's degree in technical communication; however, does this mean she knows enough to take on this initial and, for her, high-stakes writing task? In this chapter, we present a heuristic that Siena could use to determine how best to approach that initial writing task and to determine what knowledge and skills she will need for it.

This chapter includes a review of what previous research tells us about writing in the field of technical communication; a description of a heuristic for analyzing the writing requirements of a writing task; an extended example based on our study of twenty-four technical communicators' writing practices; and a summary reviewing key information in the chapter. The review of previous research may be helpful to those interested in exploring in greater detail the topics covered in the chapter—for example, new contexts of technical writing, core competencies for technical communicators, the rhetoric of technical communication, and what writing means in the field. Our heuristic is divided into six categories with several questions within each category that writers can ask about the writing tasks they may perform. This section reveals how complex the act of writing is for technical communicators, while showing how that complexity can be managed. The extended example follows a new technical writer as she begins her first job, showing how she applies our heuristic to learn about her new responsibilities and to make a potentially overwhelming experience more comfortable and manageable. The conclusion reviews key concepts about the role of writing in technical communication and suggests ways that a new technical communicator can stay current in her or his knowledge and skills in this rapidly changing field.

LITERATURE REVIEW

Scholars in technical communication have long been concerned with the skills and knowledge that technical communicators need. For example, research has focused on employment ads, the expectations of employers and managers, and the experiences of technical communication graduates (see Carliner 2001; Thomas and McShane 2007; Lanier 2009). In the past decade, however, research addressing the responsibilities and work lives of technical communicators has taken on a new urgency as the field has undergone significant change and as scholars have, increasingly, pondered the roles that those trained in technical communication (whether called technical communicators, knowledge workers, or something else—and this has been a point of debate) might play in the twenty-first-century workplace (see Whiteside 2003; Faber and Johnson-Eilola 2003; Giammona 2004; Slattery 2005; Conklin 2007). Many of these studies are concerned with identifying the new modes and contexts of practice emerging in workplaces because of changing technologies and evolving organizational structures. Conklin (2007), for example, explores the increasing importance of cross-functional teams and how work processes continually flow and adapt to changing needs, making interpersonal and project management skills vitally important (see also Anschuetz and Rosenbaum 2002; Kim and Tolley 2004; Rainey, Turner, and Dayton 2005; Ford 2007).

A number of the studies cited above focus on management of documentation processes and projects, which involve both interpersonal and technological skills. Most of these scholars focus on the "core competencies" that technical communicators should have. Slattery (2005, 354) argues that "information technologies appear to be the primary medium through which these competencies are enacted." Giammona (2004, 350), who interviewed and surveyed individuals regarded as leaders in the field, found that writing was the most important skill for technical communicators: "But the one common denominator was writing—everyone agreed that a technical communicator must, at the core, be able to write." Whiteside (2003), Hart-Davidson (2001), Hart (2000), and Hayhoe (2000) also emphasize the importance of writing and the ability to communicate. Hayhoe (2000, 151) stresses that writing is what distinguishes us in our profession.

The studies we have cited generally regard writing as a skill that technical communicators use extensively. Most of these studies also acknowledge that technical communicators use writing in combination with a complex and varied mix of additional skills, competencies, and knowledge sets. In many of these studies, the power and complexity of writing as a literacy practice sometimes seems to be in the background, or regarded as no more important, powerful, or complex than other aspects of technical communication. Yet writing may be the one competency that really binds together the array of practices we call technical communication. Other than writing, no particular set of practices seems to be constant in technical communication; rather, they vary from context to context. Writing, on the other hand, seems to work in relation to the other practices so fundamentally that, without it, the remaining set would be something quite different from technical communication.

While writing is almost always examined in relation to other skills and practices in technical communication scholarship, a few studies have focused on writing more exclusively. For example, Farkas (1999) develops a set of rhetorical principles for writing procedural instructions. His study distinguishes between human actions and system actions, and he provides several alternative models for procedures (50). Isakson and Spyridakis (1999) investigate the influence of semantic (meaning-making) and syntactic (grammatical and structural) features of a text for helping users remember information. They make suggestions about sentence structure and the placement of key information that can help readers use a text more effectively in interacting with technologies and following procedures. Schneider (2002) develops guidelines for helping writers avoid ambiguity and for determining what "clarity" really involves in particular writing situations.

Other scholars have sought to answer the question, "what counts as writing?" in technical communication. They suggest that "writing" does not necessarily look like what we usually mean by writing in some contexts. Mirel (1996), for example, has examined the rhetorical strategies that make data reporting effective in database output. She found that the classical elements of rhetoric—invention, arrangement, and delivery—are essential factors to consider in "writing" with data. She also found that structuring and organizing data in ways "that support readers' interpretive strategies" is key to effective data communication (102). Finally, Winsor (1992) questions several assumptions we commonly make about the nature of writing—for example, that it involves free creation of meaning, that a human being must be immediately present when writing occurs, and that writing requires the use of words. Her consideration of the writing that accompanies and facilitates many engineering activities reveals that none of these assumptions necessarily apply and that creativity or individual choice about what or how one writes is limited, and sometimes not even possible.

All of the studies we cite have helped with understanding emerging trends and needs in technical communication, with defining further research, and with developing curricula and courses. However, what they have not provided, as Hart and Conklin (2006) suggest, are detailed insights into the day-to-day writing practices of technical communicators—insights into the perspectives of technical communicators as they write in a variety of settings. The research study that gave rise to this chapter helps meet this need.

Specifically, our research focused on technical communicators in actual workplace settings. We constructed a questionnaire asking respondents how much and what they write; how they write; their perceptions of what writing means and entails in their work; and their perceptions of the skills needed, and the relative importance of those skills, to write effectively. We sent the questionnaire to thirty practitioners, including technical communication managers, writers in industry, contractors, and writers in consulting organizations. Our respondents were geographically dispersed through the midwest, northeast, and southwest United States. We received completed questionnaires from twenty-four practitioners, who, on average, had been in their current positions for three years and in the field for eight. Sixteen had earned graduate degrees in technical communication or in a related field. We analyzed their responses both quantitatively, by counting instances of things mentioned, and qualitatively, by looking closely at the rich explanations respondents provided in answering our questions.

More specifically, for our quantitative analysis, we tallied responses to every question—for example, types of documents produced, time spent writing, number of projects worked on. Our qualitative analysis focused on the narratives respondents provided. Our questions were open ended, and respondents were encouraged to tell us, for example, not just whether, but also how and with whom they collaborate, as well as how they go about planning and developing documents. We began by identifying broad themes that ran through these narratives, and then we developed more specific categories within the themes that we used to code the responses. The narratives provide rich detail to support our heuristic.

HEURISTIC

Our research, along with prior studies, points to a number of questions that new technical communicators can ask to determine what they will need to know and do within their work contexts. We present in this chapter a heuristic that groups these questions into six categories:

1. amount and quality of writing entailed and expected,
2. nature of the writing,
3. genres and rhetorical strategies,
4. approaches to and processes for writing,
5. knowledge and skills, and
6. personal traits and qualities.

Each of these categories gets at different aspects of the writing technical communicators do and the skills and qualities they need to do that writing. Answering the questions within each category can assist technical communicators, especially those new to the profession, with determining what might be expected and needed from them in their work contexts. In this section, we present and briefly explain the questions that make up each of the six categories.

CATEGORY 1: AMOUNT AND QUALITY OF WRITING ENTAILED AND EXPECTED

The amount of writing technical communicators do involves two aspects of their work: their job description and the specific tasks and projects they work on. Although job descriptions do not typically state, for example, "A technical communicator in XYZ Corporation will spend at least 85% of her time in writing tasks," the typical duties of technical communicators in any organization may involve a fairly consistent amount of writing. That amount, however, may vary from one organization to another. In our study, the amount of time writers spent on writing tasks ranged from somewhat less than half of their work time to nearly all of it. Specific tasks are also likely to influence the amount of time spent writing. This could mean that a technical communicator writes a lot but that all of her writing tasks involve brief documents. It could also mean that a person spends considerable time on tasks that do not involve what we often think of as writing. Instead, a technical communicator may spend many hours meeting with team members, talking to subject-matter experts (SMEs), carrying out research, and so on.

Knowing how much you will write as a technical communicator is important for several reasons, not the least of which is determining how best to manage your workload, time, and resources. Technical communicators need to plan and make informed decisions about managing competing demands, satisfying managerial and employer expectations, and, most importantly, meeting deadlines. As a result, our questions for this first category of our heuristic include the following:

- How much time will I spend writing?
- How many documents will I write at one time? How many in a year?
- How important will it be to write well? And what does it mean to write well in my industry, field, and company?

The final questions relate to the quality of one's writing and how important that is, in the context of the organization in which the technical communicator works, in relation to a particular workplace task, and in the context of the larger industry or field. Quality can involve a range of concerns, from deadlines (How much time do I have to write a document?), to what is at stake in the writing (Could readers be physically hurt if they do not understand my document?), to audience (Is it internal or external to the organization?), to the value placed on writing and other forms of documentation by the organization. How quality is defined may also vary depending on the project, its circumstances, and the organization—for example, does it mean mechanical correctness, technical accuracy, rhetorical effectiveness, or some combination of these? Further, does it mean that all documents must meet certain standards for usability or does it refer primarily to things like readability, visual appeal, and conformity to stylistic or design standards?

CATEGORY 2: NATURE OF THE WRITING

The second category of our heuristic is concerned with determining the kinds of writing one will be asked to do as a technical communicator and what that writing will entail. The kinds of documents you are assigned, and where and how they originate, can greatly determine how you write. Therefore, we recommend these questions for this category:

- How much of the content for my writing will I have to research and develop from scratch? How much will I take or borrow from elsewhere? And what, then, will I need to do with it?
- What will be involved in writing original documents in my organization? Will I need to locate and interview subject-matter experts; locate in-house source documents; locate outside sources, such as books, research studies, or Internet sources?
- What will be involved in reusing or repurposing existing documents in my organization? Will I need to know where and how to locate such documents? Will I be provided the relevant documents at the start of a project? Will I need to verify the completeness or appropriateness of the documents with which I am provided? Will I need to conduct additional research similar to what I would do when writing from scratch?

Another common practice in technical communication is for writers to work in teams to develop larger documents that are assembled and disseminated in various ways. Knowing which of these practices you will be engaging in will have a significant impact on the tasks you undertake as a writer, and on how much time and what resources those tasks require. All of these approaches to writing are likely to involve research, but they may differ in the kind of research, sources, and skills required. The nature of the writing you do can also influence the tools and technologies you use, the way you organize your work, the amount of time you allocate to various tasks and projects, and the amount of control or ownership you will have, ultimately, over the documents you write.

CATEGORY 3: GENRES AND RHETORICAL STRATEGIES

The third category in our heuristic is closely connected with the previous one. New technical communicators need to know what kinds of documents they will produce and what the requirements and conventions are for those documents. Technical communicators need to know a variety of genres. They also need to be able to move easily between genres, and they need to understand the conventions of various genres and why those particular conventions exist. Technical communicators need, as well, to be prepared to produce new kinds of documents, since needs and expectations evolve within most work contexts and with new technologies.

Technical communicators need, in essence, to know how versatile and flexible they will need to be, both in regards to the kinds of documents they will be asked to write and in regards to the rhetorical strategies they will need. The key questions for this category include the following:

- What kinds of documents will I write and in what situations?
- What genres do I need to know and understand?
- What are the conventions for those genres?
- What rhetorical skills and strategies will be most helpful to me overall and for the particular genres and documents I will need to produce?
- How will I learn about my audience? What will I need to know about it?
- How will I determine my purpose(s) in writing? How will that purpose (or those purposes) influence the documents I produce?

The questions in this category encompass the various rhetorical concerns inherent in the work of technical communicators—concerns with purpose, audience, persuasion, and so on. Such concerns always need to be at the forefront for writers, which scholars have long stressed. Rainey, Turner, and Dayton (2005, 323), for example, found the "ability to write clearly for specific audiences directed by clearly defined purposes" to be one of the most important competencies for technical communicators. Similarly, Kim and Tolley (2004, 382–383) found that rhetorical skills and knowledge of audience are essential for technical communicators. We also believe that technical communicators need to be diligent in seeking and obtaining sufficient knowledge of their audiences, and of the rhetorical contexts of their work more generally. In short, rhetorical skill and competency remain central in the field.

CATEGORY 4: APPROACHES TO AND PROCESSES FOR WRITING

As the three previous heuristic categories show, there is no single approach to writing; it depends not only on individual preferences and skills, but also on the project, company, type of document, technologies used in documentation processes, and so on. Further, writing processes, for our purposes, encompass a full range of tasks, including research, planning, drafting, reviewing and editing, revising, proofreading, and publishing. We recommend these questions for this category, and there are a lot of them.

- How do/will I write?
- What might influence how I write (e.g., individual preference, genre, organizational context, industry, tools, work environment, project complexity, deadlines)?
- What research skills will I need for my work? Or even for a particular project?
- What will I need for a project in terms of tools, skills, resources, information, and time? (This question speaks to being able to break down a project.)
- Will I write alone or as a part of a team of writers?
- What will I need to know about reviewing and editing? Will I have to review my own work? Will I review the work of others?
- Will I be open to having my own work reviewed and edited? Who will review my work? What will they focus on?
- How will I assure the technical accuracy of my work?
- How will I make sure the reviews I receive are useful?

Technical communicators need both an awareness of themselves as writers and an understanding of how the work they do—and for whom and with whom they do it—may influence their writing process. Writing processes vary considerably from one organizational setting to another. Significant variations in processes can be connected to any of several factors: individual preferences and differences; types of documents; the industry; tools; the job setting and work environment; and the specific requirements of the project, including its deadlines.

CATEGORY 5: KNOWLEDGE AND SKILLS

In addition to knowing oneself as a writer, technical communicators also need to possess technical skills and knowledge. What this encompasses is, again, highly variable, depending on such factors as organizational context, industry (e.g., finance, transportation, telecommunication, health care), position, responsibilities, and so on. The questions to ask in this category include the following:

- What technologies will I have access to in my workplace? What or how much will I be expected to know about those technologies? And what technologies, more generally, will I need to know (hardware, software, digital communications technology, new media, etc.)?
- What will I need to know about the industry for which I write? Also, will I be expected to understand the industry for which I write when I begin, or can I learn about it on the job?
- What will I need to know about the subject about which I write? Will I be expected to be an expert on the subject matter about which I'm assigned to write? If not, will I be expected to know how to find the information on my own?

As a field, we have long debated the importance of skill with and knowledge of technology, especially relative to other knowledge and skills. Some argue that such skill and knowledge are essential and primary—technology is, after all, what we're about as a field. Others, however, argue that such knowledge is secondary—that knowing how to write, for example, is much more important. Many recent discussions place the importance of knowing technology somewhere in the middle, arguing that such skill is important but no more so than putting it into a larger context of other knowledge. This is what we do. Our findings, on which we elaborate further in our extended example, come down to this: technical communicators need to understand technology, and this means they need an aptitude for learning technology. Hart (2000, 291) says, "Most experienced technical communicators have yet to encounter software we couldn't begin using productively within a day, and become skillful within about a week. Mastery can certainly take far longer, but most of what we do doesn't require that level of mastery." Technical communicators certainly need technological skills; more importantly, however, they need the aptitude to learn and begin using new technologies as needed in their work.

CATEGORY 6: PERSONAL TRAITS AND QUALITIES

The final category of our heuristic concerns the personal traits and qualities that can help technical communicators with their writing. Our questions for this category are as follows:

- As a technical communicator, will I primarily be expected to work alone or closely with others?
- Will I be expected to plan my own work processes, or will I have projects mapped out in detail by a supervisor or team leader?
- How adaptable will I need to be? How open-minded?

- What will it mean to be adaptable and flexible in my organizational context?
- How important will learning and acquiring new knowledge be in what I do?

As an example of the importance of this category, much of the research we have cited about technical communication competencies supports our findings that interpersonal skills are essential. Hart (2001, 73) says that such skills include being willing and able to interact face-to-face and often across professional, cultural, and linguistic boundaries: "*Communication is about contact between two people, not simply an exchange of words*" (emphasis in original). Interpersonal skills—the ability to listen and ask questions, in particular—are also essential to writing and to carrying out research for one's writing.

Equally, technical communicators need an interest in and passion for learning as well as an ability to adapt easily to change. Giammona (2004, 354) quotes Jack Molisani, who says, "Today, I would say the ability to learn quickly and adapt, a tolerance for change, hands-on technical skills appropriate to what you are documenting, experience in the industry in which you are writing, and communication skills are key."

HEURISTIC SUMMARY

Technical communicators can analyze writing tasks in terms of six categories, or aspects, of writing situations. First, different tasks and contexts will require different amounts of writing and different definitions of and expectations with respect to the quality of that writing. Next, writing can involve very different characteristics from one situation to another, for example, writing from scratch, repurposing existing text, writing alone, or writing in a team. Third, the situated nature of writing calls for different genres—reports, instructions, proposals, help systems, or web pages, just to mention a few possibilities. These genres and situations call for different rhetorical strategies—persuasion and argumentation, carefully documented factual presentation, formal or informal style, technical or less technical language, all depending upon factors such as audience, purpose, and what is or is not at stake with the document. Fourth, different tasks and projects require various approaches or processes: research, collaboration, review. Approaches and processes may also be governed by standards within an organization or industry; they may be determined as well by the technologies to which a writer has access. Fifth, writing tasks and situations call upon various kinds of knowledge and skills. Although job descriptions and interviews are written to help employers screen prospective employees based on the education and skills they bring to the job, a technical writer will have to assess the particular competencies that a task will demand and to continually acquire new content knowledge as well as develop new skills for emerging technologies and technical processes. Finally, different situations and tasks demand different personal qualities. Some jobs can be done by working alone; most require a great deal of interpersonal interaction, whether for close collaborative teamwork or for engaging in tasks such as interviewing and reviewing. They may also involve varying amounts of technical aptitude. One requirement of most positions in technical communication is the ability to learn quickly and independently. While occasionally workplaces may be highly structured and routine, most often the twenty-first-century corporate settings of technical communication require flexibility and adaptability.

EXTENDED EXAMPLE

In order to see how our heuristic might be applied in the workplace, let's return now to our writer, Siena, as she starts her first technical writing job. We will follow her in this example as she asks questions from our six-part heuristic and learns what writing involves in the company where she will be working. We will also share what the writers from our research had to say in relation to our six categories.

AMOUNT AND QUALITY OF WRITING ENTAILED AND EXPECTED

As Siena enters the field, she wonders first just how much of her professional time will be devoted to writing. Some of the questions she has for her supervisor include

- How much time will I spend writing?
- How many documents will I write at one time? How many in a year?
- How important will it be to write well? And what does it mean to write well in my industry, field, and company?

Siena's supervisor, Allie, will likely answer her questions about the amounts of writing she will do the way most of our respondents did: most technical communicators, especially those recently hired, spend the majority of their time writing. Eighteen of our respondents said they spend at least one-quarter of each day on writing or writing-related activities. Sixteen (almost 66%) said that writing is what they do, primarily, in their jobs. All but two said that at least 25% of their jobs entail writing.

Allie tells Siena that she will start with just one project, but that she can expect to be working on additional projects very soon. Some will be short, but others will involve months of work. Allie cannot give Siena an exact number, but she guesses that Siena could easily complete "20 or more" writing projects in a year. Overall, our respondents reported working on an average of 4.3 projects at a time and 29 projects in a year. As an example, Roberta, a medical writer in an advertising agency, said she typically juggles 4 projects at one time and completes 18 to 20 in a year. Olivia, who works for a consulting company, estimated that 80% of her work day involves writing. She said she often juggles 8 to 10 projects at one time.

Writers also often need to make decisions about quality. Siena becomes concerned about this as she thinks about juggling several projects. Allie tells her that projects occasionally have different levels of importance, depending on factors such as audience, purpose, and different stakeholders. Related to this, our findings suggested the importance of understanding just "how good" one's writing needs to be in any situation. We were initially surprised that some of our respondents said that writing skill and quality were not the most important things for them. For example, Madeline, a documentation manager, said, "Even without stellar writing skills, if you care about the user's experience, your documentation will have value." She added, "I don't consider perfect writing skills to be the most important skill, at least in our organization." Claire, a proposal writer, said, "Writing ability is necessary, but if I didn't possess the top three skills [interviewing, time management, and industry knowledge], Pulitzer Prize–winning writing skills would be useless in this position."

Most of our respondents, however, ranked quality in writing high in their work. Most said the ability to write well was essential, both in obtaining and in advancing in their jobs. One of these, a writer in a contract organization, said, "When I first began in this type of work, the ability to write and coherently construct a document was critical to my success in the position." Another, who now manages other writers, said, "Writing and editing skills—this is still number 1 for me, primarily because I cannot teach it. And the strong need for these skills is what makes me require a BA in tech writing." This respondent added that writers need not "excel at *all* [her emphasis] aspects of writing . . . as long as they're enthusiastic about having someone pitch in where they have weaknesses." Further, the ability to write well was defined broadly by most respondents—as encompassing, for example, stylistic and mechanical accuracy, sensitivity to audience and purpose, rhetorical skill, editing, clarity, and conciseness.

NATURE OF THE WRITING

In addition to how much writing she will likely do in her job, Siena wants to know the kinds of writing she will be asked to do and what that writing will entail. She asks Allie, "Will I need to create documents entirely from scratch, or will I mostly repurpose existing documents, for example, for the purpose of single-sourcing?" Allie, as a manager, is happy to hear Siena ask these questions because they are important ones for new writers. Siena needs to know how to approach the writing tasks she's assigned. She needs to understand what, precisely, she's being asked to do and why, where and how all of it will fit within the larger context of the department and organization, and so on. By understanding which tasks might be original and which might entail repurposing previous work, she can do a better job managing projects and balancing tasks. Allie might say, "You will do some work from scratch, but often your work will involve reworking existing documentation. However, most important will be making sure you know which you'll be doing before you even begin."

Based on our findings, we decided that the most important questions for this category in our heuristic are

- How much of the content for my writing will I have to research and develop from scratch? How much will I take or borrow from elsewhere? And what, then, will I need to do with it?
- What will be involved in writing original documents in my organization? Will I need to locate and interview subject-matter experts; locate in-house source documents; locate outside sources, such as books, research studies, or Internet sources?
- What will be involved in reusing or repurposing existing documents in my organization? Will I need to know where and how to locate such documents? Will I be provided the relevant documents at the start of a project? Will I need to verify the completeness or appropriateness of the documents with which I am provided? Will I need to conduct additional research similar to what I would do when writing from scratch?

Twenty of our respondents (more than 80%) said they spend at least part of their time creating documents from scratch and that doing so is central to their roles; however, twenty of them, some the same and some different, also said that they spend at least part of their time rewriting or repurposing existing documents. Char, a technical writing manager at a security software company, talked about how her work encompasses both kinds of writing: "Writing means creating the user documentation. We do repurpose most of our guides and quick-start cards, updating information for each release. There are always new products to document, so that requires creating new documentation." Char talked about updating existing documentation for new releases as well as about how existing products may also require entirely new documentation: "Sometimes, based on feedback from the Consulting Engineers, we create new documents (offshoots) for existing products."

In technical communication, repurposing documents typically involves preparing them for delivery in multiple formats. Since single-sourcing is now so common, many writers repurpose documents for this reason. Madeline, a documentation manager in a software company, said, "I spend time thinking about how to structure information so that it can be reused . . . When you move towards single-sourcing, you have to think about how to modularize information as well as how to set up the underlying template structure so that the content outputs appropriately for different types of deliverables." For other respondents, repurposing meant reusing existing text as a way to save time. For example, Claire, a proposal writer, said, "Since many of the same topics are frequently discussed, my department maintains a library of standard, or boilerplate, text that is available for use as is or customizable."

GENRES AND RHETORICAL STRATEGIES

Building on the previous questions, Siena also asks Allie if she can give her some idea of the types of documents she will be developing, including whether a new version of an existing document will be the same type of document as the original. At this point, Siena also should begin considering the larger rhetorical context of her work: Who will her audiences be for her writing? How will they read and use what she produces? What will be the purposes of the documents she produces? And finally, in what ways, if any, should her writing project an image of the company or the product? Allie may well tell her that in some situations these aspects of a project are spelled out very precisely; for many projects, however, the writer, or the team, ends up working and reworking these issues throughout the project.

In relation to the questions for this category of our heuristic, our research suggested that technical writers produce a variety of documents. When we reviewed the completed questionnaires, we counted fifty different types, which we ultimately grouped into thirty categories. The largest category, as might be expected, was that of manuals, guides, instructions, tutorials, and job aids (there were forty-one mentions of these). There were twenty-one mentions of documents such as newsletters, newsletter articles, articles for other kinds of publications, press releases, press kits, and blogs, and about twelve mentions of reports, product reviews, and minutes.

On average, each of our respondents writes eight kinds of documents.

Many are genres we commonly associate with the field, although new types of documents associated with new media are increasingly being added to these lists. As an example of the variety we found, Cecelia, one of the writers at a tax and accounting software company, listed the following: "Getting-started guides, installation instructions, walk-throughs/tutorials, user bulletins, report samples, online help (WebHelp), Captivate sequences (animated demos), training guides, conversion/comparison guides (guides for transitioning from one product to another), status reports, meeting minutes." Lists like these suggest that technical communicators need to know a variety of genres. They also suggest that writers need to be able to move easily between genres, and they need to understand the conventions of various genres. Technical communicators also need to be prepared to produce new kinds of documents, because needs and expectations evolve within most work contexts and with new technologies.

Finally, in relation to our questions about audience and purpose, we found that a common guiding principle for most technical communicators is that everything is driven by the needs of audiences. So how do technical communicators learn about their audiences? Diane, who writes for a financial company, relies on a variety of resources—her manager initially, but primarily the users themselves through interviews, follow-up queries, and observations. In our study, five respondents said they employ user interviews, and three of these also talked about observing the user with the product.

It did surprise us, however, that only six of our respondents (a quarter) talked about having direct contact with members of their audiences. The remainder (three-quarters) talked about having to rely on others in their organizations, on clients, on product documentation, and even on intuition. Within their organizations, audience information came from managers, SMEs, editors, sales representatives, and client representatives. While people in these roles may be familiar with users, accepting their assessment of audiences for technical documentation purposes presumes a familiarity with audience, especially as a rhetorical concept, that they very likely do not possess. What we conclude from these findings is that technical writers need to be diligent in seeking and obtaining sufficient knowledge of their audiences, and of the rhetorical contexts of their work more generally, employing a full range of strategies. In short, a writer like Siena may realize that she needs to begin thinking about her audiences as soon as she is assigned a project, and that she needs to be strategic in learning about and considering how best to address them. Directing questions about audience and purpose to her manager might be just one of multiple strategies Siena uses to establish an effective course for her work.

APPROACHES TO AND PROCESSES FOR WRITING

The next category of our heuristic concerns how technical communicators actually write. Siena, in all likelihood, learned in school that there is no single writing process that works in every situation. She probably also developed confidence in her ability to write well. However, she wonders if she can count on that confidence in her new situation. Some questions she will ask Allie include

- Are any processes already set in place that I will be expected to follow, or can I work according to my preferences?
- Are the genres of writing specific to the company or to the industry?
- How complex will my projects be, and how firm are project deadlines?

In response to these questions, Allie informs her that the company has "SOPs—standard operating procedures"—for different documents, but that standards need to be adapted to particular situations because every project is unique. As she considers Allie's response, Siena also likely realizes that it often may be up to her to figure out the best process for each project.

Once Siena is assigned her project, she will likely have to determine, first, how best to research it. As most of our respondents acknowledged doing, she will need to consider questions like these:

- Who are the appropriate SMEs, and how accessible are they?
- Are there documents already existing that support or explain the technology?
- Can I get access to and use the technology?

After Siena has researched the product and its users, she will begin, at some point, putting words, images, or multimedia together to compose the document. She will then need to know the answers to such questions as

- What tools and skills are required for the project?
- What will be the best way to break the project down into stages of development?
- Will I need to team up with other writers on some stages of the project?

This is where writers put their rhetorical knowledge to work in the service of the actual writing. As suggested previously, effective writers have a repertoire of rhetorical strategies they can draw on. More than a few of our respondents, for example, talked about planning their documents very deliberately, often by using outlines and templates. For these writers, organization of the document is a primary concern. Others talked about creating planning documents to move into composing, and some talked about diving right in and drafting.

Finally, Sienna should understand the review and editing process in her company. She will want to know

- Will I edit my own work?
- Will I edit the work of others?
- Who checks the technical accuracy of my work?

Several of our respondents stressed the importance of the review and editing stages of the writing process, emphasizing the contributions that others can make to one's writing. In fact, more than half said that editing their own and their colleagues' work is one of the tasks they do most often. Maureen said that in her organization, which is focused on marketing, "Everything is reviewed by someone else." Further, she said, "account managers assign projects to me, and then review what I've written to double-check I've covered the client's requirements AND adhered to brand standards."

The review processes our respondents described were almost as varied as their writing processes. Susan, the manager at the company specializing in tax and accounting software, described a review process that involves a range of constituents: "When the SME is happy with the doc (or, often, when they've run out of time to hone further), the doc is routed for formal review to multiple departments: Development (which includes the SME), Support, Training, at least one other Tech Comm staff member, and other interested parties (e.g., Sales) as required." The process Susan described is a complex one that involves negotiation, interaction, and sometimes even office politics.

KNOWLEDGE AND SKILLS

Technical communicators like Siena also need to possess technical skills and knowledge. This certainly involves skills with the technologies needed to write, design, and edit documentation. It also includes knowledge of, or ability to learn about, the technologies they will write about: for example, finance, transportation, telecommunication, health care, and so on. Siena, therefore, wonders how prepared she is for the demands of a real professional workplace. The questions she'll want to ask herself—or her manager—include the following:

- What technologies will I need to know (hardware, software, digital communications technology, new media, etc.)? How much skill will I need with these technologies?
- What will I need to know about the industry for which I write? Also, how important will it be to be familiar with the industry for which I will be writing?
- What will I need to know about the subject about which I write? Will I be expected to be an expert on the subject matter about which I'm assigned to write? If not, will I be expected to know how to find the information on my own?

Siena needs to be flexible in learning and using technology, both that which is new to her as well as that which she may already know but may be using in new ways or for new purposes.

Our respondents certainly agreed that technological skill is important: nineteen ranked it among the top skills they themselves have. Fourteen also said that knowledge of technology was essential for obtaining their jobs. Maureen, a writer in a marketing agency, summed it up: "The computer is king. If you can't use it, you're dead. I work from my home office, and spend 95% of my work time in front of the computer." Susan also stressed the importance of technology, but said that for her it is not number one: "Obviously, we need writers who are comfortable with the considerable technical aspects of the job and who won't panic when the software they're documenting crashes repeatedly, as software under development so often does. And it's definitely an advantage if they've had some experience in the actual tools that we use. That said, I never hire based on tools expertise because (a) they change all the time, and (b) we can teach these skills to a new writer." On the topic of writers learning tools, Susan had this to say: "I want writers to become experts on the product they document, not on a certain type of deliverable and the tools used to develop it. So everyone in TC needs to learn most of these tools. This makes for some variety in their work, but it means they have to be flexible and quick learners."

When asked what they use in their everyday work, respondents identified sixty-four different tools. We were also able to pull from our data several areas of specialized knowledge that relate to technology. These include knowledge of document design, web design, project management, multimedia design, content management, editing, single sourcing, and computer programming. Of course, mastery of all of these tools and areas of knowledge would be impossible, which is, again, why so many of the respondents stressed the ability to learn new tools. Further, outside of

Microsoft Office and a few other tools—namely Dreamweaver, Photoshop, FrameMaker, and RoboHelp—most tools were mentioned by only a few respondents, suggesting that there is great variety in what writers in the field are using.

PERSONAL TRAITS AND QUALITIES

As she gains experience, Siena increasingly realizes that she is going to have to deal with many different personalities and work styles. Her co-workers also will tell her that projects seldom proceed as anticipated, and, as a result, she will often have to respond to contingencies. The questions she will want to ask herself, and perhaps her manager, include the following:

- As a technical communicator, will I primarily be expected to work alone or closely with others?
- Will I be expected to plan my own work processes, or will I have projects mapped out in detail by a supervisor or team leader?
- How adaptable will I need to be? How open-minded?
- What will it mean to be adaptable and flexible in my organizational context?
- How important will learning and acquiring new knowledge be in what I do?

All of the writers who responded to our questionnaire identified traits and qualities that they have found important in their work. Interpersonal skills, for example, were viewed by most of our respondents as essential. Few jobs exist in which technical communicators work alone, with little or no need to talk to and negotiate with other people. Time and again, previous studies and our own research have made it clear that the technical communicator needs to be a "people person," outgoing, good at oral communication as well as writing, and adept at working effectively with a variety of people. As mentioned previously, technical communicators also need an interest in and passion for learning as well as an ability to adapt easily to change. Our research supports this. In short, if our writer, Siena, loves learning and has a desire to learn—if she's someone who is not afraid to learn new things—she will likely do very well as a technical communicator.

CONCLUSION

The twenty-first-century technical communication workplace is not monolithic. Your formal education is a vital foundation for a career as a technical communicator, but you also need to continue learning, and this starts the first day on the job. It is common for technical communicators to change jobs, often because they want the challenge of working in a new industry or of doing different kinds of writing. Many also take on added responsibility or new roles within their organizations (e.g., as managers or supervisors). As our writer, Siena, advances in her career, she will very likely internalize the heuristics we have made explicit in this chapter, but she will continue to seek answers to the questions about writing in whatever new situation and context she finds herself.

In particular, she will look for indications of what is expected in terms of good writing and how it is typically accomplished for particular tasks. She will want to see how genres are adapted to workplace contexts, subject matter, organizational goals, user needs and expectations, technologies and media, and other factors specific to a situation. She will expect to have to learn new subject matter, new writing and editing technologies, and new project and content management tools. She may also have to learn to work in organizational structures that are different from any in her past experience. She may have to work with colleagues, clients, or users whose cultural or national backgrounds are different from her own, or who speak English differently than she does. But one thing will probably remain the same for Siena—she will enjoy and welcome the challenges of new technologies; of working with a variety of people; of figuring out organizational processes, structures, and cultures; and of developing communication products and processes that truly connect with others to help them do their work.

DISCUSSION QUESTIONS

1. Our research suggested that knowing your audience is one of the most important things in technical communication. What are some possible approaches to learning about audience? What can you do when you aren't able to talk directly with the members of an audience? What might be some other ways to find out about the people who will use what you write?
2. How would you describe yourself as a writer? What do you believe are some characteristics or qualities you possess that will assist you with the writing you will do in your career?
3. You probably have learned a lot of research skills in school. You may also have done client-based projects in some of your technical communication courses, where you had to apply these skills to the projects that the clients asked you to do. Based on your own experience and the findings from our research, what research skills do you think will serve

you best as you begin your work in the field? How do you anticipate using those skills?

4. Susan, the technical communication manager at a tax and accounting software company, said, "I want writers to become experts on the product they document." Assuming you do not have such expertise when you are hired, what strategies might you use to become an expert, as she suggests? How much time do you think you would need to acquire the expertise you need?
5. How important do you believe it is to be familiar with the industry for which you will be writing? How might you acquire that familiarity? Also, do you think it will be more important to know about writing and to have the skills for being effective as a writer? Explain your perspectives on this.
6. Arrange to interview a technical communicator in a local industry or organization. Plan your interview to focus on a sampling of questions from at least three of the heuristic categories in this chapter. Report to your class on how closely the technical communicator's perspective aligns with what we've reported from our research in relation to these categories.
7. How might the technology used in your professional workplace compare with or differ from the kinds of technology you learned in school and the ways you used it in school? How can you best prepare yourself for using the technology that you may end up using in your work?
8. Find a technical document for a technology that interests you. Assume that you work for the company that developed this technology and that you have been asked to write another document about the technology for a different purpose or audience. Outline a plan for completing this task, using some of the relevant heuristic questions in this chapter.
9. Do you believe that your education has prepared you adequately for the writing you will do in your job? In what specific ways might that writing differ from the writing you did for your courses, both within and outside of technical communication? You might also contact a technical communicator in the workplace and ask what courses they recommend you take and what they believe you should know before completing your program and entering the field.
10. Working with another student, find technical communication job ads that interest each of you. Then take turns interviewing each other for the job you chose. As the interviewer, plan interview questions based on the ad but also using ideas from the heuristics in this chapter. Do not tell your partner what the questions will be prior to the interview.

As the interviewee, prepare for your interview by relating the heuristics in this chapter to the requirements of the ad. What qualities, knowledge, skills, and experience do you have that you think will make you a good candidate for the position? What requirements do you lack? Can you present yourself in ways that could compensate for what you lack? Do the heuristics help you do that?

WORKS CITED

Anschuetz, Lori, and Stephanie Rosenbaum. 2002. "Expanding Roles for Technical Communicators." In *Reshaping Technical Communication: New Directions and Challenges for the 21st Century*, edited by Barbara Mirel and Rachel Spilka, 149–163. Mahwah, NJ: Lawrence Erlbaum Associates.

Carliner, Saul. 2001. "Emerging Skills in Technical Communication: The Information Designer's Place in a New Career Path for Technical Communicators." *Technical Communication* 48:156–175.

Conklin, James. 2007. "From the Structure of Text to the Dynamic of Teams: The Changing Nature of Technical Communication Practice." *Technical Communication* 54:210–231.

Faber, Brenton, and Johndan Johnson-Eilola. 2003. "Universities, Corporate Universities, and the New Professionals: Professionalism and the Knowledge Economy." In *Power and Legitimacy in Technical Communication: The Historical and Contemporary Struggle for Professional Status*, edited by Teresa Kynell-Hunt and Gerald J. Savage, 209–234. Amityville, NY: Baywood.

Farkas, David K. 1999. "The Logical and Rhetorical Construction of Procedural Discourse." *Technical Communication* 46:42–54.

Ford, Julie Dyke. 2007. "The Convergence of Technical Communication and Information Architecture: Creating Single-Source Objects for Contemporary Media." *Technical Communication* 54:333–342.

Giammona, Barbara. 2004. "The Future of Technical Communication: How Innovation, Technology, Information Management, and Other Forces Are Shaping the Future of the Profession." *Technical Communication* 51:349–366.

Hart, Geoff. 2000. "Ten Technical Communication Myths." *Technical Communication* 47:291–298.

———. 2001. "Conquering the Cubicle Syndrome." In *Writing a Professional Life: Stories of Technical Communicators on and off the Job*, edited by Gerald J. Savage and Dale L. Sullivan, 69–73. Needham Heights, MA: Allyn and Bacon.

Hart, Hillary, and James Conklin. 2006. "Toward a Meaningful Model for Technical Communication." *Technical Communication* 53:395–415.

Hart-Davidson, William. 2001. "On Writing, Technical Communication, and Information Technology: The Core Competencies of Technical Communication." *Technical Communication* 48:145–155.

Hayhoe, George F. 2000. "What Do Technical Communicators Need to Know?" *Technical Communication* 47:151–153.

Isakson, Carol S., and Jan H. Spyridakis. 1999. "The Influence of Semantics and Syntax on What Readers Remember." *Technical Communication* 46:366–381.

Kim, Loel, and Christie Tolley. 2004. "Fitting Academic Programs to Workplace

Marketability: Career Paths of Five Technical Communicators." *Technical Communication* 51:376–386.
Lanier, Clinton R. 2009. "Analysis of the Skills Called for by Technical Communication Employers in Recruitment Posts." *Technical Communication* 56:51–61.
Mirel, Barbara. 1996. "Writing and Database Technology: Extending the Definition of Writing in the Workplace." In *Electronic Literacies in the Workplace: Technologies of Writing*, edited by Jennie Dautermann and Patricia Sullivan, 91–114. Urbana, IL: National Council of Teachers of English.
Rainey, Kenneth T., Roy K. Turner, and David Dayton. 2005. "Do Curricula Correspond to Managerial Expectations? Core Competencies for Technical Communicators." *Technical Communication* 52:323–352.
Schneider, Barbara. 2002. "Clarity in Context: Rethinking Misunderstanding." *Technical Communication* 49:210–218.
Slattery, Shaun. 2005. "Technical Writing as Textual Coordination: An Argument for the Value of Writers' Skill with Information Technology." *Technical Communication* 52:353–360.
Thomas, Shelley, and Becky Jo McShane. 2007. "Skills and Literacies for the 21st Century: Assessing an Undergraduate Professional and Technical Writing Program." *Technical Communication* 54:412–423.
Whiteside, Aimee L. 2003. "The Skills That Technical Communicators Need: An Investigation of Technical Communication Graduates, Managers, and Curricula." *Journal of Technical Writing and Communication* 33:303–318.
Winsor, Dorothy A. 1992. "What Counts as Writing? An Argument from Engineers' Practice." *Journal of Advanced Composition* 12:337–347.

KAREN SCHRIVER

16 What Do Technical Communicators Need to Know about Information Design?

SUMMARY

As a technical communicator, you will need to develop your expertise not only in writing, but also in information design. Whether you are designing reports about the environment or manuals for smartphones, you will need to be concerned with how to present your messages visually and verbally. As organizations around the globe expand the audiences for whom they create products and services, the demand for well-designed text and graphics has become increasingly important. Although most technical communicators have honed their skills in verbal expression, they often lack knowledge of information design and have few strategies for integrating visual and verbal content. They may not be familiar with the growing research on information design that could help them make decisions about designing content spatially and typographically. This chapter offers a heuristic for structuring content visually and for making the structure salient through grouping, organizing, and signaling.

INTRODUCTION

Miguel recently acquired a new job with a government agency within the U.S. Department of Agriculture. His department's mission is to spread the word about healthy eating habits to families across the United States. Miguel's supervisor started him off in his new position with two tasks. The first task is to revise a flyer for parents about healthy foods and their benefits. The flyer consists of a list of foods that research has shown to be nutritious, and gives reasons why they are nutritious. Miguel's second task is more demanding. His supervisor asked him to create the first in a new series of documents to inform parents about issues related to nutrition—issues such as obesity, nutritional supplements, weight gain, natural cures, and myths about healthy eating. The supervisor expects that Miguel's document will serve as a template for the series. He told Miguel that the research on healthy eating has changed substantially since the department last printed its information, rendering the previous content obsolete. Miguel must begin with a careful review of the current research on nutrition. He recognizes that to do the task well, he must synthesize the most recent nutritional research and then design the content from scratch.

Miguel's overall goal for these tasks is to create easy-to-understand texts that can be presented on paper or on the department's website. He begins by searching many nutrition-related databases with an eye toward compiling the available data that could be used to make recommendations for healthy eating. Initially, he thinks the task is obvious and straightforward, but he begins to feel a bit overwhelmed when he finds that the amount of available research is enormous, and that it is scattered across professional and popular literature in print and on the web. Moreover, much of the information he finds is contradictory, making it hard to understand and to draw conclusions from.

What approach could Miguel take to meet these rhetorical challenges? How should he proceed in making decisions about the content? Once he gathers the appropriate content, how can he design the information so that it will be easy for any parent to understand? What are the alternatives for organizing the material visually and verbally? What are the options for displaying the content spatially? What sorts of graphics could he use? Could his choice of typography encourage reading?

This chapter presents a heuristic that Miguel could use for addressing these rhetorical challenges or other complex problems in technical communication. In particular, the heuristic explores three information-design principles for making one's thinking visible to readers: (1) grouping content rhetorically, (2) organizing content visually, and (3) signaling structural relationships. Each of these principles will be explored through practical examples that illustrate how they can guide technical communication. Taken together, the ideas presented in this chapter can help you place more emphasis on designing visually—attending carefully to issues of graphics, spatial display, or typography—and focus more attention on making the structure of your content visible to readers. When technical communication skills are combined with sophistication in information design, it can lead not only to more effective professional communications, but also to the development of the communicator's expertise (Schriver 2012). This chapter will help you expand your repertory of strategies for solving communication problems by drawing on evidence-based principles of information design.

WHY RESEARCH ON INFORMATION DESIGN IS USEFUL

As the above scenario from the Department of Agriculture shows, technical communicators could benefit from research about strategies for visual design. In this way, they could both manage their own creative process and do a good job in persuading supervisors that their design choices are sound. Research on information design helps technical communicators understand how different people typically engage with visual and verbal content. It provides insight into the kinds of textual or graphic "moves" that tend to make content clear and understandable—promoting comprehension, satisfaction, and usability. Moreover, information-design research shows us how visual design may influence the persona projected by the content and how design can influence whether readers believe in the credibility of the content and the trustworthiness of the authors (Schriver 1997).

WHAT IS INFORMATION DESIGN?

Information design is the art and science of integrating writing and design so that people can use content in ways that suit their personal goals. Information design involves making communication artifacts by shaping verbal language and visual language (and increasingly, in many technical and scientific contexts, mathematical language). A fundamental goal for information design is to enable and enhance relationships among stakeholders for an artifact—that is, among the variety of audiences, clients, critics, readers, listeners, users, and viewers who have a stake in the content. The field of information design focuses on devising novel ways to enable relationships among people through the effective design of content (Frascara 2010). By drawing on principles of information design, technical communicators can design more rhetorically effective communications.

As technical communicators shape their communications for stakeholders, they need to orchestrate word and image in order to achieve the optimal selection of content, the most appropriate organization, the best level of detail, and the best mix of media. Of course, achieving the optimum is not easy. Research tells us that technical communicators need to acquire sophisticated skills in modeling stakeholders' processes of interpretation (Schriver 1989, 1997). When research is available that can help them make better choices, experienced technical communicators opt for writing and design strategies that have been evaluated for their effectiveness with readers.

Experienced technical communicators recognize that they may perceive content differently than their audiences, who typically bring different knowledge, background, experience, or culture to bear during interpretation. Experts pay close attention to the findings of empirical research in order to understand the kinds of writing or design choices that may help or hinder readers as they attempt to make sense of text and graphics.

The stakes are high because bad information design can leave stakeholders with a lasting negative impression, inviting them to ignore a message, misunderstand it, poke fun at it, or simply give up and stop reading. Technical communicators need to worry about people's cognitive and emotional responses to their content.

INFORMATION DESIGN EMPHASIZES THE VISUAL

The emphasis on designing content visually stems from information designers' long-standing belief that the appearance of a communication influences whether people will want to read it. Studies of reading show that people who are confronted with content begin to interpret that content immediately (Anderson 2009), and that the visual display of that content can help or hinder people's interpretation. Research suggests that people may form immediate opinions of the visual display of content. For example, website users found it easy to rate a website as attractive or unattractive, and surprisingly, they could make such judgments reliably and consistently based on their looking at a webpage for only one-twentieth of a second (Lindgaard et al. 2006).

Moreover, researchers have shown that people remember content presented visually more easily than content presented verbally (Paivio 1969). Importantly, research finds that a careful integration of both words and pictures engages people more effectively than either alone (Sadoski and Paivio 2001). When designers provide access to their content through both visual and verbal means—what psychologists call "dual coding"—readers will have two ways of understanding the content and are more likely to remember it (Paivio 1990). Studies show that people tend to remember more when they acquire new content visually and verbally, rather than just visually or just verbally.

Because the human eye is hardwired to interpret and organize what we see, people tend to make immediate inferences about what they see in their field of vision. This aspect of human interpretative processes has implications for information design. On one hand, if the visual display meets the readers' expectations, readers will recognize the structural cues, be able to navigate the content, and be more likely to sustain their reading. On the other hand, it also means that if the visual display of the text does not attract viewers or confuses them, reading may never begin or will stop before much of the content is considered.

While research on information design is broad and diverse, technical communicators can most productively start with three areas central to displaying content visually:

- grouping content rhetorically,
- organizing content visually to show contrasts, and
- signaling structural relationships.

GROUPING CONTENT RHETORICALLY

Technical communicators already know how important it is to shape their writing with a reader or viewer in mind. Taking an audience-centered perspective applies to grouping content as well. Information designers are concerned with making sure the visual display of the content is rhetorically effective. This means formatting the content into meaningful groups that readers will notice, expect, and appreciate.

Research suggests that visual grouping gives readers a sense of the overall structure (Tullis 1997). When text and graphics are organized into meaningful semantic clusters, it makes it easier for readers to chunk the content (Kahn, Tan, and Beaton 1990). Grouping can also reduce cognitive load by helping readers remember content, which can make the content seem less complex, leading to fewer errors and increased satisfaction (Niemela and Saarinen 2000).

Grouping Can Be Visual or Verbal

Technical communicators are highly adept at grouping verbal content according to its purpose. For example, in authoring user assistance, writers separate overview information from procedural information. In constructing a report about a scientific research study, the writer purposely separates information relevant to the literature review from material about the study proper. In fact, a good part of technical communicators' work involves grouping content strategically, striving to make implicit structures explicit for readers.

Technical communicators can make verbal groups more evident by employing visual devices such as sidebars, itemized lists, boxes, shading, color, and white space. When content is formatted consistently (e.g., all procedures use enumerated lists with a short line length while all overview text is formatted with a longer line length), readers can readily perceive intended relationships among the content elements.

In addition to technical communicators' developing skills in visually structuring verbal content, it is also important for them to develop strategies for grouping visual content spatially; for examples, see Brumberger and Northcut 2012. Structuring visual content involves grouping visual materials to make an argument or to tell a story, creating purpose-driven groups from source materials such as photos, technical illustrations, digital art, tables, charts, diagrams, and data displays.

Grouping Can Have Cognitive and Affective Benefits for Readers

When content is grouped in ways that allow readers to form meaningful relationships among the elements, readers can often make connections across the content that they might miss otherwise. Grouping content spatially makes the content more coherent, allowing readers to recognize how the pieces of the message fit together. In this way, grouping helps make what otherwise might be invisible structures apparent to the reader. Grouping not only organizes the content, it also renders it visually conspicuous—quite important for busy readers, impatient readers, less-able readers, and people who are reading in a second language.

How the content is grouped may also influence readers' first impressions of the message (Lindgaard et al. 2006), setting in motion positive or negative attitudes about the content (Schriver 1997). This makes it important to catch the reader's attention and to make a good impression at first glance.

Grouping Can Simplify Complex Content

Technical communicators are often faced with the task of reorganizing lengthy content into meaningful groups, particularly as they help their clients move content from paper to the web. Breaking the text into short paragraphs (e.g., one to three sentences) promotes faster reading, and importantly, the shorter length influences readers' sense of how much effort it will take to read. Researchers at the Poynter Institute for Media Studies used eye tracking to investigate how people read print and online newspapers. They found that readers skim and scan the content, following the modular clusters of the newspaper's layout (Stark Adam, Quinn, and Edmonds 2007). In a related study, readers tended to give stories with short paragraphs twice as much attention as those with longer paragraphs (Outing and Ruel 2004).

Grouping Can Show Semantic Relationships

By clustering content in ways that make the text simple and inviting, technical communicators can make reading less effortful. We can also use grouping to show how the content is logically and semantically related. For example, we can use white space to group semantically related content. When groups of content are positioned in close proximity, whether on a page or screen, readers can easily infer their relatedness. In this way, proximity is a powerful grouping tool.

Grids (discussed later in this chapter) are ways to structure space. They are modular visual design systems for organizing pages, screens, and three-dimensional spaces. They can be powerful tools for showing semantic relationships among elements of information spaces. Technical communicators can benefit from gaining familiarity with concepts related to the design of typographic grid systems—concepts such as layering, grouping, separating, zoning, and highlighting (Müller-Brockmann 1985; Keyes 1993; Samara 2002). Some of these concepts, mainly developed by design practitioners, have been tested; for example, research finds that by aligning content elements and by consistently positioning them, designers can help viewers find information more quickly (Parush, Nadir, and Schtub 1998).

The overall visual look, including the layout, significantly shapes users' perceptions of consistency and their satisfaction in browsing tasks (Ozok and Salvendy 2000). When readers can form judgments about the nature of the content quickly and see how elements are related, they are more likely to continue reading. Studies of how people read on the web suggest that getting people to sustain their reading represents one of the biggest challenges for authors of websites, and that people expect content that is organized to fit their unique reasons for coming to the content (Schriver 2010). Getting people to read and keep reading is an important benefit of grouping content rhetorically.

ORGANIZING CONTENT VISUALLY TO SHOW CONTRASTS

In the early part of the twentieth century, Gestalt psychologists systematically studied how the properties of the visual world shape our perceptions (Wertheimer 1922; Köhler 1947). One of the earliest discoveries of Gestalt psychology was that the way things look depends not just on the properties of their elementary parts, but also, and more importantly, on their organization. They pointed out that *contrast is fundamental to human perception* and that the human eye is attracted to areas of high contrast (dark-light, large-small, thick-thin, saturated color–unsaturated color).[1] In designing content, contrast can be rendered graphically or typographically.

Enhancing Contrast

Technical communicators can create graphic contrast by juxtaposing changes in size, shape, color, weight, saturation, and position (Mullet and Sano 1995; Tufte 1983). When used purposefully, contrast can reveal the architecture of the content, making the hierarchy visually present and helping readers see relationships among the parts (Ivory, Sinha, and Hearst 2001). Contrast also helps people as they search for content. It highlights key information on a page, segregating main points from minor ones (Jenkins and Cole 1982; Scharff, Hill, and Ahumada 2000). Contrast can be achieved in many ways: for example, through integrating pictures within pictures, text within text, and pictures in relation to texts. When the eye is confronted with contrast, it provokes curiosity and invites the reader to look closer to see what is going on.

The best typographic contrast is achieved by using black type on a white background. In every study in which this aspect of the text has been studied, the same result was obtained: the darker the type and the whiter the background, the better the legibility of the text (Scharff, Hill, and Ahumada 2000; Muter and Marrutto 1991). Moreover, studies of low-vision readers suggest that contrast is a fundamental perceptual aid to their seeing the text at all. Reece (2002) found that low-vision readers rely heavily on typographic contrast to see the text, adding extra reason for technical communicators to concentrate on building effective typographic contrast into the overall design of their content.

Researchers have studied the role of typographic features in text design since the 1800s. Consequently, most of what we know about typographic design has been derived from the way people read on paper. In the past decade or so, researchers in journalism, information design, usability, and human factors have conducted studies of online typography. Some findings of the typographic research apply to both print *and* online information design, while other findings apply particularly to print or to electronic displays (computer screens, tablet displays, smartphones).

Typeface: Serif or Sans Serif?

Research that applies to both print and online applications tells us that typographic *contrast* is a critical feature of good design. Good typography clearly signals the hierarchy and structure of content. When the contrast is working well, readers or viewers can easily make discriminations about how the content is organized. Moreover, good contrast facilitates search and rapid retrieval of information. Typographic contrast can be signaled in many ways, including contrasts of light to dark, thick to thin, roman type to italic, uppercase to lowercase, serif to sans serif. Figure 16.1 shows the difference between a serif typeface (16-point Palatino) and a sans serif typeface (16-point Myriad Pro).

In the research on typography, one finding appears repeatedly in the literature. Studies comparing serif and sans serif faces find that readers pay more attention to the amount of contrast among styles within a typeface (e.g., light, medium, bold, extrabold, black) than they do to the distinction between serif and sans serif (Schriver 1997; Spencer, Reynolds, and Coe 1974).

Research shows that when the typographic resolution is excellent, serif or sans serif typefaces are equally legible and equally fast to read. Thus, when text is printed on paper at high resolution, the text will be

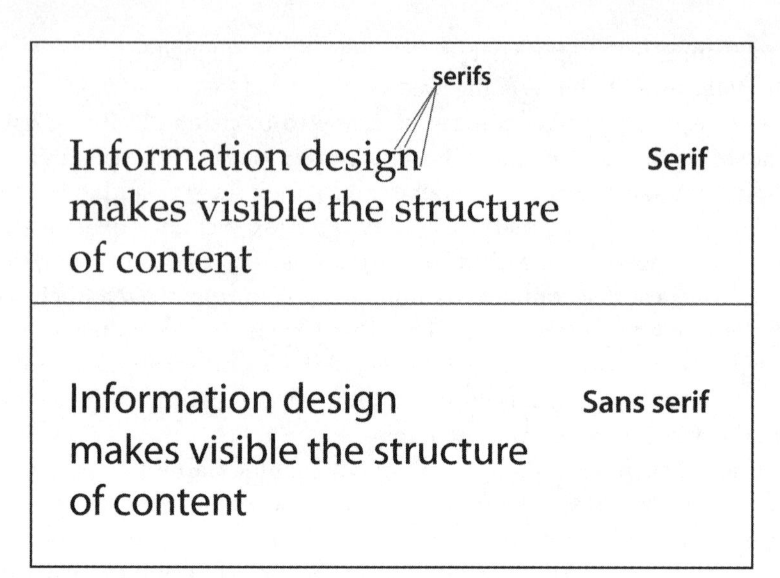

Figure 16.1. Differences between serif and sans serif typefaces

equally legible whether the type is serif or sans serif. However, when the resolution is average or poor, sans serif is more legible (Bernard et al. 2001). The legibility of either serif or sans serif typefaces at a certain point size may differ, for example, depending on the resolution of the computer monitor, smartphone screen, tablet computer, or video projector; it is common for type displayed on high-resolution screens to appear smaller but crisper while type displayed on low-resolution screens appears larger but fuzzier. Legibility matters a lot when busy readers must distinguish between pairs of characters such as *o*'s and *e*'s, *8*'s and *6*'s, *o*'s and *o*'s. Practical situations in which readers must make rapid discriminations among numbers or characters include e-mail addresses, URLs, credit card numbers, serial numbers, order numbers, and prescription numbers.

Research on paper documents showed that when Americans engage with continuous text, such as a short story, they preferred serif faces, but when reading more technical prose, such as found in instruction manuals, they preferred sans serif (Schriver 1997, 289–303). Anecdotal reports suggest that Europeans are as comfortable with sans serif typefaces for reading continuous text as they are with serif faces. Researchers conclude that people tend to prefer what they are accustomed to reading.

Even though there is no difference in the legibility of serif and sans serif type when the screen resolution is good, people still have preferences for typefaces. And whether they are young or old, as they read online, most people prefer sans serif.

In a number of studies, low-vision readers (that is, severe loss of vision even with glasses) and readers with "normal" vision (that is, 20/20 with or without glasses), were found to prefer sans serif to serif type in computer displays 87–95% of the time (Reece 2002). Readers also preferred roman (nonitalic) type to italic type (67–82% of the time). Overall, Reece's study underscores the importance of typographic contrast for low-vision readers and suggests that both reduced-vision readers and normal readers show strong preferences for sans serif typefaces online. Theofanos and Redish (2005) remind us that it is important to think about low-vision users when we design text for online displays because vision impairments are more prevalent than we think and the number is growing as the baby boomer generation ages.

Type Size

Generally speaking, research on type size has investigated the impact of point size on reading speed, reading accuracy, and reader preference. Most studies have concentrated on the point sizes that are most legible for body text: 10-point, 12-point, and 14-point type (Bernard and Mills 2000; Bernard, Liao, and Mills 2001; Dyson 2004). The results of these studies are fairly consistent, leading researchers to draw these conclusions.

- Ten-point type is read more slowly, but more accurately, than 12-point.
- Twelve-point is read faster, but less accurately, than 10-point.
- Smaller typography slows reading, but tends to be read with better accuracy (perhaps because more concentration is required to see it).
- Most readers prefer larger type (12-point to 14-point) rather than smaller type (10-point).
- Children (ages nine to eleven) prefer 14-point sans serif.
- Older readers (ages seventy and above) prefer 14-point sans serif.
- Partially sighted or visually impaired readers prefer 14-point to 16-point sans serif.

Uppercase versus Lowercase Type

When sentences and paragraphs are displayed using upper-case type (all capital letters), they can be hard to read (Vartabedian 1971). When text is set in all uppercase, the text is rendered a uniform rectangle, without the distinctive word shapes that allow for rapid reading. In fact, when the body text is displayed in all capital letters, reading speed can be slowed about 13 to 20% (Breland and Breland 1944).

Recent discussions of typographic case remind us that there is nothing inherently unreadable about uppercase (Weinschenk 2009). Rather, readers' familiarity with upper- and lowercase may underlie the superior reading speed associated with mixed case. In short, we read typographic displays more quickly when the text is formatted in ways that are more familiar, which for readers in Western cultures means type set using upper- and lowercase letters. Vartabedian's research also shows that when uppercase is employed sparingly, it can make it easy to locate those items because they stand out due to the contrast.

Reading speed is optimal when uppercase and lowercase letters are used together (Rickards and August 1975). The variation in character height among upper- and lowercase letters allows rapid discrimination of letters and facilitates word recognition (Paterson and Tinker 1946). Thus, a combination of upper- and lowercase makes the reading process more smooth and efficient, enabling the eye to track the text more easily. When extra emphasis is needed, bold has been found to be a better cue than uppercase (Coles and Foster 1975).

SIGNALING STRUCTURAL RELATIONSHIPS

Technical communicators signal the content's structure in two primary ways: visually (through layout and typography) and verbally (through cues such as topic sentences, logical connectives, and overviews).

Signaling Visually

To signal the structure visually, we can use a variety of graphic and typographic techniques—size, position, weight, style, repetition, and alignment. For instance, suppose a technical communicator wanted to show that three textual elements were of equal importance in the hierarchy of a communication. It would be appropriate to signal the elements by (1) aligning them vertically as columns of equal size, and (2) giving each column a heading using the same typographic styles. These visual signals would show the parallel nature of the elements within the structure.

Size and position are primary visual cues to signal importance. When a message is dominated by a few large elements (e.g., photos), their size tells the reader they have priority in the message structure. Alternatively, when textual content is placed in a focal position for the reader (such as on the first page of a website), that position indicates the element's significance within the context.

Signaling the various levels of the text's structure can be accomplished through choices in typography. When well chosen, the typographic treatments allow readers to distinguish among, for example, first-level, second-level, and third-level headings. By signaling the structure conspicuously, readers are more likely to notice important relationships in the content and to recognize thematic continuities.

In cases in which encouraging the reader to notice is critical, such as a warning label on a medicine bottle, designers often employ *double signaling*—that is, using more than one cue to draw attention to the message. For example, signals could involve size, weight (e.g., boldface), color, or position. However, it is important to be judicious in signaling with typography, as too many signals (e.g., too much boldface) or too much repetition of a particular cue (e.g., too many itemized lists in a row) can flatten the visual hierarchy, making it hard to tell what is important because nearly everything is signaled.

Designers strive to make their layouts show at a glance the rhetorical relationships among message elements. They do so by using features such as grids, colors, shapes, backgrounds, orientation, and size; for a case study of forms redesign, see Schriver 2011. Research suggests that these dominant features are perceived immediately, making it important that they are well designed and that they cue the structure effectively (Malamed 2009).

Signaling Verbally

Lengthy or complex content may be signaled by layering the content into levels, with more abstract or general content at the top level and more specific content at lower levels of the hierarchy. Technical communicators aim to make the structure explicit through careful composition of verbal cues, including previews, summaries, sidebars, headings, subheadings, topic sentences, advance questions, logical connectives (e.g., *and*, *or*, *consequently*, *next*), structural cues (e.g., *first*, *second*; *on one hand*, *on the other*), metadiscursive cues (*recall my first point*, *imagine the opposite interpretation*), pull quotes, legends, and captions (Spyridakis 1989a, 1989b).

Many print and online documents have a hierarchical structure, signaled by making broader topics (or more important topics) more prominent than specific topics (or less important topics). Subtopics are typically cued by subordination (Farkas 2005). For example, a website's home page is at the top of the hierarchy, its links provide branches to content at a more specific level, and these branches are split again at each level of the hierarchy. While simple content often has a structure only two levels deep, more complex content often requires a four-level or five-level structure.

Designers can improve the quality of their content by (1) organizing it into logical and semantically related rhetorical groups, (2) thinking carefully about how many levels of content are needed, (3) determining how the groups should be organized to orchestrate a message, and (4) using verbal devices such as labels, names, captions, and headings that enable readers to make quick judgments about the nature of each group.

USING INFORMATION DESIGN TO SHAPE CONTENT: A THREE-PHASE HEURISTIC

As a way of summarizing what technical communicators need to know about information design, figure 16.2 presents a heuristic for structuring content visually. It assumes that a thorough analysis of stakeholders' needs and expectations has been conducted and that the communicator has some content to work with. As shown, the heuristic has three interactive phases: grouping, organizing, and signaling. Technical communicators can use the heuristic to remind themselves of considerations they need to take into account as they plan the display of their content.

PHASE 1: GROUP CONTENT INTO RHETORICAL CLUSTERS

The first phase of the information-design heuristic involves grouping content into rhetorical clusters. A *rhetorical cluster* is a collection of coordinated text elements (visual, verbal, or mathematical) that is designed to guide the reader in engaging with the content. A key feature of a rhetorical cluster is that its elements are interdependent, dynamically shaping the reader's interpretation (Schriver 1997, 343). For example, a newspaper

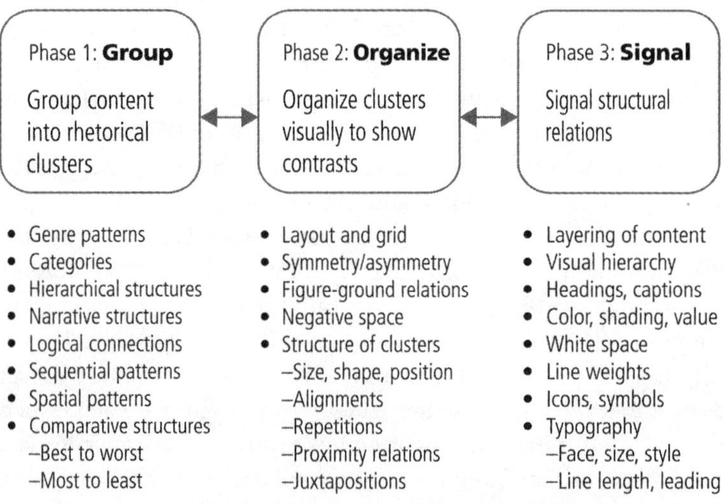

Figure 16.2. A three-phase heuristic for structuring content visually, depicting recursive activities for making structure salient through information design

story might be composed of text, a few levels of headings, a photograph, and a caption. Readers who examine a newspaper might be attracted first to the photograph. Many readers would then inspect the caption in order to interpret the photograph. This coupling during interpretation makes the photograph and the caption work in concert as a rhetorical cluster. The idea is to guide interpretation and to arouse the reader's curiosity.

Rhetorical Clustering versus Chunking

A rhetorical cluster is not a chunk, a commonly used term in technical communication to describe written text that has been segmented into short self-contained topics suitable for presentation on a screen. Chunks described by technical communicators do not correspond to those discussed by psychologists who study human memory. For psychologists, a chunk represents a package of related information in long-term memory (Miller 1956). When technical communicators characterize chunking, they mean introducing white space to break up large sections of text, or designing itemized lists or tables in lieu of lengthy paragraphs. Chunking can help busy readers see a lengthy text as more manageable. It can also facilitate readers' scanning, particularly when each chunk has a clear heading or caption.

Chunking is usually considered an editing strategy for tightening the writing in order to make the chunks of content stand on their own while fitting them into a larger text structure. In contrast, rhetorical clustering is an invention and planning strategy for writing or design, encouraging audience-centric groupings of content for maximum rhetorical impact. While chunking focuses on breaking up an existing text into smaller topics, rhetorical clustering organizes visual and verbal content in ways that allow readers to see important relationships among textual and graphic elements, for example, hierarchical relationships, logical connections, semantic or temporal relations. It is a good idea to consider these relationships prior to considerations of chunking, but certainly they interact.

Devising Preliminary Groupings for the Content

Before choosing a scheme for grouping the content into rhetorical clusters, consider readers' likely purposes and goals for using the information. Draw on what you know about their knowledge, beliefs, expectations, culture, and reading level as you devise a preliminary grouping. Make use of familiar genre patterns (such as problem-solution, if-then, or cause-effect) as you think about optional designs. Sketch a few structures for the content, exploring different ways of grouping the key ideas.

Figure 16.3 shows how a technical communicator's goals for the content can be paired with familiar patterns for organizing ideas. It presents a few typical goals for content and some attendant grouping techniques.

- Explain: flowcharts and hierarchy maps
- Unfold in layers: pyramids—iceberg and inverted pyramids
- Narrate: temporal sequencing
- Instruct: topical progressions or action/step sequences
- Compare: itemized lists, matrixes, tables
- Illustrate part-to-whole: picture-in-picture, hierarchical flow
- Detail cause and effect: node networks (or concept maps) and flow diagrams

These grouping techniques offer ideas for organizing an overall structure or for nesting different structures. Using multiple types of rhetorical clusters in the same artifact adds visual contrast and interest to the content, especially to those typically text-heavy genres such as letters, instructions, reports, and proposals. Figure 16.3 shows only a few of the many grouping techniques available to technical communicators.

Good communicators know that the same content can usually be presented in many ways. But even skilled technical communicators may fall short of planning innovative designs because they ignore the generative power of "trial balloons" en route to a good prototype. Because they work under the pressure of tight deadlines, technical communicators may rely on the first organizational pattern they think of, or merely dump their content into predetermined slots of a template.

Rhetorical clustering is a practical way to try out different structures early in the writing and design process before too many commitments are made. By exploring alternative organizational patterns, technical communicators are more likely to generate a better structure for their content.

Rhetorical clustering allows communicators to form a visual roadmap, helpful in planning an artifact's big picture as well as its little picture. For example, a technical communicator may organize a website using the iceberg approach, also called progressive disclosure or staged disclosure (Lidwell, Holden, and Butler 2003; Nielson 2006). In using the iceberg approach, the technical communicator strives to simplify complex content by organizing it as layers, deferring details or advanced topics to lower in the structure (see figure 16.3, "unfold in layers"). Layer 1 is the top level, or tip of the iceberg. For websites, this is the home page, and here technical communicators must integrate a number of key rhetorical clusters—such as one devoted to organizational identity (logo, name, taglines), another to products or services, and others to navigation and structure. When these

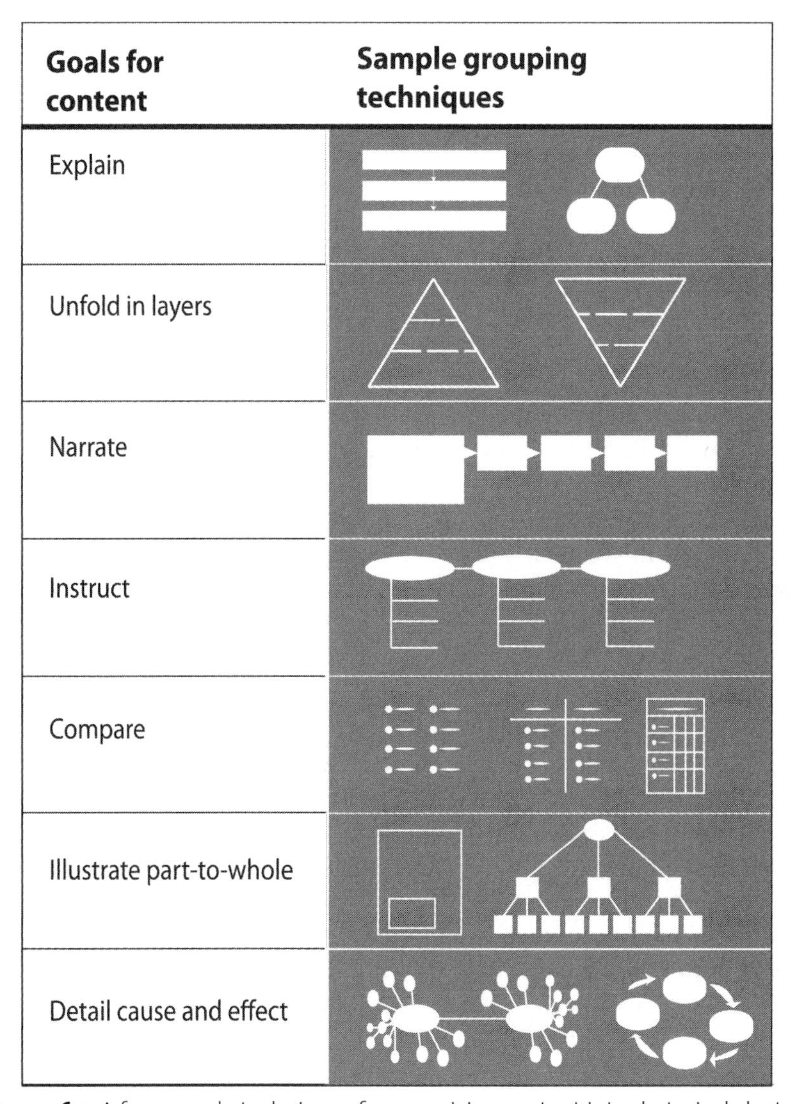

Figure 16.3. A few sample techniques for organizing content into rhetorical clusters

rhetorical clusters are both well designed and carefully integrated, the content provides an explicit trail of what to expect, what some have called good "information scent" (Pirolli and Card 1999), making viewers want to engage with the message, following the trail of cues to lower layers of the message. Layer 2 in the iceberg approach usually refers to the body of the content, and for websites, this is associated with attractive and easy-to-use interior pages. Layer 3 is the detail level, and with websites, this usually means supplementary information (e.g., audio, video, PDFs) and links to related content.

In contrast to the iceberg approach is the inverted pyramid, another technique for layering content, typically used by journalists (see figure 16.3). With this method of structuring content, the most important or newsworthy information is presented first (represented by the widest part of the pyramid). The key rhetorical cluster for the beginning of a news story—the lead—concisely capsules who, what, when, where, why, and how. Level 2 of the news story provides the details, often with rhetorical clusters of text, photos, charts, and graphs. And the third level offers more background information that completes the story. As we can see, the iceberg approach and the inverted pyramid can make it easier to organize the text at the global level and to construct reader-oriented goals for each rhetorical cluster.

PHASE 2: ORGANIZE CLUSTERS VISUALLY TO SHOW CONTRASTS

The second phase of the heuristic (shown in figure 16.2) points to the need to organize rhetorical clusters visually, with the goal of making visible the intended contrasts among elements. Research reminds us that readers tend to form first impressions of content quickly, cuing on the structure and the tone. It is important to organize the visual display of the text and graphics by considering the order in which readers (or viewers) may look at them. This means striving to guide the reader's eye through the content by making the structure visibly clear. The reader's priorities for using the content should dictate the overall spatial display and the need for predictability in the content's positioning.

Some of the time, it will be useful to visually organize the layout by using a grid—a modular system for spatially displaying the content in columns and rows. Grids allow consistent positioning of textual and graphic genre elements (e.g., all first-level headings of a report will fit on one line spanning columns 1 and 2; if the headings need more horizontal space, the designer will shorten them to fit the grid). At other times, the content needs to stand out from the rest of the text or graphics (the surround), and the grid will need to be broken. Providing contrast sometimes means interrupting the rhythm of a sequence. For example, we can break the flow of well-organized but heavily textual content with dramatic changes in size and proportion by interspersing occasional full-page graphics. Sometimes no grid is preferable. Space limitations for this chapter prohibit an extended discussion of layout and grid design, but technical communicators can profit from exploring the graphic possibilities afforded by designing grid systems. Technical communicators need practice in trying out alternative grids for their projects, in learning when to make purposeful deviations from a grid, and in realizing when to abandon a grid entirely; for a discussion of making and breaking of grid systems, see Müller-Brockmann 1985 and Samara 2002.

In organizing rhetorical clusters visually to show desired contrasts (e.g., differences in hierarchy, function, logic, semantics, time), it is important to try out different spatial displays and gauge their relative impact. Once a few alternative displays have been fleshed out, it is a good time in the design process to assess people's interpretations of the prototypes. Obtaining early feedback offers a crucial sense of how people (outside of the design team) may construct the persona projected by the information design (e.g., serious, playful, manipulative, corporate, condescending). Feedback from stakeholders can help technical communicators reconsider issues such as the shape, size, position, and proximity of clusters, as well as about how different clusters interact when juxtaposed (Schriver 1997).

PHASE 3: SIGNAL STRUCTURAL RELATIONS

Once the content has been grouped and meaningful contrasts have been established, technical communicators need to ensure that readers will readily grasp the structure of the message. This requires revisiting stakeholders' likely purposes for engaging with the content, to determine which content elements should be signaled explicitly. The third phase of the information-design heuristic (shown in figure 16.2) asks technical communicators to signal structural relations. The signaling phase operates recursively with the processes of grouping and organizing—each set of decisions dynamically influences the others. For example, the choices a technical communicator makes in designing the page or screen using typographic cues may lead to a reconsideration of proportions for the grid design. A change in the shape of a text element might suggest a change of color to make that shape recede or become more prominent. Indeed, every information-design decision can have an impact on every other.

The technical communicator's goal in carrying out the third phase of the heuristic is to guide the reader's eye through the content by carefully choosing signals that establish the structure and tone. This can be accomplished, for example, through strategic use of white space, color, shading, typography, line length, or other signals. As we saw in the research reviewed earlier, signaling can be visual or verbal—carried out graphically, typographically, or linguistically. Above all, choose graphic and typographic cues that reveal meaningful relationships in the hierarchy of the content.

If the message is primarily textual, strive for typographic cues that emphasize contrasts that bring the structure to the fore. Let us take an

Abstract Reading online is a complex interaction among people, technology, text, and graphics. This four-part article reviews the research literature from 1980 to 2010 about how people read online. Part one investigates the purposes that people bring to reading online, exploring how differences in goals, expectations, and reading skill influence what people do. Part two explores the impact of computer and mobile technologies on reading, asking how technologies enable and constrain reading. Part three integrates the research on good writing and focuses on the text features that help people to understand, remember, and appreciate online content—from words to whole-text considerations (e.g., noun strings, voice, headings, structure, and text density). Part four examines the research on the visual display of content and consolidates the research literature on graphic issues from typography to overall visual impression (e.g., typeface, line length, grouping, hierarchy, and contrast). This article consolidates key issues of interest for information designers and summarizes what we have learned about reading online in order to more effectively fulfill our goals as advocates for readers.

Figure 16.4. The original version of an abstract

example. Figure 16.4 is the original version of an abstract for a scientific paper. The content is formatted as a single block of text. It displays the text in the serif font Times Roman, using a justified format (sometimes called justified right). This format gives the text a solid rectangular shape. Notice that the typographic treatment renders the text a uniform shade of gray and the only prominent visual cue is the bold italics employed for the word *abstract*.

In evaluating such a text with an eye toward creating meaningful typographic contrast, it is important to identify the structure of the message and how to signal it more explicitly. In lengthy structures, this entails selecting a heading style for each level of the text (for example, major headings, minor headings, and subheadings). For each level of the hierarchy, choose a typeface (font), font size, style (light, regular, bold, italic, black, extrablack), and a color. Keep in mind that readers notice the differences among the styles within a typeface (e.g., light, regular, bold, black) as well as whether text is displayed in sans serif or serif. In choosing typography, it is important to apply styles purposefully and consistently. Choose a typeface with excellent contrast among the styles within the face (that is, it has good contrast between the boldface and the regular style). In general, limit the number of typefaces to two that contrast well, such as one sans serif and one serif.

Figure 16.5 shows a second iteration of the abstract presented in figure 16.4. The technical communicator reformatted the abstract to display the text in ragged-right format by removing the justification of the type. She also introduced vertical line spacing to designate six chunks, a visual cue to make the organization of the abstract more apparent. Even so, the abstract could profit from more effective signaling of the structure.

In figure 16.6, the technical communicator modified the design a bit further by introducing a sans serif type (Frutiger bold) to highlight the linguistic cues designating the four parts of the article (e.g., "Part one").

Abstract Reading online is a complex interaction among people, technology, text, and graphics. This four-part article reviews the research literature from 1980 to 2010 about how people read online.

Part one investigates the purposes that people bring to reading online, exploring how differences in goals, expectations, and reading skill influence what people do.

Part two explores the impact of computer and mobile technologies on reading, asking how technologies enable and constrain reading.

Part three integrates the research on good writing and focuses on the text features that help people to understand, remember, and appreciate online content—from words to whole-text considerations (e.g., noun strings, voice, headings, structure, and text density).

Part four examines the research on the visual display of content and consolidates the research literature on graphic issues from typography to overall visual impression (e.g., typeface, line length, grouping, hierarchy, and contrast).

This article consolidates key issues of interest for information designers and summarizes what we have learned about reading online in order to more effectively fulfill our goals as advocates for readers.

Figure 16.5. An initial revision of the abstract in figure 16.4

Abstract Reading online is a complex interaction among people, technology, text, and graphics. This four-part article reviews the research literature from 1980 to 2010 about how people read online.

Part one investigates the purposes that people bring to reading online, exploring how differences in goals, expectations, and reading skill influence what people do.

Part two explores the impact of computer and mobile technologies on reading, asking how technologies enable and constrain reading.

Part three integrates the research on good writing and focuses on the text features that help people to understand, remember, and appreciate online content—from words to whole-text considerations (e.g., noun strings, voice, headings, structure, and text density).

Part four examines the research on the visual display of content and consolidates the research literature on graphic issues from typography to overall visual impression (e.g., typeface, line length, grouping, hierarchy, and contrast).

This article consolidates key issues of interest for information designers and summarizes what we have learned about reading online in order to more effectively fulfill our goals as advocates for readers.

Figure 16.6. A second revision of the abstract in figure 16.4

In figure 16.7, the final iteration of the abstract, the technical communicator shifted the visual display from a series of indented chunks to an itemized list. Notice that this display explicitly nests the four parts of the review within the body of the abstract, making them a visually distinct group yet part of the same rhetorical cluster.

As we can see, each successive version of the revised format introduces more visual segregation among the content elements. The final version of the abstract makes it easier for the reader to rapidly accomplish an important goal: to determine if the content of the scientific paper is of interest.

Let us consider a second example that emphasizes the need for graphic and linguistic cuing rather than typographic cuing. Imagine a technical communicator who is designing a customer catalog of laptop computers and smartphones that range in price from $100 to $3,000. Because the message must be conveyed primarily through graphics, it will be important to consider the size and position of each graphic and whether any layers of content are needed. The process might start with a collection of images of laptops and smartphones of many different sizes.

The technical communicator's first task would be to devise ways to make all images of laptops a similar size, and then repeat the process for the smartphone images. When same-size images are displayed in the catalog, they instantly cue potential buyers of the parallel nature of those products. The next consideration would be to devise a way to display items in parallel fashion, for example by their functionality and price. The technical communicator could use shading or bounding boxes to help readers see relationships among items (such as orienting the laptops left to

Abstract Reading online is a complex interaction among people, technology, text, and graphics. This four-part article reviews the research literature from 1980 to 2010 about how people read online.

- **Part one** investigates the purposes that people bring to reading online, exploring how differences in goals, expectations, and reading skill influence what people do.
- **Part two** explores the impact of computer and mobile technologies on reading, asking how technologies enable and constrain reading.
- **Part three** integrates the research on good writing and focuses on the text features that help people to understand, remember, and appreciate online content—from words to whole-text considerations (e.g., noun strings, voice, headings, structure, and text density).
- **Part four** examines the research on the visual display of content and consolidates the research literature on graphic issues from typography to overall visual impression (e.g., typeface, line length, grouping, hierarchy, and contrast).

This article consolidates key issues of interest for information designers and summarizes what we have learned about reading online in order to more effectively fulfill our goals as advocates for readers.

Figure 16.7. A final revision of the abstract in figure 16.4

right and putting a gray box behind all laptops that are $500 or less). The idea is to use graphic cues such as size, position, alignment, background, and color to make relationships salient for readers. In this way, readers can glance over the content and quickly determine how items are related, which ones look interesting, and which items to skip.

Readers of such catalogs would expect that similar products would have descriptive blurbs of about the same length. Having text that is visually parallel shows readers that they are scanning a family of competing products. By drawing on both textual and graphic cues, the technical communicator could call attention to particular products that have special features, such as smartphones that have GPS tracking. The technical communicator might decide to use a large table format with images on the top row and blurbs on the second row, enabling readers to compare each product picture by picture and feature by feature.

As we can see, the writing and design processes interact recursively. While sometimes technical communicators start their work with mainly images, at other times they start with mainly text. As they execute their plan for a communication artifact, the writing may influence the design or vice versa. It is important to allow for this healthy symbiosis between writing and design by thinking about the integration of the visual and verbal during the invention and planning phase of technical communication.

Consider Preferences for and Familiarity with Cues

Most people prefer sans serif typefaces as they read online—whether they are children, college students, senior citizens, or people with low or corrected vision. A sans serif face for electronic communication will likely satisfy the preferences of most online readers (Schriver 2009). Children, older adults (average age seventy), and readers who are dyslexic have been found to benefit from reading typefaces (either serif or sans serif) set between 12 points and 14 points, with most people preferring 14-point type over 12-point type (Bernard, Liao, and Mills 2001; Wilkins et al. 2009; O'Brien, Masfield, and Legge 2005). However, as we consider people's preferences for typographic cues, it is important not only to consider the point size of a typeface, but also its *x-height*, a key factor in typeface identification and legibility (see figure 16.8).

The x-height refers to the height of a lowercase *x* in the alphabet of a given typeface; it is the distance between the baseline of a letterform (bottom of lowercase *x*) and the mean line of a letterform (top of lowercase *x*). Figure 16.9 illustrates how x-heights may vary from typeface to typeface. X-heights vary not only in the amount of vertical space they occupy, but also in the amount of horizontal space they require. In choosing

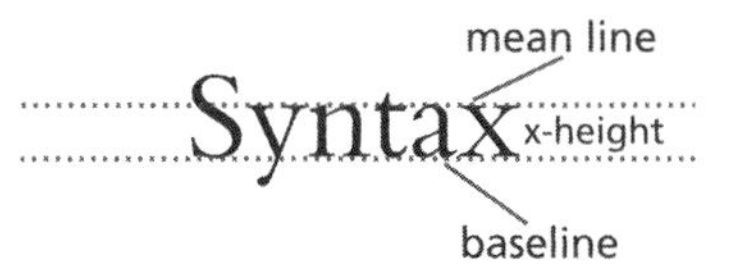

Figure 16.8. An example of x-height; the typeface is Bembo (24 point).

Figure 16.9. How x-heights vary for different typefaces set at the same point size

a typeface, then, it is important to compare competing typefaces at the particular point size deemed appropriate for the task (e.g., what would be required for displaying paragraphs or captions). Typefaces with small x-heights tend to make the text look smaller than one would expect for a given point size (e.g., 12-point type may appear to be 10-point). In such cases, it makes sense to increase the point size to enhance its legibility. Conversely, typefaces with large x-heights may make the text look crowded and dense, requiring the designer to increase the leading (the space between the lines of type) in order to increase the legibility.

Anecdotal reports suggest that most readers prefer what they are accustomed to. North American readers tend to prefer serif typefaces for the body copy of lengthy text presented on paper. In contrast, because sans serif faces were popularized in European countries, Europeans may exhibit a higher tolerance for reading lengthy texts displayed using sans serif faces. Even so, many European books and newspapers employ serif typefaces for the body copy. Because attitudes about typography are evolving, it is important to study the culture of the people who will engage with the artifact and attempt to employ cues they will prefer, find comfortable, and negotiate easily.

One thing we can say with certainty is that people from every culture appreciate good contrast between the headings and the body copy. This can be achieved by using one of the following: (1) sans serif type for headings and serif type for body text, (2) a serif type for both headings and body text, as long as it has good contrast within the face, or (3) a sans serif face for both headings and body text, again, as long as it has good contrast within the face.

Assess the Quality of the Design and the Intelligibility of the Cues

Even when one has a good idea of the reader's likely strategies for using the content, technical communicators must evaluate the goodness of their predictions by evaluating the content with readers. For example, it may be that people read typefaces set at smaller point sizes (such as 10-point type) more slowly because they are less legible, but readers may also remember more about what they have read and construct a more accurate understanding of the text because they are spending more time interpreting it (Dyson 2004). Usability testing can alert us to such speed-accuracy trade-offs, forcing us to make decisions about which is more important for the situation. Moreover, usability testing can guide technical communicators in revising their overall design as well as their visual and verbal systems for signaling the structure. The findings of usability testing can alert us to visual or verbal cues that are missing, incomplete, inconsistently executed, or just plain unclear (Schriver 1991, 1992). As it turns out, unclear content and omissions make people shut down and stop reading. When technical communicators combine the use of evidence-based design heuristics with user-experience testing, they increase the likelihood that their designs will be read.

EXTENDED EXAMPLE

To illustrate the heuristic above, I return to my introduction, in which Miguel was assigned two tasks: designing the first in a series of documents about healthy eating, and revising the visual design of a flyer that Miguel's supervisor thought could be used in the series. Let's look at how Miguel was able to draw on information-design research and the three-phase heuristic in his redesign of the flyer and design of the prototype.

TASK 1: REDESIGN A FLYER

Miguel began his revision task with the original flyer, shown in figure 16.10. The supervisor felt the design looked a little dull, but thought the content was good. The original flyer extended over two pages and was intended to function as a summary of the foods that research had found

HEALTHY FOODS

Almonds and walnuts (and other nuts)—Lowers both total and LDL cholesterol levels.
Apples—Low in calories, high in soluble fiber, which helps lower cholesterol.
Avocados—Rich in monounsaturated fat and fiber; source of plant sterol and antioxidants.
Blueberries (and other berries)—Great source of antioxidants and dietary fiber.
Citrus fruits—Lots of vitamin C, folate, thiamine, and potassium.
Cruciferous vegetables—Have unique compounds that are felt to be cancer protective.
Fat-free or 1% milk (and yogurt)—Excellent source of calcium.
Garlic (and onions)—Linked to anticlotting, cholesterol lowering, and cancer protection.
Legumes (including beans, peas, lentils, peanuts, and soy)—Vegetarian source of protein; low in calories and saturated fat; good source of vitamin B6, potassium, and zinc.
Melons—Good source of lycopene and vitamin C.
Olive oil (particularly virgin olive oil)—Beneficial to your health not only for its monounsaturated fat (oleic acid), but also because it is rich in polyphenols.
Red wine—Contains bioflavonoids, phenols, resveratrol, and tannins, which have antioxidant and anticlotting properties; raises HDL cholesterol.
Salmon (and other fish)—Rich in omega-3 fatty acids; great source of protein and iron.
Spinach—Source of vitamins A, K, C, and B6; riboflavin; folate; and potassium.

Figure 16.10. The original document listing healthy foods

to be the most nutritious. The first thing Miguel noticed was that the flyer was worse than dull; it was ugly. He doubted whether members of the intended audience would read it. He diagnosed the problem as one of lack of contrast, noticing that it did not signal the content very well. Miguel wanted to employ a less routine typeface than Times Roman. He felt that he could come up with a better layout that would cue the relationship between the names of the food and the explanation about why those foods were healthy choices. Miguel also questioned whether the alphabetical structure made it boring, but decided to keep that organization and focus mainly on revising the visual display of the content.

Figure 16.11 shows Miguel's revision. He employed the same content, but gave it a more explicit structure. First, he changed the visual format from a list to a table. In so doing, he removed the gray background to increase the contrast and thereby improve the legibility. Miguel suspects that shifting the display to a tabular format will promote quicker access of the content, but he decides to wait until he pilot tests the redesign before he makes this claim to his supervisor.

Second, he changed the typography from a serif face (Times Roman) to a sans serif face (Frutiger), employing mixed styles and weights within the Frutiger family for the side heads and the body copy. He used Frutiger bold condensed for the side heads "healthy foods," Frutiger light condensed for the elaboration of the foods, and Frutiger light for the explanations—the part that he felt needed typography with a larger x-height. Miguel also revised the typography for displaying the major heading "healthy foods," from 12-point Times Roman set in all capital letters to 24-point Apple Chancery using title capitalization. These typographic updates improve the attractiveness of the flyer and increase the contrast between the major heading and the list of foods, their elaboration, and the explanations. Third, Miguel altered the verbal structure by shifting from the topic-based title "healthy foods" to the question "healthy eating?"

In addition, Miguel paired the heading "healthy foods" with the scenario heading "why they're good for you." Scenario headings orient the content from the reader's perspective (Flower, Hayes, and Swarts 1983). Miguel highlighted these linguistic cues by reversing the type for the column heads (white type on gray background). Miguel's tabular format for the body copy makes referencing why a given food is nutritious easier than the original format, which made the blurbs appear as miniparagraphs. Although Miguel's revision is clearly better than the original, we can see that the text needs further modification if it is to be understood by parents. Miguel reassures his supervisor that in his next iteration, he will replace words such as *monounsaturated* and *bioflavonoids*.

TASK 2: DESIGN A PROTOTYPE

Miguel's second task was to design a paper prototype for the series on nutrition. His supervisor recommended that he start with a one-page handout, suggesting that after it met with his approval, he would assign Miguel more complex tasks. His supervisor was interested in seeing whether Miguel could draw on his original research to design the content for the prototype.

Healthy Eating?

Healthy Foods	Why They're Good for You
Almonds and walnuts	Lowers both total and LDL cholesterol levels.
Apples	Low in calories, high in soluble fiber, which helps lower cholesterol.
Avocados	Rich in monounsaturated fat and fiber; source of plant sterol and antioxidants.
Blueberries (and other berries)	Great source of antioxidants and dietary fiber.
Citrus fruits	Lots of vitamin C, folate, thiamine, and potassium.
Cruciferous vegetables	Have unique compounds that are felt to be cancer protective.
Fat-free or 1% milk (and yogurt)	Excellent source of calcium.
Garlic (and onions)	Linked to anticlotting, cholesterol lowering, and cancer protection.
Legumes (including beans, peas, lentils, peanuts, and soy)	Vegetarian source of protein; low in calories and saturated fat; good source of vitamin B6, potassium, and zinc.
Melons	Good source of lycopene and vitamin C.
Olive oil (particularly virgin olive oil)	Beneficial to your health not only for its monounsaturated fat (oleic acid), but also because it is rich in polyphenols.
Red wine	Contains bioflavonoids, phenols, resveratrol, and tannins, which have antioxidant and anticlotting properties; raises HDL cholesterol.
Salmon (and other fish)	Rich in omega-3 fatty acids; great source of protein and iron.
Spinach	Source of vitamins A, K, C, and B6; riboflavin; folate; and potassium.

Figure 16.11. A revision of the list of healthy foods in figure 16.10

Phase 1: Group Content into Rhetorical Clusters

Miguel began by thinking about his audience as he sorted through the information he had collected in his research. He had identified what he thought were the top fifty empirical findings about nutrition. Miguel was surprised that so many things he had heard about healthy eating were either only partly true or completely wrong. He felt that many people might be mystified about the science of nutrition and that his goal would be to help to set the record straight.

Miguel had also collected many photographs, drawings, and cartoons in order to illustrate healthy choices. For example, he had assembled drawings of people eating, photos of fruits and vegetables, close-ups of whole grains, cartoons of people overeating, photos of lean meat, drawings of the U.S. Department of Agriculture's food pyramid (and its revision, the food plate). This abundance of material proved a bit overwhelming, and Miguel needed a strategy for organizing the first piece in the series.

Figure 16.12 presents a progression of pencil sketches Miguel created in coming up with ideas for the prototype. He began by grouping the findings about nutrition according to topic (left side of top row), starting with types of foods: fruits, grains, vegetables, protein, and so on. He considered what the topics might look like if arrayed as a list or as links with nested content (right side of top row).

Next he organized his photos and visuals by sorting them according to size, kind of image, and color (left side of middle row). This led him to draft a few additional structures (right side of middle row). He considered a paragraph-photograph sequence arranged vertically, which he thought might serve to associate key photos with the text. Then he entertained a picture-in-picture format, with a collage of foods as a background image and several small photos that highlighted the immediate topic. He also thought about using a line chart (showing how the average American's weight had increased over the years) with a pop-out textual blurb about the problem of obesity.

Still not convinced he had a workable structure, Miguel then explored designing rhetorical clusters for the goals of (1) narrating and explaining and (2) showing comparisons and explaining (third row of figure 16.12). He first sketched a time line that integrated text and graphics to show how ideas about healthy eating had evolved (left side of third row). Then he considered using a few matrix formats to show key comparisons among pieces of the content, such as ideas about healthy eating in the 1980s versus 2010s, or things that people believe to be true about healthy eating, but that are actually false (right side of third row).

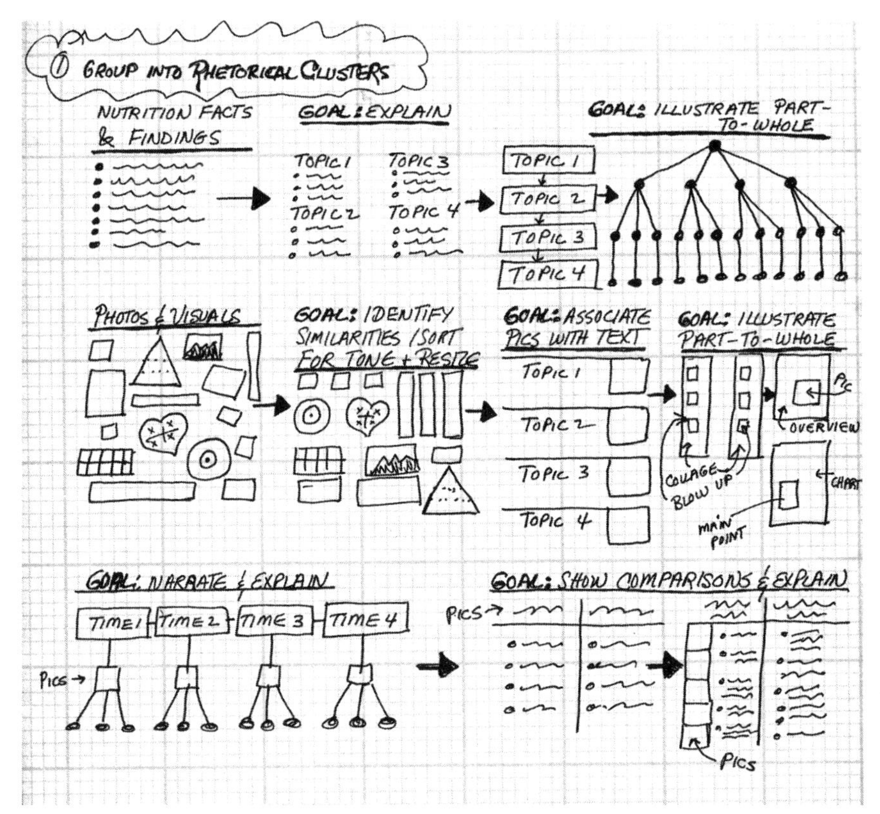

Figure 16.12. Phase 1 of the information-design heuristic, "group content into rhetorical clusters." Miguel sketches ways to group the content as he plans a prototype for a handout about healthy eating.

Phase 2: Organize Clusters Visually to Show Contrast

Next Miguel took the sketches from this first phase of the heuristic (grouping into rhetorical clusters) a step further. Drawing on phase 2, Miguel organized the clusters into a few very basic layouts, trying to get a sense of his visual options and to see which ones might present the content better to his audience. Figure 16.13 shows a series of pencil sketches Miguel made to visually organize the rhetorical clusters for contrast.

As Miguel tried out these alternative ways of viewing his content, he began to imagine the words and images he might use. He tried to predict what an audience of parents might conclude about healthy eating as they read each version. This led Miguel to scrutinize whether his goals for the project should be to explain (top sketches in figure 16.13), to narrate and explain (middle sketches), or to show comparisons and explain (bottom sketches).

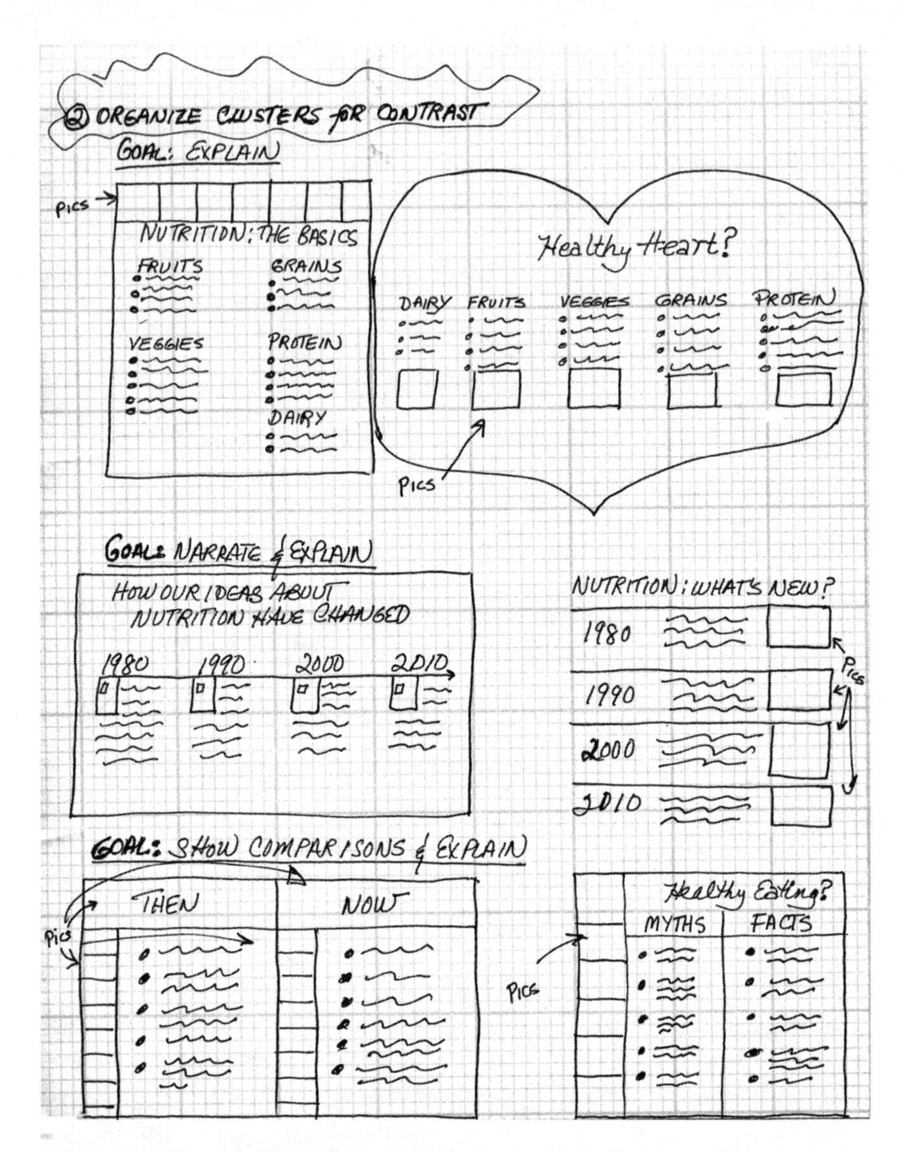

Figure 16.13. Phase 2 of the information-design heuristic, "organize clusters visually to show contrast." Here Miguel sketches a few structures in order to plan some advice about healthy eating.

Miguel concluded that for this first piece in the series, he would focus on helping parents see a few comparisons between what they think about nutrition and what research actually shows about nutrition. He decided that holding the single goal of "explaining about nutrition" was too narrow for his purposes and chose not to use the top sketches. He rejected the structures he sketched for "narrate and explain" on the basis that parents probably do not care about the history of nutrition; they care about what to do about nutrition. He felt he could make the strongest case for his content and for his readers (and to his supervisor) by using the structure shown in his last sketch, "myths and facts."

A lesson Miguel took from the sketching exercise was that seeing the content through the lens of alternative structures proved useful—helping him generate new ideas about how to present the message. It was not until he had seen the alternatives side by side that he realized how visualizing the structures helped him clarify his rhetorical goals for the project.

Feeling that he had made a good decision, Miguel turned to his computer to execute the plan. He began working on ideas for the layout. He divided his attention between thinking about visuals, the optimal typography, and how much vertical space he needed for the text. After Miguel designed a few iterations of the layout in which he placed sample text and photos, he concluded that a simple three-column grid was what he needed. His final grid is shown in figure 16.14.

Miguel chose the three-column grid because it allowed him to organize the content using column 1 for the visuals and columns 2 and 3 for the text. He decided that the space he had available in column 1 was too narrow for anything complex, and that the rhetorical function should be

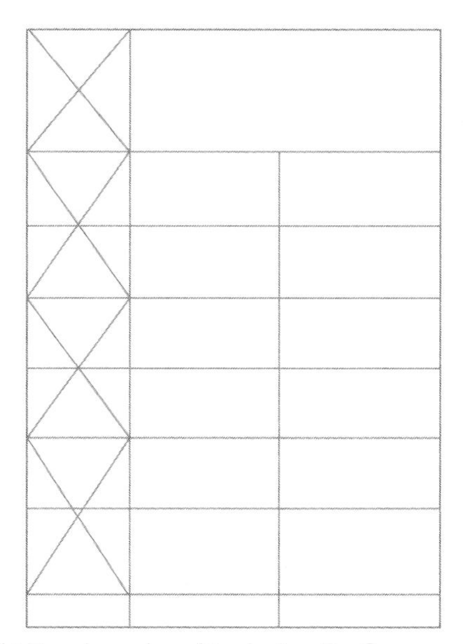

Figure 16.14. The grid Miguel employed to design the flyer presented in figures 16.15 and 16.16

to set the tone and to attract the eye to inspect the content. He sorted through the many visuals he had collected, looking for photos that had a similar look and color value, especially for realistic soft-edge photos of nutritious food. Once he made his selections, he resized the photos and cropped them to fit the grid. He wrote and edited the text for the set of "myths and facts," so that each item was roughly the same length.

Phase 3: Signal Structural Relations

Miguel then worked through the final phase of the heuristic, "signal structural relations," to produce a draft of the prototype. He attempted to integrate the overall appearance of the layout with a major heading that spans columns 2 and 3, "Healthy Eating?" (He purposefully employed the same major heading and displayed it using the same typeface as the first flyer. Miguel's intent with these repetitions was to start a document family, one with a unified visual and thematic identity.) Miguel then formatted the body text to fit columns 2 and 3 and gave each a heading, "Myths" and "Facts." He stacked the photos of fruits and vegetables in the first column and removed any white space between them. Once he completed the draft, he asked his information-design colleague Louise to look it over and offer some feedback. Miguel's initial draft and the comments from his colleague are displayed in figure 16.15.

As Louise's comments show, there were several problems with the information design of Miguel's prototype. The headings "Myths" and "Facts" that were to preview the itemized content did not function very well because their spatial position violated the Gestalt principle of proximity, which suggests that "elements that are close together are perceived to be more related than elements that are farther apart" (Lidwell, Holden, and Butler 2003, 160). The headings "Myths" and "Facts" needed to be closer to the actual content. As Louise pointed out, the "headers float within the gray bar." Similarly, a Gestalt problem of proximity was created by the horizontal position of the itemized "myths" (in column 2); the text was too close to the photos, and there was too much white space between columns 2 and 3.

Louise noticed that the heading "Facts" (column 3) was touching the major heading "Healthy Eating?," a typographic oversight on Miguel's part. Louise also mentioned that the heading "Facts" was not centered over the column, but shifted slightly to the right.

Louise's biggest problem by far with the prototype was the spatial positioning of the itemized content. The horizontal positioning of the items was unhelpful in guiding readers to make quick and appropriate comparisons between the myths and facts. That there were six clusters of myths

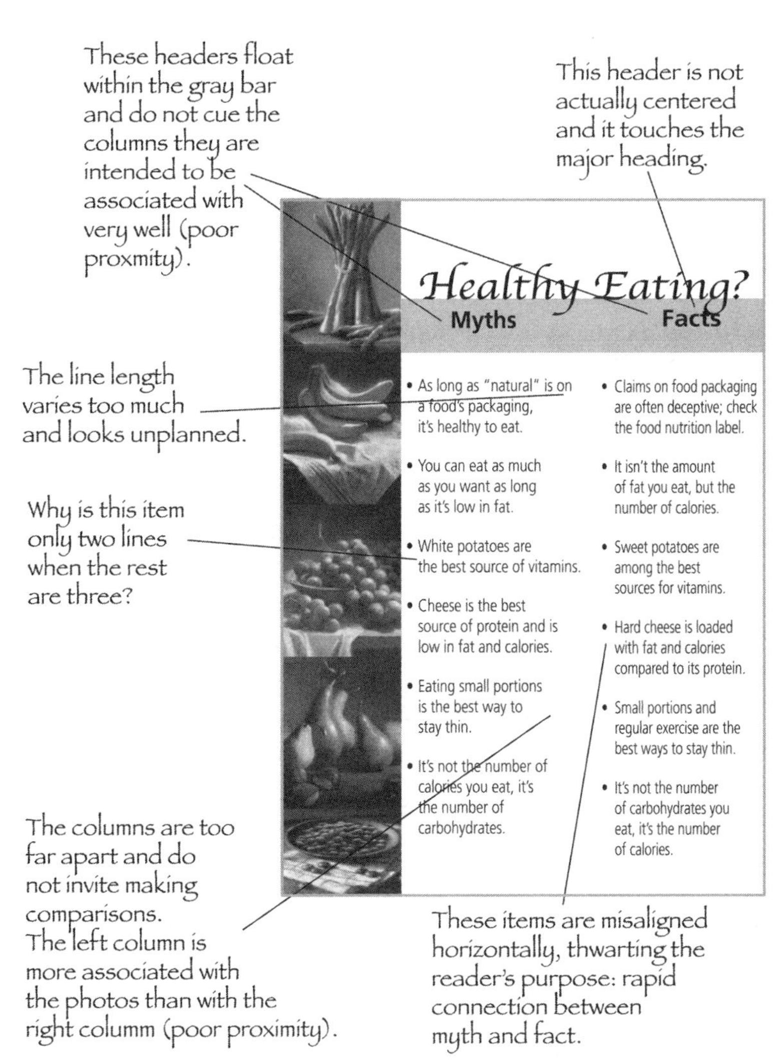

Figure 16.15. Phase 3 of the information-design heuristic, "signal structural relations." Miguel's first draft, with comments about the signaling from a colleague experienced in information design.

and facts was not immediately apparent. The columns were too far apart. While the horizontal alignment for items 1 to 3 was only slightly off, bulleted items 4 to 6 were completely misaligned. Moreover, the line length employed for the itemized content appeared to be random, with some lines jutting out much farther than others.

To improve on the next iteration, Louise suggested that Miguel try a different visual strategy for signaling the itemized content. She recommended that Miguel think of a clear way to make it obvious that there were six clusters, each with two items, a myth and a fact.

Miguel considered each of Louise's constructive criticisms. He began his revisions with the easier ones to execute. First he worked on improving the proximity relations between the headings and the itemized content. He did so by moving the headings to align with the bottom of the gray bar (at the top of the layout) and by making sure the headings were indeed centered over columns 2 and 3. Next he standardized the line length of the items, so that each had a similar visual shape and occupied no more and no less than three lines of type. This move allowed him to achieve the horizontal alignment he sought, displaying each myth on the same line as each fact. Miguel realized that while he had worked hard to make the writing for the "myths and facts" parallel, editing each item for length and syntax (e.g., see item 6), he had neglected to use clear visual signals to make those parallels evident. Miguel agreed with Louise about the need for a different strategy for signaling the itemized content. After a few more drafts, Miguel made the six clusters more visible with two additional revisions. Initially he moved the columns closer together (creating better proximal relations). Then he inserted a thin gray rule between each cluster, making the format resemble that of a table. Miguel's final version is shown in figure 16.16.

Miguel was pleased with his final result, but wanted the opportunity to test his draft before he considered it finished. He was curious to know if the design encouraged reading and whether the main points were clear. As he looked over the six clusters of myths and facts, he questioned whether it was a wise idea to include messages about only two foods—potatoes and cheese—in the context of more general points about nutrition (see items 3 and 4). He considered moving those items to another text he would create later, but ran out of time before he needed to meet with his supervisor.

Miguel's supervisor was very satisfied with his work and believed it was an excellent prototype on which they could base other short pieces. He felt it would make a good first impression with readers—that it was clear and conveyed a positive persona—and would show that the Department of Agriculture cared about helping citizens understand healthy eating. Of course, he noticed the advice about potatoes and cheese. He said that including these items made the content's level of abstraction inconsistent across the list. But he thought the advice was good and that these foods had been connected to obesity. His final call was to keep the spotlight on potatoes and cheese. He also approved Miguel's request to test the prototype and suggested that Miguel create a plan to assess the quality of the entire series.

Healthy Eating?

Myths	Facts
• As long as "natural" is on a food's packaging, it's healthy to eat.	• Claims on food packaging are often deceptive; check the food nutrition label.
• You can eat as much as you want as long as it's low in fat.	• It isn't the amount of fat you eat, but the number of calories.
• White potatoes are the best source of vitamins.	• Sweet potatoes are among the best sources for vitamins.
• Cheese is the best source of protein and is low in fat and calories.	• Hard cheese is loaded with fat and calories compared to its protein.
• Eating small portions is the best way to stay thin.	• Small portions and regular exercise are the best ways to stay thin.
• It's not the number of calories you eat, it's the number of carbohydrates.	• It's not the number of carbohydrates you eat, it's the number of calories.

Figure 16.16. The final version of Miguel's prototype flyer about healthy eating

CONCLUSION

In this chapter, we have seen that by drawing on principles of information design, technical communicators can improve the quality of the content they create. When the structure of a message is made prominent visually and verbally, readers are more likely to attend to the message and respond positively to it. Technical communicators need not be graphic artists in order to improve the visual display of their content. Rather, they need to be willing to devote the time it takes to excel in three key activities:

- grouping content rhetorically,
- organizing content visually, and
- signaling structural relationships.

The research presented in this chapter provides very strong evidence that each of these activities can help improve readers' engagement with content. These activities are especially powerful when they are employed iteratively and when they are combined with usability testing. Content that has been shaped by these activities and followed up with usability testing is likely to be easier to understand, more used, better liked, and better remembered. Moreover, each component of the information-design heuristic is supported by empirical evidence.

As the case study of Miguel's design process shows, using the heuristic helped him consider different visual structures for his content. His pencil sketches were a valuable catalyst to his design thinking. Seeing the options, even in very rough form, allowed him to imagine the resulting documents more fully and eliminate those designs that would likely be ineffective. And the feedback from his information-design colleague helped him recognize that he needed to worry about the consistency of visual cues just as much as consistency in writing and editing.

Taken as a whole, the research tells us that information design is neither decoration nor artifice. Rather, good information design supports readers' cognitive and emotional interactions with content. In this way, information design is about enabling communication. My research on expertise in information design suggests that as technical communicators become more sensitive to the graphic, spatial, and typographic possibilities for their content, they become more flexible and more skilled in designing messages (Schriver 2012).

The expansion of technical communicators' concern from focusing mainly on verbal aspects of messages to both visual and verbal aspects is important because many messages are never read. At first glance, people judge them to be too complicated, boring, or ugly. As technical communicators develop their expertise in information design, they will be better able meet their readers' cognitive and emotional needs for content. They will also find that information design is not about the details, such as choosing between serif and sans serif typefaces (although the details are important), but rather about the big picture—that is, building bridges to make sure people engage with the message. That means understanding the content people need, structuring it in ways that invite engagement, and envisioning novel ways to shape the message. The hardest part of the work lies in building those rhetorical bridges.

DISCUSSION QUESTIONS

1. Take another look at figure 16.2, the heuristic for structuring content visually. In your opinion, which of the three phases will be the most difficult for students whose educational background has focused mainly on writing? What aspects of visual design do you think will be most challenging for you? What aspects interest you most?
2. People may respond emotionally as well as cognitively to the communications they read. Think about the last time you reacted emotionally to a document or a website. How did you respond? Excited? Enthusiastic? Annoyed? Frustrated? What was it that made you react as you did? What were the characteristics of the message or its presentation that triggered your response? What role did good or bad information design play in how you felt?
3. Look at the seven grouping techniques presented in figure 16.3. Think of a topic in technical communication that could be organized in various ways. Choose two of the grouping techniques and describe how the topic could be organized using each technique. In your opinion, which one provides the best structure for the content? Why?
4. Find a short example of poor information design (two pages or two screen captures). Make an electronic version of the example so you can display it for the class. Prepare a short presentation in which you first analyze the stakeholders for the example and their likely purposes for using the communication. Then draw on information-design research and the heuristic in figure 16.2 as a framework for evaluating why the information design is inadequate. As you discuss the example, identify the specific visual or verbal features that make the design poor.
5. Collect a set of four articles written on a single scientific or technical topic (e.g., nanotechnology, sleep disorders, cloud technology, electric cars). Two of the articles should be written for a lay audience (e.g., from *Science News*) and the other two for professionals in the field (e.g., *Science*). Analyze the articles for the ways they organize the content. Are similar structures employed for expert and novice audiences? What are the key differences among the visuals included? Are they effective? What can we learn from these practices?
6. Choose a local nonprofit agency and collect a sample of their paper and electronic documents. Create screen captures of their website and

print out representative pages that typify the agency's design moves. Analyze the documents for their writing and visual display of the content. What tone and persona do the designs project? Do paper and online versions give similar impressions?

7. Gather the paper documents and the printouts of the website you collected from the nonprofit agency. Lay out the document set on a large table, side by side: paper and online documents. Assess the impression the document set creates "as a family." Is it easy to tell that all of the documents are designed by the same agency? Are there elements that make the organizational voice inconsistent? Create a class presentation in which you detail your analysis, pointing out recommendations that you would make to the agency for improving the coherence and consistency of their communications.
8. Collect an example of the front page of your favorite electronic newspaper or service. Capture the image electronically and then import it to a design program, such as Adobe Illustrator. "Select the entire page" and use the software to "gray out the transparency of the page" so you can still see the shapes of the textual and graphic elements, but not read them. Then, use the software's tools to trace the grid of the page and make a grid overlay. Pay attention to the underlying structure of the body copy and graphics. Save the original page in full color and the page with your grid overlay. Bring the files to class and be prepared to describe how the content is displayed. How many columns and rows are employed in the design? What are they used for? Why is the information grouped as it is?
9. Choose a commercial organization with an online presence that has online competitors (e.g., banks, software manufacturers, airlines, rental car agencies, hotel chains, consumer product makers). Identify three websites that represent the competition (e.g., three different airlines). Collect screen captures from their websites. Then select two key features that are relatively similar across the websites that can be compared. For example, for airlines you could assess the booking information or trip-planning tools. Compare and contrast the three sites by making a series of tables that highlight the similarities and differences in their design strategy. Prepare a short report that discusses your findings.
10. Look for a document that presents an information design you admire. "Reverse engineer" the document—that is, try to identify the visual and verbal strategies that make it tick. Drawing on the research in information design, discuss why you think the document is effective.

NOTE

1. For a discussion of Gestalt principles applied to information design, see Schriver 1997, 303–326.

WORKS CITED

Anderson, John R. 2009. *Cognitive psychology and its implications*. 7th ed. New York: Worth Publishers.

Bernard, Michael L., C. H. Liao, and M. Mills. 2001. The effects of font type and size on the legibility and reading time of online text by older adults. In *Proceedings of the Special Interest Group on Computer-Human Interaction Extended Abstracts '01*, 175–176. New York: Association of Computing Machinery. http://psychology.wichita.edu/hci/projects/elderly.pdf.

Bernard, Michael L., and M. Mills. 2000. So, what size and type of font should I use on my website? *Usability News* 2 (2). http://www.wsupsy.psy.twsu.edu/surl/usabilitynews/2S/font.htm.

Bernard, Michael L., M. Mills, M. Peterson, and K. Storrer. 2001. A comparison of popular online fonts: Which is best and when? *Usability News* 3 (2). http://psychology.wichita.edu/surl/usabilitynews/3s/font.htm.

Breland, K., and M. K. Breland. 1944. Legibility of newspaper headlines printed in capitals and in lower case. *Journal of Applied Psychology* 28:117–120.

Brumberger, Eva, and Kathryn Northcut. 2012. *Designing texts: Teaching visual communication*. Amityville, NY: Baywood.

Coles, P., and J. J. Foster. 1975. Typographic cuing as an aid to learning from typewritten text. *Programmed Learning and Educational Technology* 12:102–108.

Dyson, Mary C. 2004. How physical text layout affects reading from screen. *Behaviour and Information Technology* 23 (6): 377–393.

Farkas, David K. 2005. Explicit structure in print and on-screen documents. *Technical Communication Quarterly* 14 (1): 9–30.

Flower, Linda, John R. Hayes, and H. Swarts. 1983. Revising functional documents: The scenario principle. In *New essays in technical and scientific communication*, edited by P. V. Anderson, R. J. Brockmann, and C. R. Miller, 41–58. New York: Baywood Press.

Frascara, Jorge. 2010. Data, information, design and traffic injuries. *VideoLectures.net*, March. http://videolectures.net/aml2010_frascara_diti/.

Ivory, Melody Y., Rashmi R. Sinha, and Marti A. Hearst. 2001. Empirically validated web page design metrics. In *Proceedings of the 2001 Human Factors in Computing Systems Conference*, edited by Michel Beaudouin-Lafon and Robert J. K. Jacob, 53–60. New York: ACM.

Jenkins, S. E., and B. L. Cole. 1982. The effect of the density of background elements on the conspicuity of objects. *Vision Research* 22:1241–1252.

Kahn, M., K. C. Tan, and R. J. Beaton. 1990. Reduction of cognitive workload through information chunking. In *Proceedings of the Human Factors and Ergonomics Society 34th Annual Meeting*, edited by D. Woods and E. Roth, 1509–1513. Santa Monica, CA: Human Factors and Ergonomics Society.

Keyes, E. 1993. Typography, color, and information structure. *Technical Communication* 40 (4): 638–654.

Köhler, W. 1947. *Gestalt psychology*. New York: Liveright.

Lidwell, William, Kritina Holden, and Jill Butler. 2003. *Universal principles of design*. Beverly, MA: Rockport Publishers.
Lindgaard, Gitte, Gary Fernandes, Cathy Dudek, and J. Brown. 2006. Attention web designers: You have 50 milliseconds to make a good first impression! *Behaviour and Information Technology* 25 (2): 115–126.
Malamed, Connie. 2009. *Visual language for designers: Principles for creating graphics that people understand*. Beverly, MA: Rockport Publishers.
Miller, G. A. 1956. The magic number seven, plus or minus two: Some limits on our capacity for processing information. *Psychological Review* 63:81–97.
Müller-Brockmann, J. 1985. *Grid systems in graphic design*. Revised ed. New York: Hastings House.
Mullet, Kevin, and Darrell Sano. 1995. *Designing visual interfaces: Communication oriented techniques*. Mountain View, CA: SunSoft Press.
Muter, P., and P. Marrutto. 1991. Reading and skimming from computer screens and books: The paperless office revisited? *Behaviour and Information Technology* 10:257–266.
Nielson, Jakob. 2006. Progressive disclosure. *Alertbox: useit.com*, December 4. http://www.useit.com/alertbox/progressive-disclosure.html.
Niemela, M., and J. Saarinen. 2000. Visual search for grouped versus ungrouped icons in a computer interface. *Human Factors* 42 (4): 630–635.
O'Brien, Beth A., Stephen J. Masfield, and Gordon E. Legge. 2005. The effect of print size on reading speed in dyslexia. *Journal of Research in Reading* 28 (3): 332–349.
Outing, Steve, and Laura Ruel. 2004. Eyetrack III: Online news consumer behavior in the age of multimedia. http://poynterextra.org/eyetrack2004/index.htm.
Ozok, A. A., and G. Salvendy. 2000. Measuring consistency of web page design and its effects on performance and satisfaction. *Ergonomics* 43 (4): 443–460.
Paivio, A. 1969. Mental imagery in associative learning and memory. *Psychological Review* 76:241–263.
———. 1990. *Mental representations: A dual coding approach*. New York: Oxford University Press.
Parush, A., R. Nadir, and A. Schtub. 1998. Evaluating the layout of graphical user interface screens: Validation of a numerical computerized model. *International Journal of Human-Computer Interaction* 10 (4): 343–360.
Paterson, D. G., and M. A. Tinker. 1946. Readability of newspaper headlines printed in capitals and lower case. *Journal of Applied Psychology* 30 (April): 161–168.
Pirolli, Peter, and Stuart K. Card. 1999. Information foraging. *Psychological Review* 106 (4): 643–675.
Reece, Gloria Ann. 2002. Text legibility for web documents and low vision. PhD dissertation, University of Memphis.
Rickards, E. C., and G. J. August. 1975. Generative underlining strategies in prose recall. *Journal of Educational Psychology* 67:860–865.
Sadoski, Mark, and Allan Paivio. 2001. *Imagery and text: A dual coding theory of reading and writing*. Mahwah, NJ: Erlbaum.
Samara, Timothy. 2002. *Making and breaking the grid*. Beverly, MA: Rockport Publishers.
Scharff, Lauren, Alyson Hill, and Albert Ahumada. 2000. Discriminability measures for predicting readability of text on textured backgrounds. *Optics Express* 6 (4): 81–91.
Schriver, K. A. 1989. Evaluating text quality: The continuum from text-focused to reader-focused methods. *IEEE Transactions on Professional Communication* 32 (4): 238–255.
———. 1991. Plain language through protocol-aided revision. In *Plain language: Principles*

and practice, edited by E. R. Steinberg, 148–172. Detroit, MI: Wayne State University Press.

———. 1992. Teaching writers to anticipate readers' needs: A classroom-evaluated pedagogy. *Written Communication* 9 (2): 179–208.

———. 1997. *Dynamics in document design: Creating texts for readers*. New York: John Wiley and Sons.

———. 2009. Using design to get people to read and keep reading. *Health Literacy Out Loud Podcast (HLOL)*, December 12. http://www.healthliteracyoutloud.com/?s=schriver.

———. 2010. Reading on the web: Implications for online information design. Ljubljana Museum of Architecture and Design Lecture Series on Visual Communications Theory: On Information Design. http://videolectures.net/aml2010_schriver_rotw/.

———. 2011. La retórica del rediseño en contextos burocráticos (The rhetoric of redesign in bureaucratic settings). In *Information design*, edited by Jorge Frascara, 156–165. Buenos Aires: Ediciones Infinito.

———. 2012. What we know about expertise in professional communication. In *Past, present, and future contributions of cognitive writing research to cognitive psychology*, edited by Virginia W. Berninger, 275–312. New York: Psychology Press.

Spencer, H., L. Reynolds, and B. Coe. 1974. Typographic coding in lists and bibliographies. *Applied Ergonomics* 5:136–141.

Spyridakis, J. H. 1989a. Signalling effects: Part I. *Journal of Technical Writing and Communication* 19 (1): 227–239.

———. 1989b. Signalling effects: Part II. *Journal of Technical Writing and Communication* 19 (4): 395–415.

Stark Adam, Pegie, Sara Quinn, and Rick Edmonds. 2007. *Eyetracking the news: A study of print and online reading*. St. Petersburg, FL: Poynter Institute for Media Studies.

Theofanos, Mary Frances, and Janice Redish. 2005. Helping low-vision and other users with web sites that meet their needs: Is one site for all feasible? *Technical Communication* 52 (1): 9–20.

Tufte, E. R. 1983. *The visual display of quantitative information*. Cheshire, CT: Graphics Press.

Tullis, T. S. 1997. Screen design. In *Handbook of human-computer interaction*, edited by M. Helander, T. K. Landauer, and P. Prabhu, 503–531. New York: Elsevier Science.

Vartabedian, A. G. 1971. The effects of letter size, case, and generation method on CRT display search time. *Human Factors* 13 (4): 363–368.

Weinschenk, Susan. 2009. It's a myth that all capital letters are inherently harder to read. *What Makes Them Click: Applied Psychology to Understand How People Think, Work, and Relate*, December 23. http://www.whatmakesthemclick.net/2009/12/23/100-things-you-should-know-about-people-19-its-a-myth-that-all-capital-letters-are-inherently-harder-to-read/.

Wertheimer, M. 1922. Untersuchungen zur Lehre von der Gestalt: 1. Prinzipielle Bemerkungen. *Psychologische Forschung* 1:47–58.

Wilkins, Arnold, Roanna Cleave, Nicola Grayson, and Louise Wilson. 2009. Typography for children may be inappropriately designed. *Journal of Research in Reading* 32 (4): 402–412.

ANNE FRANCES WYSOCKI

17 What Do Technical Communicators Need to Know about New Media?

SUMMARY

Like other communication technologies, new media—networked digital communication technologies—entwine with how we act together socially, culturally, economically, politically, and in the workplace. Technical communicators thus need to stay aware not only of new software and hardware but also of how we can use and so shape new media to support our work as communicators. By offering heuristics for using a rhetorical approach for analyzing technical communication contexts that engage with new media, this chapter suggests how the rhetorical approaches many technical communicators already use can also address projects using new media. This chapter—and the rhetorical heuristic—also suggest how to consider shifts in living and working environments that seem solely technical but that are also rhetorical and political.

INTRODUCTION: ONE POSSIBLE FUTURE

On a morning in 2025 (in Vernor Vinge's 2006 novel *Rainbows End*), three high schoolers walking to school decide to duck (without paying) into the local amusement park—Pyramid Hill—to play a little Cretaceous Returns, one of the virtual games offered in the park. The high schoolers are "wearing": their clothing is wired, as are their contact lenses, so they can interact—with no need for other hardware and carrying nothing in their hands—with the public and, in this case, private networks overlain on the hills, gullies, and scrub of northern San Diego County. Because, as Vinge describes, these networks spread throughout the countryside and because people are wearing the digital devices that allow them to see in the environment what we now see only on computer screens or book pages, these students, like the people managing the environment to deal with flooding, can see and interact with visual representations, hanging in the air, of the County Flood Control's tunnels and pipes; they can also see maintenance records for that infrastructure of tunnels and pipes and can see instructions for tending the infrastructure. (The students use this infrastructure to sneak into the park.) Similarly, people in this imagined future can send representations of themselves (accurate representations or modified representations that allow one to appear as, say, a rabbit) to interact with others at great distances, either through talking or through "silent messaging" (where one's text messages hang silently in the air before others, typed not with a keyboard but through subtle body movements; there is no longer a need for cell phones).

In line with the other chapters in this volume, I could now shape this chapter by posing the following problem: Imagine a technical communicator in the future, Cassandra Fortuna, who is responsible for developing training materials for the people charged with maintaining the County Flood Control's networked visual overlays (the overlays described above, the ones illegally brought up by the students); what sort of materials do you think Fortuna should develop? Through what sorts of process should she develop them? With whom should she work?

In addition, however, I add a further challenge: How should Fortuna prepare herself as a technical communicator as our media are shifted, redesigned, and presented to us as "new media"?

If new media *were* only about software or genres of electronic texts, the answer to the final question above would be a list recommending that you learn the most commonly used software packages for creating webpages; help systems; blogs and other social-networking sites; instructional video games, software, or videos; and single-source and content management systems. But because the production, distribution, and consumption of such texts can differ considerably from the production, distribution, and consumption of "traditional" print texts, technical communicators engaged with new media need to engage not only with software and genre features but also with larger concerns such as the expectations that audiences have for new media and the new sorts of relations we establish with each other as we choose among available new communication technologies.

In this chapter, I therefore discuss ways in which our relationships with each other and our texts are shifting because of how we use new media. Out of those discussions, I build a heuristic for thinking rhetorically when you work on projects that engage with new media in the future—and now. I will apply that heuristic to an extended example that grows out of the possible future predicted by *Rainbows End*. I situate new media work within an understanding of how texts of all kinds are designed to persuade us toward particular actions or beliefs by being designed to address particular audiences operating within particular times and places. I want to help you consider how these shifting contexts in which technical communication texts are produced—contexts that swirl together the technological and the social—also shift what it is to write, to be a writer, to be an audience, and to make judgments about the efficacy of texts.

LITERATURE REVIEW

The following sections review the literature around new media and technical communication by considering, first, a definition for "new media" and, second, how the aspects of media that make them "new" affect our understandings of what is possible with our media and so with our relations with others.

DEFINING "NEW MEDIA"

For many, "new media" are simply any communication produced, distributed, and/or consumed through computers: in this view, "new media" are nothing more than blogs, online newspapers, digital help systems, computer games, podcasts, text messaging, Facebook, GPS systems, MP3s, interactive Flash poetry, wikis, content management systems, Twitter, digital television recorders, or (you can add to this list with further examples). In their production, distribution, and consumption, these "new media" stand in contrast to the media with which we were familiar leading into the late twentieth century: books, newspapers, magazines, radio programs, television shows and commercials, and movies. This definition—"any communication produced, distributed, and/or consumed through computers"—separates the *what* of a communication (what we read, see, or hear) from the *how* of the communication, implying that it doesn't matter whether (for example) I write a report by hand or with my keyboard because what I write is separable from the medium in which I produce it and through which you, as reader, see it. Starting in the late nineteenth century, however, theorists and scholars started questioning whether we can separate the content of a message from how we produce, distribute, and consume it, and such analyses of the relations among our communication technologies and our culture, politics, economics, and even sense of self continue into the present (see, for example, Deibert 1997; Eisenstein 1979; Kittler 1999; Saenger 1997; or Winner 1986). In defining any new medium, then, we need not only to differentiate which of its characteristics are indeed new but also to stay alert to how those characteristics interact with other aspects of our lives together.

As you consider the characteristics I list below, then, ask yourself whether I have captured what makes new media new and whether my descriptions help you stay mindful of how new media entwine with how we live and work together. Following these characteristics, I will investigate how they shape new communicative possibilities; based on the characteristics and their possibilities, I will present a rhetorical heuristic for technical communicators working with new media.

1. *New media result from digitization.* With digitization, "graphics, moving images, sounds, shapes, spaces, and texts . . . become computable; that is, they comprise simply another set of computer data" (Manovich 2002, 20). When they no longer exist as different materials but are stored in computer code, media such as video, sound, pictures, and text can be blended together in texts; think of how television, radio, music, video, movies, newspapers, and magazines are no longer separate media coming to us from different sources or through different devices but are all accessible and mixed together on single devices, a process referred to as media convergence. In addition, in ways much more easy than with print media, digitized media can be copied, modified, reused, and mashed together, resulting in what is often referred to as remix culture. Explorations of remix culture acknowledge that anyone—professional or not—with access to a networked computer can build new cultural objects out of existing cultural objects, publish the new objects, and hence participate in larger cultural and political discussions; such explorations also acknowledge that the ease of copying and reusing existing texts does not sit well with current copyright laws and conceptions of intellectual property, which were formulated within the structures of print cultures. See, for example, copyright lawyer Lawrence Lessig's book *Remix* (2008) on how remix culture and current copyright laws are at odds.

2. *New media entail using code to control the presentation and distribution of media.* Reading a newspaper on a laptop or cell-phone screen means not only that the newspaper's words and pictures have been digitized but also that they have been shaped by programming code to appear as they do; they have also been shaped by code to travel through digital networks. Coding also is used to determine what you are allowed to see by media owners: for example, do you have to register and create a password to read a particular newspaper?

3. *New media depend on digital networks.* Without the wireless and wired networks linking our computers, we could not share data. Webpages would be meaningless if we had to print and mail them to each other; Google would not be possible if Google's programmers were not able to set up code that searches through the world's networked hard drives in order to index the webpages stored on them; collaborative writing environments like wikis wouldn't exist because we would be unable to share, at a distance, the documents on which we work together. *Interoper-*

ability is the term often used to name how different systems—computer hardware and software, differing networks—can be networked to enable such work.

4. *New media are faster than print media.* A text to your friend can bring you a response much faster than paper-and-envelope mail would, and it is not unusual for an afternoon's surfing on the Internet to bring us information from across the globe as events happen. The number of calculations your computer carries out as you surf the Internet, do your taxes online, or even play as simple a game as Tetris is considerably more than any human could carry out on paper in a day.

5. *New media enable different kinds of interactivity than print media.* Print media has its own forms of interactivity: novels engage our imaginations to dream up differing futures; we do not have to read books from front to back but can dip in at any page to read. Because of digitization and coding and their speed, however, new media can change in response to our actions as print cannot: in addition to how a story online can spark our imaginations or how we can click the links of an online magazine in any order we want, an online test can be programmed to change in response to our answers, such that it can—without making us aware of it—take us through a review of the information if we are struggling or to harder questions if we do well. My actions in a computer game are very likely to cause different results than yours. In addition, networking means that I can have collaborative or shared interactivity with others—as in massively multiplayer online games like World of Warcraft, whether in the same room or across the country.

6. *New media are becoming ubiquitous.* As the physical components of computing devices shrink, new media can appear across a wider range of environments than print texts. You probably have a cell phone and perhaps a separate MP3 player, and your car may display its minute-by-minute mileage. You may have a digital video recorder attached to your television, or perhaps for her last birthday you gave your mother a digital picture frame that cycles through family photographs. As Byron Hawk and David M. Rieder (2008, xiv) note in writing about "small tech," our uses of these smaller devices is "signaling a move from traditional, visual software interfaces to the ad hoc interfaces created by gestural objects in material ecologies": we use our hands not only to call others on our phones, but also (for one example), while out walking in the city, we can point the phone's camera at restaurants we pass to learn how others rate the restaurants, as the Urban Spoon application on iPhones can do. Such interactions among small devices, our bodies, and our environments are spreading as we develop and use devices for isolating and studying DNA, work with ambient video, or

experiment with using sound projections to create learning environments for people moving through historical sites. Hawk and Rieder's arguments imply that technical communicators need to develop different and focused strategies for learning about and using different small tech devices, should they be appropriate for a particular audience and context, because each device fits into our social and cultural lives differently.

Why do the above characteristics of new media matter? In describing the characteristics, I have noted some implications for technical communicators; in the next section, I tease out these characteristics still further to show how they have been used to shape new communicative possibilities, possibilities that in turn shape how technical communicators can think rhetorically about developing effective new media texts in line with their purposes and values.

NEW MEDIA DEVELOPMENTS THAT AFFECT THE WORK OF TECHNICAL COMMUNICATORS

The characteristics I laid out above for defining new media texts play out in different ways as those who develop new digital technologies and texts emphasize and combine various aspects of the characteristics. Below I examine—from simplest to most complex—how the characteristics listed above are changing the designing, planning, and publishing landscape for technical communicators.

Multimodality

Sometimes people use *new media* and *multimodality* interchangeably, for multimodality is one of the first features of online texts noted by communicators working on computers; new media is not the same as multimodality, however. Multimodality is just one result of the first characteristic of new media noted above—the digitality of new media—and is a way of naming our ability easily to mix pictures, sounds, animations, video, and alphabetic text on digital screens. (Each of those kinds of ways of presenting information—pictorial, audible, alphabetic, etc.—is a communication mode.) Multimodality existed before the computer—you experience it every time you see a page in a book that mixes alphabetic text with photographs or you see a movie with a sound track—but digitization made such mixing much easier than it was with print technologies. At least three aspects of digital multimodality require heightened attention from technical communicators.

First, audiences using computers regularly expect multimodality. An audience might simply expect that texts demonstrate more attention to the visual layout of information on printed pages, by using color and photographs or giving more attention to typography. But audiences can also expect that such texts—especially when delivered on digital screens—will range across available modes, mixing (for example) video with animation and incorporating sound. R. J. Jacquez, senior product evangelist at Adobe Systems for Technical Communication and eLearning, notes this when he claims that online experiences "have raised users' expectations" and that "users' expectations should drive the direction we take as technical communicators": he argues that we should therefore make our communications "engaging" for audiences, using "video, hands-on interactivity, blended learning and the idea that we learn by doing" (quoted in Ellison 2009, 11).

Second, as implied by what I have written in the preceding paragraph, technical communicators can now choose among a range of modes for shaping information. Will it make more sense to present instructions step by step in video or on paper? Will presenting instructions spoken by a gentle-voiced reader make the instructions more effective than if the reader reads them quietly? (See Kim et al. [2008] for a thoughtful discussion on shaping multimodal documents for particular audiences and contexts.)

Finally, because audiences for certain technical communications now expect more engaging texts and because technical communicators can choose from a range of modes, the design process can ask for more creativity from technical communicators. As S. Scott Graham and Brandon Whalen (2008, 66) argue in an article about the complexities of designing new media texts, "designing and developing new media communication can be a dynamic, creative, intuitive, nonlinear (and sometimes childlike) process, which might explain why so much of new-media communication is dynamic, creative, intuitive, nonlinear (and sometimes childlike)." Technical communicators may need to learn to use design strategies from the creative professions for framing approaches to communication situations and for developing creative directions to address those situations. (For assistance in starting to learn such approaches, see John Chris Jones's *Design Methods* [1992] and Bryan Lawson's *How Designers Think: The Design Process Demystified* [2009].)

Single-Sourcing and Content Management

When different kinds of information are digitized, they can all be treated alike in storage and display systems. For example, a company can have a database in which it stores all the alphabetic text written for its documentation systems, with the text separated into various chunks, each chunk indexed by the topics it covers; all digitized photographs, videos, and charts can be similarly indexed and stored. To make a new document, a communicator can pull up all texts (alphabetic or not) indexed for a particular term, to see what already exists and so avoid duplication of efforts. Once information is digital, indexed, and stored as just described, it "can automatically and simultaneously appear in user manuals, help files, and press releases that can in turn be automatically altered to appear in print, on the Web, or on mobile devices. Once initial designs are created, fonts, colors, and layout are added on the fly for the specifics of each genre and/or medium, and with, for example, a simple change to a style sheet, aesthetic changes can easily be applied to past as well as future documents, making it easy to maintain organizational consistency" (Clark 2008, 36). Under such systems, webpages can be built in immediate response to user queries, as happens every time you search Amazon.com or the Internet Movie Database. This ability to use the same information in different documents is called single-sourcing; managing the information that is single-sourced is called content management.

Single-sourcing and content management enable efficient use and reuse of information by communicators across a company's divisions, but they require those who index the information to imagine index terms for future uses for the information. They also require decisions about just how small the stored chunks of texts should be: should information be indexed as pages, as paragraphs, as sentences, or as some combination? In addition, those composing new content for such databases must seek standardized features—such as tone of voice or level of complexity—for *all* their writing.

For technical communicators, content management—because it separates text from its presentation—creates particular challenges. Technical communicators have been accustomed to developing individual documents to address the needs of particular audiences in particular contexts; content management systems that enable communicators to work with existing chunks of information require, at their most stringent, that only existing information be poured into a template with no modifications for who will use the information or where. While this can work exceedingly well for organizations like Amazon.com and the Internet Movie Database, it can be trouble for a document intended to address audiences with differing or very specific needs. Content management systems can also make writing seem to be nothing more than the creation of bits of contextless content rather than the construction of complex and rhetorically aware documents. (See Anderson 2008, Clark 2008, or Hart-Davidson et al. 2008 for further discussions about single-sourcing, content management, and the roles of technical communicators.)

Web 2.0

Quoting from *Wikipedia* (2010b) seems appropriate for defining what happens when software designers experiment with interactivity: "The term 'Web 2.0' (2004–present) is commonly associated with web applications that facilitate interactive information sharing, interoperability, user-centered design, and collaboration on the World Wide Web. Examples of Web 2.0 include web-based communities, hosted services, web applications, social-networking sites, video-sharing sites, wikis, blogs, mashups, and folksonomies. A Web 2.0 site allows its users to interact with other users or to change website content." Applications that enable users to interact over networks set up conditions that should engage technical communicators in at least three ways: collaborative writing environments, user-generated information, and, simply, the existence (and continuing development) of applications that require rethinking a communicator's relationships with audiences.

Technical communicators can use wikis and other collaborative writing environments to develop documents together at close or far distances. In either case, collaborating communicators must learn how to negotiate the purposes, arrangements, and presentations of texts. This is no different than any collaborative writing situation and is an aspect of any distributed working environment; the challenges of such collaborations are well known and discussed in technical communication (see, for example, Spinuzzi 2007).

What is potentially new to technical communicators, however, is collaborating not with each other but with their audiences. As Society for Technical Communication Fellow Rich Maggiani (2009, 20) writes, "A collaboration through social media, properly undertaken, results in the truest form of audience-centered content. And isn't that what technical communication is all about?" Maggiani describes how a technical communicator might use online forums, wikis, or blogs to "to capture the collective knowledge of the community" (20), and, in the process, Maggiani acknowledges that such ways of working shift the position of a technical communicator from a generator of content to a facilitator or moderator. Similarly, R. J. Jacquez of Adobe Systems discusses using Twitter to communicate with Adobe product users, and—based on his experiences—considers a future role of "content curator" for technical communicators who would range across social-networking sites, "constantly finding groups and information, organising data and then pulling out the most relevant content on a specific issue and distributing it to the members of that group" (quoted in Ellison 2009, 12). When technical communicators use social-networking software to take advantage of their audiences' knowledge, not only do communicators and audiences collaborate, but the role of technical communicator can change to managing user-generated content.

For such management to work, technical communicators need to know the social-networking software their audiences are likely to use, and, importantly, they need to know how to use the software rhetorically so that users want to contribute and want to interact with technical communicators and each other—and technical communicators will also need to accommodate the desire for instantaneous response that such software encourages.

Gaming and Immersive Environments

Digital systems' speed and interactivity enable games and simulations, for pleasure as well as education and training; on networked digital systems, the same game or simulation can be experienced by multiple people at great distances. Games can be so immersive that players' sense of where they are physically is replaced by the play environment; they experience physical and emotional responses as though they were bodily within the environment. Gaming's engagements and immersive qualities offer technical communicators study possibilities and challenge them to incorporate the possibilities of gaming—and of audience expectations around gaming—into technical communications.

In an article on how groups work together in a massively multiplayer online game, Lee Sherlock (2009) examines how players used both in-game chat features and "out of game" writing environments—blogs, wikis, FAQs, message boards—to learn about and organize game play. Sherlock uses the game as an environment for addressing technical communication theorist Clay Spinuzzi's questions about how digital work systems can support workers' innovations. In considering the relations among players' freedom to shape the game and the actions of the game's publisher to control the game, Sherlock notes that "what appears to be an open system of information generated by players to assist in gameplay activity is tied to a complex constellation of legal, economic, and political factors" (283), although players were able to use the range of writing systems available to them to develop, critique, and incorporate innovations. To Sherlock, players' tactics in using the available writing systems suggest possibilities for worker-empowering digital working environments and for the design of technical communications.

Because games provide varying learning environments for players, research into the learning possibilities of games has grown exponentially. James Gee (2003), for example, argues that "good video games" take advantage of thirty-six learning principles, such as *identity* (players take on appealing characters or build them for themselves), *self-knowledge* (players learn not only what is being taught by the game but also about themselves and their own capabilities), *psychosocial moratorium* (players fail with few consequences and so are motivated to risk and explore), and *discovery* (learners are not told what it is they are to learn but can experiment and discover). Gee argues both that games are useful for education and that education should be more like good games—a call that technical communicators should be exploring.

Similarly, some who design digital interfaces for online professional environments argue that such interfaces should take advantage of the principles of immersive gaming. Marc Sasinski (2009), for example, an experience designer, argues that "everyday users are now bringing mental models" from gaming to their expectations about all digital interfaces and that designers need therefore to incorporate principles from gaming in their "day-to-day design work." Although the techniques Sasinski recommends overlap with those recommended by Gee, Sasinski recommends them not only because of the learning they encourage but also because of the emotional connections they encourage. These observations about games encourage technical communicators to consider what expectations their audiences bring about gaming to any technical communication situation and so to design communications that take advantage effectively of gaming's potential to meet those expectations.

Technical communicators considering gaming approaches for communications should also be aware of "alternate reality gaming" (ARG), which "uses the real world" rather than computer screens for play (*Wikipedia* 2010a): designers of such games can make use of cell phones, e-mail, blogs, or any other media to set up puzzles or other playing situations that extend across neighborhoods, cities, or countries. An early ARG, a 2004 promotion for the Halo 2 game, is I Love Bees, whose playing and purposes have been described by its lead designer, Jane McGonigal (2008). She sees such gaming as helping players experience and learn how to participate in a changing media environment where decisions and knowledge are constructed through collective actions, as in *Wikipedia* or the World Economic Forum's Global Risks Prediction Market. Such sites, McGonigal claims, "use digital networks to connect massively multi-human users in a persistent process of social data gathering, analysis, and application. Their goal: to produce a kind of collectively generated knowledge that is different not just quantitatively, but also qualitatively, in both its formation and its uses" (199); participants understand themselves as "playing a singular, meaningful role in the network, with valuable individual micro-contributions to make to the massively scaled effort" (201). Although many ARGs, like I Love Bees, have engaged players in solving fictional problems for commercial ends, there is a developing genre of "Serious ARGs": World without Oil, for example, helped players imagine a future without oil, and The Black Cloud helped high school students learn about indoor air quality (*Wikipedia* 2010a).

Another serious game—showing possible alternatives to ARGs—is Foldit. Foldit is a website developed and supported by a wide range of scientists at the University of Washington, the U.S. Defense Advanced Research Projects Agency, the National Science Foundation, and Microsoft, and enables anyone with access to the web to solve puzzles related to how protein molecules fold. According to the website, "Foldit attempts to predict the structure of a protein by taking advantage of humans' puzzle-solving intuitions and having people play competitively to fold the best proteins" (Foldit 2011). Players come up with novel protein structures, which scientists then analyze for their usefulness in (for example) disease eradication. In 2011, players helped determine—in ten days—the folds of a monkey virus that causes AIDS; researchers had been struggling with this problem for fifteen years. Such games challenge technical communicators not only to consider such involved gaming situations for large-scale communication contexts but also to consider how collective intelligence as a concept and a practice is shifting audiences' understanding of themselves.

Technological Decisions That Are Rhetorical Decisions

In a book about computer games and simulations, Ian Bogost (2007, 5) observes that "computers run processes that invoke interpretations of processes in the material world." For example, in a game in which players run an international hamburger chain, the game's procedures limit what players can do as they manage "the third-world pasture where cattle are raised as cheaply as possible; the slaughterhouse where cattle are fattened for slaughter; the restaurant where burgers are sold; and the corporate offices where lobbying, public relations, and marketing are managed" (28). Bogost argues that this game "makes a procedural argument about the inherent problems in the fast food industry, particularly the necessity of overstepping environmental and health-related boundaries" (31): because players play bound by the game's procedures, they learn how these procedures work in the world as well as on the computer. Bogost argues that these observations about what he calls "procedural rhetorics" apply to all games, and that attention to procedural rhetorics can help us develop stronger games as well as analyze the persuasion of games. Bogost asks us to attend not simply to what we see of the game on a screen if we wish to understand how the game functions rhetorically; we must also attend to the procedures that the game's designers have chosen to represent through the game's coding. In playing any game, Bogost argues, the procedures that quietly structure game play act on us, persuading us toward a particular view of the world's workings. In asking us to see that there is rhetoric in what might seem technical, Bogost's efforts parallel those of other writers and scholars.

Copyright lawyer Lawrence Lessig (2006), for example, compares decisions made by two universities about access to their networks. Lessig argues that a decision at one campus to allow open and anonymous access was based in First Amendment thinking; the other campus required users to register their computers first, with no room for anonymous contribution and with, potentially, full tracking of a user's actions. For Lessig, these two networks "differ in the extent to which they make behavior within each network regulable. This difference is simply a matter of code—a difference in the software and hardware that grants users access" (34). The coding of any technology thus embeds values and so encourages certain kinds of attention and behavior. (For variations on this discussion, and elaboration on how code, hardware, and software are not neutral but embody values, see Chun 2008 or Galloway 2006.)

Obviously, technical communicators need to be alert to the values and related rhetorical effects of their technical decisions—and this gives them another responsibility, as technical communicator Rebekka Anderson (2008) describes in writing about content management. Anderson argues that while content management systems "may streamline information development and communication processes, they will not therefore necessarily improve the information products and communications since the needs of the producers and consumers of the information are subordinated to the process of information development" (68). When decisions about content management are made solely for the sake of efficiency, that is, "the focus is on the information, not the people who might actually wish to use it" (74). For Anderson, this situation requires more than technical communicators advocating for users and documents designed for particular purposes; instead, she argues that "as long as technical communication scholarship lacks visibility and accessibility, focuses exclusively on end users and rhetorical problems, and fails to make strong business arguments for rhetorical work, those making critical business decisions will continue to view [content management] as a technical solution to the sociotechnical and rhetorical challenges of empowerment, collaboration, quality, usability, and technology adoption" (63). Anderson thus argues for technical communicators who work with new media to understand what they do within the large contexts of culture, society, and work—and to advocate on behalf of users over technology.

A RHETORICAL HEURISTIC FOR TECHNICAL COMMUNICATORS WORKING WITH NEW MEDIA

Rhetoric is a method for understanding how any communication does what it does. Rhetoric is for analyzing texts, to help you name and consider the strategies used in a text and whether you wish to be persuaded by it. Rhetoric is also a method for helping those who compose texts consider the audiences for whom a text is made, the contexts in which the text and audiences circulate, and the purposes for which the text is designed; with such understandings, technical communicators can decide what strategies to use in shaping any text.

Figure 17.1 diagrams a rhetorical understanding of a technical composing process; the quoted words come from the U.S. Department of Labor's description of what technical communicators do ("What's the Difference" 2009). I use this diagram to provide a three-part framework for the questions I am about to offer, questions technical communicators can ask while working with new media; I use it also to support the claims I have been making that working with new media is not simply about learning any new software but is, in addition, being alert to how new communication technologies shift relations among technical communicators, the documents and other texts they produce, and their audiences. Keep in mind that the questions I offer here supplement the questions and heuristics offered in this book's other chapters; for example, questions about ethical considerations in working with new media should be supplemented by the questions in the chapter on ethics.

Note, too, that the design process sketched in this figure may look linear and imply that questions are asked once; instead, composers move back and forth among the process's stages, using information gained at one stage to rethink what they learned at other stages. As you move through the questions below, you might note that some questions repeat

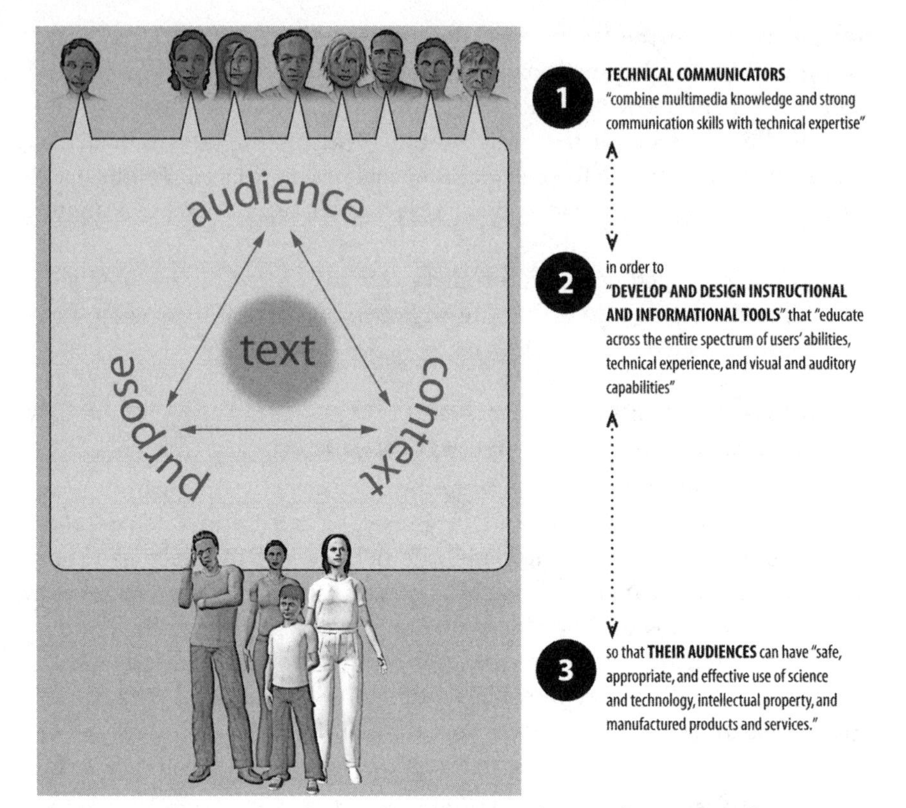

Figure 17.1. A rhetorical understanding of text production. (Quotations from "What's the Difference" 2009.)

concerns from other areas; this redundancy is intended, to keep you from fixing too early on approaches that will benefit from further consideration.

QUESTIONS FOR TECHNICAL COMMUNICATORS WORKING WITH NEW MEDIA

1a. Can one person carry out all the requirements of a new media project—or is a team needed? Most new media projects require the following: rhetorical knowledge; multimodal and interactive design abilities; writer/content developer; instructional designer; subject-matter expert; project management; attention to the social needs of the team; technical expertise for media production; testing; someone who can advocate, in the sponsoring organization, for the rhetorical work of the project in the ways described by Anderson (2008) above.

1b. If you are working at a distance from team members or the organization sponsoring the work, how will you develop comfortable working relations? How will you develop decision-making processes? What software will you use for collaboration, and what sorts of collaboration does the software encourage?

1c. Given that the abilities needed on a project may not be as clearly differentiated as they first seem, how will you negotiate who makes the final decision about the domain under which questions and problems are addressed? For example, who will decide whether a question about the functioning of a webpage is technical or rhetorical?

1d. How will you negotiate developing style sheets for the documents you produce? (A style sheet, guide, or manual is a set of standards for the writing and design of documents; the standards address grammatical, mechanical, and visual design conventions to be used by all writers on a project or in an institution so that all documents are consistent and represent the values of the project or institution.)

1e. How will you establish processes for tracking changes, storing texts, and making backups of work?

1f. Will you be developing new content for this project, or using existing content? Do you have the resources for developing new content? Do you need to supplement existing content?

1g. If you are working with existing templates for producing texts, how will you establish a sense of ownership and pride in the work? What leeway is there to modify the templates for your particular purposes?

1h. What support do you need from the organization(s) with which you work in order that you stay knowledgeable about new media technologies and their possibilities for technical communication? What responsibilities do you have to yourself to stay knowledgeable about new media possibilities, and what resources and time are reasonable for you to commit?

QUESTIONS FOR DEVELOPING AND DESIGNING INSTRUCTIONAL AND INFORMATIONAL TOOLS

Developing instructional and informational tools requires much conceptual work, as technical communicators research and think through the needs, knowledge, and contexts of those for whom they are developing communications and as they fine-tune their understanding of their communicative purposes for their audiences.

Considerations of Audience

Later I will ask you to consider how you might work with the real people who use the documents you produce, but the considerations immediately below are about how you, as technical communicator, need to conceive of your audience's qualities. As you move through these questions, keep in mind this chapter's arguments about audience expectations for digital communications to be multimodal as well as aesthetically and emotionally engaging.

2a. What relations do you want audiences to establish with the documents and the technology being used? For example, do you want audiences to be able to personalize the documents by making choices about interface colors or an avatar? Should the documents and technology provide users with immediate response, or will they need to encourage audiences toward more patient uses?

2b. What emotional relations do you want to establish between the audience and the documents? Should the documents seem friendly, authoritative, playful, challenging, or . . . ?

2c. What sense of the person or people behind the documents do you want the audience to develop? Do you want the audience to see you or the company responsible for the text as distant but friendly authorities, people who are similarly trying to learn, funny know-it-alls, or . . . ?

2d. How might your documents empower users rather than placing them in a passive relation with the technology being used or taught?

2e. How might your documents encourage audiences to engage and collaborate more with others? Should the documents be designed to be used by one person at a time or several?

2f. Should the documents you produce engage audiences through their bodies and movement (such as suggested by small tech or ARGs, as described above) or primarily through the visual interfaces of screens?

2g. Should your documents address your audience anonymously or as avatars with chosen names, or should they address your audience using their real names and characteristics?

2h. What sorts of help will the audience need in learning what you are teaching? Do they need explicit instructions up front, or can you embed context-sensitive help into the documents?

Considerations of Purpose

2i. What genres of new-media texts—blogs, games, help systems, social-networking systems like Facebook, and so on—might help you achieve your purposes?
2j. What arrangements within the documents will most support your purposes? Should your audience move through by means of a decision tree, a menu, or an index?
2k. What are the emotional purposes of your project? What emotional relations should your project encourage between audiences and the documents and technologies you are using?
2l. What aesthetic considerations will help your documents be engaging?
2m. How will you describe your purpose? Although many communicators conceive their purpose as "The audience will learn software package X," a more useful statement of purpose is "Our audience, completely new to software X and shaky about using computers, will gain a comfortable and friendly initial competence with using features A, B, and C of software X through a gentle and playful approach that will put them at ease and help them understand they cannot make any mistakes." Notice how such a statement of purpose immediately starts to suggest the look and feel of the documents to be developed.
2n. Does your statement of purpose help you stay within the project's temporal and financial constraints? Do you/your team have the abilities to develop the project?

Considerations of Context

2o. Where, with whom, and using what device(s) will your audience encounter the documents you produce? What do these contexts tell you about your audience's mood and attention level at the time they use the documents?
2p. Will the documents be used in multiple contexts by the same audience? For example, in the scenario opening this chapter, those tending the flood-control infrastructure need documents to support both routine and emergency maintenance. Be sure to brainstorm a range of contexts in which your audience will use the documents.

2q. Will the documents be used across contexts by different audience members? For example, will people be in different countries or states, at homes or offices, late at night or midday, on cell phones or desktop computers? What do these different possible contexts tell you about an audience's different needs and expectations?

2r. What does the place of use tell you about available technologies? For example, if you are developing documents for other countries, do you know the level of cell-phone use or computer access? What does this tell you about the size of screen for which you need to develop, or audience expectations about interactivity?

Considerations of the Text Itself

2s. Are you considering what you are producing as a document or a simulation? As a game or a video? As an experience or an environment?

2t. Will you use one, two, or more modes?

2u. Can your audience access the documents easily, both in terms of having access to the necessary technologies and in terms of knowledge for using the documents?

2v. If it suits the documents' purposes, can they be shared easily among different audience members?

2w. Will these documents be used once or repeatedly?

2x. What is the maximum size of the documents, in terms of download and storage, so that you will not overstretch your audiences' patience or technological limits?

2y. Do you have the resources—human, technical, monetary, temporal—to produce what you envision?

2z. Can you justify producing digital rather than print documents, in terms of time, cost, and environmental impact?

QUESTIONS FOR ENGAGING WITH THE "REAL" AUDIENCE

The "real" audience is people who live and breathe and who will use your documents. If you can interview them or a subset of them before and during the project, and engage them in collaborating and testing, your project is more likely to succeed.

3a. What can you learn from your audience to help you in designing?

3b. What sorts of audience research—interviews, surveys, observations of their technology uses—will most help you learn what you need to design effective documents?

3c. Where and how will you engage the audience in your designing and planning?
3d. How can your communication help people use new media to "serve their own cultural, political and social visions" (Srinivasan 2006, 497)?

APPLYING THE HEURISTIC

Having analyzed how communicating with new media entwines us with more than just new technologies, let's return to the view of the future with which I opened this chapter, to use the heuristic I offered: Imagine a technical communicator in the future, Cassandra Fortuna, who is responsible for developing training materials for the people charged with maintaining the County Flood Control's networked visual overlays (the overlays described above, the ones illegally brought up by the students); what sort of materials do you think Fortuna should develop? Through what sorts of process should she develop them? With whom should she work?

Imagine that Fortuna is part of a team that includes a technical specialist who knows the county's drainage system and the state requirements for such systems, a programmer, a support person who will track process documents, and a project manager (who is also managing four other projects). In an initial face-to-face discussion, the programmer suggests that it would be easy simply to make accessible to those working in the field the already existing documents about the technical aspects of the flood control system; in fact, the programmer says, she just finished work on another project in which all existing technical support documents were indexed into one system, so that it will be easy to enable field technicians to call up the index and so access all existing documents virtually, using the systems embedded in the work clothing required by the county. The project manager quickly agrees, saying that the indexing project was undertaken precisely to save the company money by making all documents accessible in this way; the training the team should develop, the manager says, is simply the training on accessing and using the index.

Fortuna agrees that the field technicians need to learn to access and use the new index, and that the team should develop such training—but she also wonders (realizing that she needs to take on the role of rhetorical advocate in this project) whether other kinds of training and perhaps some new documents might be necessary, given the contexts of the field and the knowledge most field technicians have. The heuristic above suggests that the team needs to learn about their audience and the contexts in which the audience works; the project manager agrees to provide Fortuna time to learn more about the field conditions and the field technicians' knowledge.

Over the next month, applying heuristic 3a, Fortuna talks with several field technicians—both experienced and new—to get a sense of their working conditions and of the knowledge they bring, and she develops a survey to learn more specifics (heuristic 3b); because Fortuna learns just how independently the technicians in the field work and the differing conditions they encounter, she includes in the survey the question, "What are the oddest circumstances under which you've had to make fixes in the field?"

From carrying out the survey with all field technicians, Fortuna can report to the project team that (1) the experienced field workers' rich knowledge of the drainage system is not incorporated into any current documents but is shared informally; (2) work in the field is usually routine checking and maintenance, but fires, floods, and earthquakes occur regularly enough that technicians must learn to access and apply training materials under considerable stress; and (3) because all sorts of people pass through the backcountry where the most crucial parts of the drainage system are, field technicians have to be prepared to deal with and learn from odd situations (as when one worker recently stumbled on several teenagers illegally accessing the system who then lectured the technician on the security lapses and poor technical quality of the system).

Having done such research into the contexts and audiences for the training materials, Fortuna argues for the following materials and purposes and for additions to the development team. (Note how the heuristic's considerations about purpose, context, audience, and text weave together.)

Materials for learning how to access the index. Given that the field technicians need to access the index quickly and frequently, Fortuna argues (applying heuristics 2a and 2b) that the training for using the index should consist of short videos the technicians can access from anywhere at anytime. (She argues that video makes more sense than print because the workers will learn best from demonstrations of processes rather than having to read; in addition, the videos can be presented in a quick but friendly tone so that technicians think positively of the index.)

A social-networking site for the field technicians. Because the experienced field technicians have been sharing information informally and because this information is so rich, it should be made available to all the technicians (heuristics 2d and 2v). In addition, drawing on heuristics 2a, 2e, 2k, and 3d, Fortuna proposes that, because this information changes and develops in response to what happens in the field, it also needs to be available in ways that technicians not only can access but can add to. Similarly, because the technicians are proud of this knowledge, the social-networking site should be "owned" by the technicians, so that they will continue to use and add to it: the technicians should be involved in the design of the site and should choose who among them will advocate for the site and keep it active.

A simple game simulation of how the system works when there is fire, a flood, or an earthquake. Although developing a simulation will be expensive (both because developing a game is expensive and time consuming and because it requires [heuristics 1a and 1h] adding a game-design specialist to the production team), a simple simulation will pay for itself through future savings (heuristic 2z). An incident that came to light through the survey (in an example of heuristics 2o and 2p) demonstrates the possibilities of future savings: during the last fire and flood season, a quarter-mile section of piping had to be replaced because a technician, acting quickly after the fire, made a choice that led to the pipe being destroyed during subsequent flooding; the cost of that replacement alone would cover the cost of developing the game. In addition, the game could be marketed to other flood-control districts, turning it into a potential revenue generator.

A social-networking site for the public to give feedback and advice to the county. Although a site for public feedback and advice is technically not training materials for the field workers, it could help the county develop better relations with the people it serves as well as potentially learn from others. This site could be linked to the technicians' site mentioned above, so that the technicians' knowledge base (about what the public observes and what the technicians learn about the public) could expand (heuristics 2d and 2e).

The project manager approves the first two kinds of training because they can be developed within the current budget and schedule; the manager requests further research from Fortuna into the actual costs and development time of the game simulation, so the manager can request those resources from higher up.

CONCLUSION

Being a technical communicator in our time of new media means being an advocate for the rhetorical conditions under which technical communication is produced and used. The heuristic I offer here (used with this book's other heuristics) should help you think about new media rhetorically in work and other contexts—while you also consider the expectations audiences have for new-media interactivity and engagement and how new media shape our cultural, social, and political lives together. It is impossible, of course, to list all possible questions you should ask as you pursue projects and help your audiences become more confident and competent users of technology, but the more you apply the questions offered here, the more you will learn how to ask subtle, context-specific questions.

DISCUSSION QUESTIONS

1. Pick any technological object you use frequently: a cell phone, a laptop computer, an MP3 player. What different kinds of knowledge do you need to use the object? To use the object well? (Think about lightbulbs: you need to know where to buy them, how to screw them into a socket, when to replace them, how to turn them on; you also need a minimal understanding of electricity and wiring, of money, of the uses of light, of how stores work, etc.) List all the social systems—like those of power companies, stores, and so forth—to which the use of the technology ties you.
2. Pick any technological object you use frequently: how would your life be different if the object didn't exist?
3. Pick any technological object you use frequently: how would you redesign the object for people who use senses differently than you do? For example, redesign your cell phone so that someone who didn't see well could use it, or redesign your laptop so that people who do not have use of their hands could use it.
4. Through short e-mail interviews with employees or a site visit, research an institution in which you might work as a technical communicator in the future. What technological knowledge do you need to bring to the institution? What new media does the institution currently use, and how do those technologies contribute to the working environment of the institution?
5. How will you stay informed about new digital communication technologies in the future? Do you think you or institutions for which you work should be responsible for keeping you informed about them and trained in their use?
6. Chat with a younger sibling or friend about their attitudes toward and uses of cell phones or other digital communication technologies. How do the younger person's attitudes and uses differ from yours?

Compare your observations with those made by two other people in class; working with your two colleagues, prepare and present a short presentation in which you contrast your attitudes and uses with those of the younger people and in which you consider how those attitudes and uses prepare the younger people to have relationships with their friends and colleagues that differ from yours.

7. How well are you prepared to be a facilitator of technical communication and knowledge (as described above), learning from audiences and helping them learn from each other, in addition to or instead of being the person who makes knowledge for others? What more do you need to do to help prepare for this? Do you think such facilitation is appropriate, given your own experiences using and learning about the technologies you use now?
8. Imagine that you need to create documents to help parents of children with cancer learn about a new treatment. Brainstorm the range of emotions parents might feel in such situations, and develop several different statements of purpose in which you take into consideration different emotions the parents might be feeling and how the documents you develop might address those emotions. How does imagining the different emotions they feel encourage you to imagine developing different materials for the parents?
9. Choose a location on your campus or in your town that has importance for the community. (The location could be important because of a historical incident, for example, or because it is a popular gathering site.) Imagining that you could use any available new-media technology, use the heuristic in this chapter to sketch three differing site-specific approaches you could use for informing a local, popular audience about the site. You will need to study the site by going there and observing how people use the site, and you will need to consider the kinds of emotional engagements you think the audience should develop with the site, given your purposes. Consider podcasts, touch screens, interactive projection systems, and so on.
10. Using this chapter's heuristic, design materials to support two differing audiences in learning about using new technologies. First, use the heuristic to help design materials for people in retirement communities so that they can better use their computers and other digital devices available to them to stay in touch with their families and their health-care providers; do not assume that such an audience does not know how to use computers and mobile devices. Second, use the heuristic to help people in middle schools become better researchers:

design materials that will help them understand why it is useful to research the sources behind scientific, political, or other claims and then help them learn how to find and evaluate those sources.

WORKS CITED

Anderson, Rebekka. 2008. "The Rhetoric of Enterprise Content Management (ECM): Confronting the Assumptions Driving ECM Adoption and Transforming Technical Communication." *Technical Communication Quarterly* 17 (1): 61–87.

Bogost, Ian. 2007. *Persuasive Games: The Expressive Power of Videogames*. Cambridge, MA: MIT Press.

Chun, Wendy Hui Kyong. 2008. *Control and Freedom: Power and Paranoia in the Age of Fiber Optics*. Cambridge, MA: MIT Press.

Clark, Dave. 2008. "Content Management and the Separation of Presentation and Content." *Technical Communication Quarterly* 17 (1): 35–60.

Deibert, Ronald J. 1997. *Parchment, Printing, and Hypermedia: Communication in World Order Transformation*. New York: Columbia University Press.

Eisenstein, Elizabeth. 1979. *The Printing Press as an Agent of Change*. Vols. 1–2. Cambridge: Cambridge University Press.

Ellison, Matthew. 2009. "Social Media in Technical Communication." *Communicator*, 10–12.

"Foldit: Solve Problems for Science." 2011. Center for Game Science at University of Washington. http://fold.it/portal/. Accessed December 18, 2011.

Galloway, Alexander. 2006. *Protocol: How Control Exists after Decentralization*. Cambridge, MA: MIT Press.

Gee, James P. 2003. *What Video Games Have to Teach Us about Learning and Literacy*. New York: Palgrave Macmillan.

Graham, S. Scott, and Brandon Whalen. 2008. "Mode, Medium, and Genre: A Case Study of Decisions in New-Media Design." *Journal of Business and Technical Communication* 22 (1): 65–91.

Hart-Davidson, William, Grace Bernhard, Michael McLeod, Martine Rife, and Jeffery T. Grabill. 2008. "Coming to Content Management: Inventing Infrastructure for Organizational Knowledge Work." *Technical Communication Quarterly* 17 (1): 10–34.

Hawk, Byron, and David M. Rieder. 2008. "Introduction: On Small Tech and Complex Ecologies." In *Small Tech: The Culture of Digital Tools*, edited by Byron Hawk, David M. Rieder, and Ollie Oviedo, ix–xxiii. Minneapolis: University of Minnesota Press.

Jones, John Chris. 1992. *Design Methods*. New York: Wiley.

Kim, Loel, Amanda J. Young, Robert A. Neimeyer, Justin H. Baker, and Raymond C. Barfield. 2008. "Keeping Users at the Center: Developing a Multimedia Interface for Informed Consent." *Technical Communication Quarterly* 17 (3): 335–357.

Kittler, Friedrich A. 1999. *Gramophone, Film, Typewriter*. Translated by Geoffrey Winthrop-Young and Michael Wutz. Stanford, CA: Stanford University Press.

Lawson, Bryan. 2005. *How Designers Think: The Design Process Demystified*. Fourth ed. Philadelphia: Architectural Press.

Lessig, Lawrence. 2006. *Code, Version 2.0*. New York: Basic Books.

———. 2008. *Remix: Making Art and Commerce Thrive in the Hybrid Economy*. New York: Penguin.

Maggiani, Rich. 2009. "Technical Communication in a Social Media World." *Intercom*, 18–20.

Manovich, Lev. 2002. *The Language of New Media*. Cambridge, MA: MIT Press.

McGonigal, Jane. 2008. “Why I Love Bees: A Case Study in Collective Intelligence Gaming.” In *The Ecology of Games: Connecting Youth, Games, and Learning*, edited by Katie Salen, 199–228. John D. and Catherine T. MacArthur Foundation Series on Digital Media and Learning. Cambridge, MA: MIT Press.

Saenger, Paul. 1997. *Space between Words: The Origins of Silent Reading*. Stanford, CA: Stanford University Press.

Sasinski, Marc. 2009. “Engaging the User: What We Can Learn from Games.” *Johnny Holland Magazine*. http://johnnyholland.org/2009/08/31/engaging-the-user-what-interaction-designers-can-learn-from-video-games/. Accessed January 21, 2010.

Sherlock, Lee. 2009. “Genre, Activity, and Collaborative Work and Play in World of Warcraft: Places and Problems of Open Systems in Online Gaming.” *Journal of Business and Technical Communication* 23 (3): 263–293.

Spinuzzi, Clay, ed. 2007. *Technical Communication in the Age of Distributed Work*. Special issue of *Technical Communication Quarterly* 16 (3).

Srinivasan, Ramesh. 2006. “Indigenous, Ethnic and Cultural Articulations of New Media.” *International Journal of Cultural Studies* 9 (4): 497–518.

“What’s the Difference between Technical Communicator and Technical Writer?” 2009. *Society for Technical Communication*. http://archive.stc.org/story/tc_tw.asp. Accessed March 16, 2012.

Wikipedia. 2010a. “Alternate Reality Game.” Accessed February 22, 2010.

———. 2010b. “Web 2.0.” Accessed February 22, 2010.

Winner, Langdon. 1986. *The Whale and the Reactor: A Search for Limits in an Age of High Technology*. Chicago: University of Chicago Press.

18

REBECCA E. BURNETT, L. ANDREW COOPER, & CANDICE A. WELHAUSEN

What Do Technical Communicators Need to Know about Collaboration?

SUMMARY

Collaboration is important because virtually all workplaces rely on group-based decision making and projects, often increasing creativity, productivity, and the quality of both process and product. The chapter presents a definition of collaboration as an intentional, sustained interaction toward a common goal. Topics about collaboration explored in the chapter include cognition and learning, processes used in collaboration, and technology that supports collaboration. Throughout the chapter, examples are drawn from an extended scenario about a grant-funded project at a health-care organization. The chapter introduces strategies to help technical communicators become more successful collaborators and avoid pitfalls when collaborating.

INTRODUCTION: BUILDING COLLABORATION IN THE WORKPLACE

Cassandra worked on grant-funded research projects at a health-care organization with scientists, project managers, information technology personnel, and administrative assistants. After a major project was funded, she joined a five-member group. Each member brought a different skill set and therefore assumed a different project role with different responsibilities.

The project's primary goal was to provide peer support to Korean women recently diagnosed with breast cancer by matching them with other women who had survived breast cancer. These women, working as volunteers, provided emotional support, information about breast cancer diagnosis and treatments, and help locating resources such as child care during medical appointments. Under the direction of the grant's principle investigator, Dr. Samantha Smith, the project group planned to develop culturally appropriate and bilingual (English/Korean) patient recruitment brochures, informational sheets about breast cancer, and in-take and tracking forms.

When Cassandra and Samantha learned that the project had been funded, they were both excited. As a well-known breast cancer researcher,

Table 18.1. *Members of a collaborative project that supported Korean women diagnosed with breast cancer*

Key collaborator	Project role
Samantha	breast cancer researcher
Cassandra	technical communicator
Patricia	study coordinator
John	database administrator
Diana	administrative assistant

Samantha was the subject-matter expert, while Cassandra was her technical editor. In the proposal, Samantha had specifically named Cassandra as the technical communicator, and three additional staff members: a study coordinator, a database administrator, and an administrative assistant (see table 18.1). Thirty volunteer breast cancer survivors would also be recruited to provide peer support, and a minimum of 300 patients would be enrolled during the two-year project.

Cassandra and Samantha had previously written several proposals to fund other breast cancer projects and had also collaborated on several final reports about these projects with John, the database administrator. Diana routinely provided administrative support to all staff, so Cassandra, John, and Diana had already established strong working relationships together.

The study coordinator, however, would be a new hire. This person would recruit and train the volunteer breast cancer survivors, enroll new patients, and serve as liaison between study participants and staff. She or he would need to develop strong relationships with both patients and project staff to help ensure that the staff could develop effective materials. The group hired Patricia, a bilingual (English/Korean) breast cancer survivor for the position. She brought the perspective of someone who had not only survived breast cancer but also was intimately familiar with the target culture.

Shortly after Patricia was hired, the group met in the conference room to discuss each person's role and to draft a long-term work plan. Samantha would oversee all aspects of the project; all staff would report directly to her. On the conference-room whiteboard, Diana began listing the project's major tasks and the person(s) responsible. The group would hold biweekly meetings, she explained, while individuals would also interact in the following smaller groups to help make the project more manageable:

- Cassandra, Patricia, and Samantha would jointly create the project recruitment materials.
- Samantha and Patricia would compose draft documents, which Cassandra would revise and edit.
- After recruitment was completed, John would create a patient-tracking database, and Diana would enter new patient information into the system as well as regularly update patient records.

During the biweekly meetings, each small group would report its progress to the group as a whole. The biweekly meetings would allow each group to gain a broader perspective on its work from the other group members. During these meetings, Samantha would encourage each person to offer substantive suggestions and feedback on all aspects of the project.

Samantha then addressed the software tools needed to accomplish each task. Creating some of the materials—such as the patient recruitment brochures—would require using publishing design software. Cassandra was taking a class for just this purpose. John would use the database management system that he routinely used for other projects to create data-entry screens for Diana. John would also train Diana to use the database management system.

The staff members worked in the office during regular business hours and were available for in-person or over-the-phone collaboration. Samantha, on the other hand, saw patients at her clinic several miles away three times a week. The group decided e-mail would be the best method for contacting her.

As the group discussed the project's goals, each individual's tasks and responsibilities, the available settings, and the required tools, Cassandra realized that her relationships with each of her coworkers and with Samantha in particular would change. Working with each person individually would require adjustments in communication styles. John, for example, seemed to prefer direct, businesslike interactions. Samantha, on the other hand, had previously assumed a personal style. However, Cassandra knew that Samantha now needed to assume the role of "boss," moving beyond their personal relationship and building working relationships with the other members of the group. All group members would need to make adjustments to be effective collaborators on this project.

As Cassandra's situation illustrates, many factors influence the ways in which technical communicators must approach and plan collaborative interactions with their group members. Gaining a more in-depth understanding of her situation will help you learn to work more effectively as a technical communicator on group-based projects. This chapter begins with a discussion of what collaboration is and what it is not, exploring the ways it affects technical communication practices. Next, a literature review summarizes critical research about factors that influence collaboration. Then, a heuristic provides ways to apply that research to your own collaborative projects. An extended example follows the heuristic, demonstrating the ways in which Cassandra applied it to her own situation.

LITERATURE REVIEW: WHAT YOU CAN LEARN ABOUT COLLABORATION FROM RESEARCH

As the opening scenario illustrates, many types of work involve collaboration. Collaboration will be vital to your professional life, regardless of your career. According to researchers, "recent estimates conclude that group-based work methods exist in nearly 70% of U.S. firms" (Lowry et al. 2006, 632). Not only is collaboration crucial in most workplaces, it also takes place in as much as 75–85% of workplace writing (Burnett 1991; also see Lunsford and Ede 1990).

Three assumptions underpin this chapter's discussion about collaboration in technical communication:

- *Collaboration is rhetorical.* Successful technical communicators consider collaborative contexts, identify purposes, select appropriate content, respond to audiences' needs, organize information, and support arguments. Failing to address rhetorical elements can reduce the effectiveness of collaboration, so technical communicators must collectively assess the rhetorical situation.
- *Collaboration is a process.* Successful technical communicators understand that they are part of a group whose members must cooperate and share responsibility for completing high-quality work while maintaining a schedule. Each collaborator has a dual responsibility: completing individual work effectively and helping others do the same. Failing to fulfill these responsibilities can derail the process, so technical communicators must approach collaboration strategically to avoid problems.
- *Collaboration is multimodal.* Successful technical communicators understand that their processes and products include written, oral, visual, and nonverbal communication across a range of print and digital media. Failing to recognize various opportunities and challenges inherent to different modes and media can lead to inefficiencies, misunderstandings, and technological disasters, so technical communicators must learn to use modes and media to the group's advantage.

Understanding these assumptions increases the likelihood that your collaborations as a technical communicator will avoid pitfalls.

LITERATURE DEFINING COLLABORATION

Before defining what collaboration *is*, consider what collaboration *is not*. It is not having each group member write a separate section of a long document and then cobbling the document together at the end. It is not giving a group presentation where each speaker is only responsible for discussing her or his individual contribution on the overall project. It is not simply using computer-mediated communication (e.g., social networking, file or video sharing, blogs, instant messaging, chats, and online forums). It is also not being in a group that leaves you doing all the work.

At its core, collaboration involves substantive *interactions* between and among *people* who share *goals* and exchange information as they work toward those goals in a variety of *settings* and with a variety of *tools*, either because the task size or *complexity* is too great for a single person or because the task will benefit from multiple *perspectives* (cf. Nunamaker et al. [1991], who pose a similar but less-elaborated definition). Collaboration is an intentional, sustained interaction toward a common goal, and thus it involves all of these factors, which form the vertical axis on the matrix of this chapter's heuristic:

- *Interactions*. Collaborative interaction is sustained and extended over time, anywhere from a few hours to many years. When you collaborate, anticipate the frequency, depth, and breadth of the interaction. Articulate and agree on a common understanding of the tasks and processes; such discussion and agreement virtually always ensure more effective collaboration (Barthelmess et al. 2006; Soller et al. 1999).
- *People*. You and your collaborators are committed to the same technical communication tasks, but each person brings different backgrounds and perspectives, and each person plays different roles. The people in your group are not necessarily equal in subject-matter experience, relevant expertise, or organizational rank. Working together may diminish though not necessarily eliminate hierarchical differences. Even when hierarchies aren't obvious, they probably exist and influence the interaction (Rosen 2007).
- *Goals*. Collaborators share common goals determined among the collaborators themselves or mandated by external or internal sources. Buy-in usually increases when technical communicators

articulate a shared understanding or representation of the task (Flower 1987).

- *Settings*. Collaborators work in a variety of spaces and places, in both face-to-face and distance settings (Winsor 2000). You might work with a colleague in the next office or halfway around the world. In thinking about the work environment, technical communicators consider accessibility specified in the Americans with Disabilities Act (Reece 2002) as well as factors such as convenience, safety, physical and psychological comfort, access to technology and other useful resources, sufficient and appropriate work space for varied activities, and privacy for productive work.
- *Tools*. Collaborators select tools based on the project and available resources (Hewett and Robidoux 2010). Typical tools include technology for communicating regularly (e.g., telephone, e-mail, videoconferencing), web-based applications for sharing and working on drafts (e.g., file-sharing and multiple-user editing applications), and software features that make collaboration easy (e.g., the track-changes feature in word-processing software). Collaborators also need to be creative in adapting software to improve interaction. For example, presentation software is not typically thought of as a collaborative tool, but you can collaborate by using the commenting feature or adding text boxes for annotations.
- *Complexity*. Some tasks are simply too large or complex for an individual. You should expect several types of complexity: procedural, social, and intellectual (Galegher and Kraut 1994). Procedural complexity can occur in large-scale collaborations (e.g., in high-energy physics, which sometimes has groups in the hundreds) or in small groups that have vaguely defined goals and responsibilities. Social complexity can occur in projects that require a technical communicator to gather input from a range of contributors with differing interpersonal competence or differing skills (see Cross 1994, 2001). Intellectual complexity can appear with what are called "wicked problems"—typically those so complex that an apparent solution in one area often generates new problems in other areas; they always require collaboratively generated solutions (Conklin 2005). For example, addressing climate change, managing water resources, and developing national health education campaigns to reduce sexually transmitted diseases are all wicked problems.
- *Perspectives*. Multiple perspectives and multiple voices often produce more effective solutions, which can involve what Mikhail Bakhtin (1981) refers to as heteroglossia: the presence of multiple

voices and perspectives within language that still serves authorial intentions. Different voices often function in different linguistic registers, characterized by changes in vocabulary and syntax for particular purposes, audiences, or situations. Skillful technical communicators adapt and respond to such changes.

Technical communicators can anticipate encountering each of these elements and need to consider their interplay.

Engaging in attitudes and behaviors that are likely to result in productive collaboration is essential in the workplace. Understanding the following categories, which form the horizontal axis on the matrix of this chapter's heuristic, can help you engage with your group members more effectively and productively:

- *Cognition and learning*. Collaboration involves moving beyond an individual approach to a group-based perspective, which requires cooperative learning, helping each other work toward a collective understanding and action.
- *Small-group processes*. Group members must manage factors that influence group performance.
- *Technology*. Collaboration can benefit from tools adapted for shared tasks, a subject that researchers often explore in relation to computer-supported cooperative work.

LITERATURE ABOUT COGNITION AND LEARNING

Researchers have studied the ways individuals think and learn in groups and the ways they help each other learn in virtually all professional areas.

Cognition is a collection of mental processes that constitute thinking. Effective collaboration involves thinking in ways that enable group members to construct shared meaning. Collaborators need to agree substantively on beliefs that are central to their collaborative goal, but their beliefs can never align perfectly because of differences in areas such as experience, intelligence, and values. Trust and negotiation are central to face-to-face and technology-based interactions that technical communicators use to share beliefs critical to common goals (Stahl 2005; Van den Bossche et al. 2006).

Cooperative learning involves people working together in small groups to accomplish shared goals. Small groups typically receive instruction from a teacher, trainer, or manager and then work on a task until all group members successfully understand and complete it (Johnson and Johnson 1998). No individual group member possesses all of the information, skills, or resources necessary to accomplish the goal alone. Cooperative learning depends on positive interdependence: individuals reach their own goals only if other group members do as well. One widely accepted view explaining interdependence involves Vygotsky's (1978) "zone of proximal development," the difference between what a person can accomplish individually and what the same person can accomplish when assisted by a more knowledgeable person (see also Stahl, Koschmann, and Suthers 2006). Ensuring that every collaborator in a group performs at the highest possible level increases the likelihood of reaching the goal successfully.

SMALL-GROUP PROCESSES

Research about small groups provides useful information about procedural factors that technical communicators can manage: group size and membership; time, space, place, and related resources; social loafing; conflict; and leadership.

Group Size and Membership

Decisions about productive group size should depend on the nature of the task and the available resources. When Paul Benjamin Lowry and his colleagues (2006) studied the ways group size affects communication, they drew two conclusions:

- The study's three-person groups engaged in higher-quality communication than six-person groups.
- Variables such as appropriateness, openness, richness, and accuracy were greater in three-person groups than in six-person groups.

While Lowry and his colleagues did not find that being in a three-person group significantly improved the quality of discussion, they suggested that "more complex projects may benefit from using much smaller groups" (657).

While group size is important, differences among group members can also affect the quality of collaboration. Naomi Kleid (2004) notes that classroom groups are usually carefully constructed to give students a productive experience. For example, instructors may consider factors such as ability, workplace experience, level of software proficiency, gender, and English proficiency. However, workplace managers consider different factors, such as employee availability, previous experience, and skill level. Workplace professionals' attitudes and behaviors are also influenced by recognizing that their tasks have long-term significance and that they may work together again (Kleid 2004), not always true in classroom collaboration.

Time, Place, Space, and Resources

Negotiating time, place, space, and resources to do collaborative work is critical, directly affecting people's commitment and engagement. Nearly all technical communicators have work that overflows available time, but the "duration, intensity, and continuity of involvement" of collaboration differ in classrooms and workplaces (Kleid 2004). In classrooms, task duration and complexity as well as member availability are usually controlled by the semester and assignment schedule. In the workplace, task duration and complexity are affected by more demands, including other deadlines, client/customer expectations, and limited availability of other group members, all of whom likely have competing responsibilities in multiple groups in various locations. Schedules are rarely as predictable as an academic semester: projects can last hours, days, weeks, months, or even years.

Collaborators working at a distance who report high levels of mutual involvement—more turn taking, more talking than emoting, more referencing work products—often have greater success (Healey et al. 2008). They also want to believe that all of their group members are fully engaged. For many people, "this sense of engagement breaks down . . . when it becomes apparent that . . . assumptions about the other speaker are ungrounded" (Healey et al. 2008, 174). For example, professionals in the workplace have a sense of where work-related conversations should take place. If a particular collaborator is not at a workstation but is instead driving or otherwise engaged (signaled by background noises such as a toilet flushing, traffic honking, or birds chirping), other collaborators may sense a lack of interest and commitment. In these circumstances, collaborators are not just in different physical spaces but in different mental places, which can negatively affect group efforts.

The spaces and resources that work groups use affect three aspects of collaborative interaction: formality, "presence" (i.e., "group members' recognition that other group members are actively contributing to the work" [Bemer, Moeller, and Ball 2009, 157]), and confidentiality. Amanda Metz Bemer and her colleagues (2009) argue convincingly that the physical environment—both space and resources—affects interaction, and thus productivity and attitudes.

Social Loafing

Social loafers (sometimes called free-riders) are collaborators who don't assume their fair share of the work, a problem usually affecting distance/virtual groups more than face-to-face groups. Larger groups often have more loafers than small groups because lack of engagement and productivity are more difficult to identify quickly or easily. Further, managing and coordinating the workload of larger groups requires more overall effort. Groups with undefined tasks often have more loafers than groups with defined tasks, because group members may be unsure about what they should be doing (Suleiman and Watson 2008). Social loafing has the potential to derail collaboration. If the project requires the effort of five people but two are loafers, then either some of the work doesn't get done, or the three active collaborators are overburdened and probably resentful.

Sometimes social loafing is a result of people simply not thinking for themselves and going along with whatever others say, which is called groupthink (Janis 1972). This tendency for a group to develop a singular mentality that isn't open to dissent or alternative perspectives ignores the benefits of productive conflict. If groupthink replaces productive challenges, the quality of the work suffers.

Conflict

Many collaborators want to avoid conflict, but incorporating productive conflict can be good, increasing interaction and improving the quality of deliverables. Voicing disagreement and urging collaborators to consider alternatives can result in substantive conflict if the disagreements and alternatives focus on rhetorical elements such as context, argument, and purpose. Substantive conflict focuses on the intellectual content of the collaboration. On the other hand, unproductive conflict results from interpersonal disagreements or problems with group processes.

To ensure that conflict is both substantive and productive, as a technical communicator you should only voice disagreements and offer alternatives that are serious, relevant to the task at hand, and supported by strong reasoning and evidence (Burnett 1993, 1994b). You and your collaborators should avoid unproductive conflict, which can drain resources and weaken ties among group members. As you and your collaborators interact, consider each member's attitudes, which include self-confidence and motivation as well as a sense of responsibility and receptivity to planning and collaboration. Awareness of a collaborator's self-confidence, for example, can help you gauge how assertive you should be when responding to that collaborator's ideas. You can develop the following repertoire of verbal moves:

- mildly assertive behaviors (e.g., prompting a group member to clarify and contribute information) can spur substantive conflict;
- more assertive behaviors (e.g., challenging a collaborator with critical questions or opposing viewpoints and directing a change to

an approach or artifact) sometimes work, but they can backfire if the collaborator isn't receptive (Burnett 1994a).

With strategically chosen verbal moves, you can help your collaborators improve their technical communication (Burnett 1993). Minimizing unproductive conflict and encouraging productive conflict take conscious effort.

Leadership

Collaborative interaction can be nonhierarchical, involving a flat organization with no identified facilitator or leader; however, lack of hierarchy and defined leadership often leads to problems because no one is in charge. Effective leaders encourage a unified effort, facilitate interaction, and encourage collaborators toward a common goal. In general, collaborative groups function better with facilitators and leaders, who may be the same people, but don't need to be.

- Facilitators manage group processes. They schedule meetings; establish and maintain technologies for interaction (e.g., e-mail lists, file-sharing sites); collect, maintain, organize, and disseminate work documents (e.g., meeting minutes, document drafts); recommend policies for group interactions, decision making, and conflict resolution; and monitor participation.
- Leaders influence the efforts of a group working toward a common goal. They motivate the group, ascertain the needs of group members, and understand their working context (Denmark 1993). They plan and organize activities, functioning as bridge builders by helping collaborators with strategies that include "keeping agreements, telling the truth, showing respect, and demonstrating a commitment to the relationship" (Bell and Patterson 2006, 14). Effective leaders help a group articulate and accomplish a shared vision. The outcome of this collaborative effort is "greater than that which could be accomplished by any of the individuals acting alone" (Hartwig 2008, 3).

Facilitating and encouraging collaborators toward a shared goal is called knowledge leadership. Knowledge leaders engage everyone in the group, encourage and offer constructive feedback, create a working environment that expects mutual trust and respect, model productive behavior, and encourage initiative (Hardy and Connect 2008).

Is leadership the same in classrooms and workplaces? Not often: "Undergraduate students are uncomfortable with uncertainty, lack of structure, and negotiation (as are many people), but professionals know they must manage uncertainty, create structure, and negotiate differences" (Kleid 2004). Thus, some students look for leaders who reduce uncertainty, dictate explicit organizational structure, and simplify negotiation, but these behaviors may not be productive for group interaction or decision making.

In both classrooms and workplaces, leadership is often gendered (Yoder 2001)—women and men may have different styles of communicating and leading groups. Social constructions of gender influence many women to assume conventionally female styles of leadership and many men to assume conventionally male styles of leadership (Matusak 2001). Women's gendered style, while critical for getting work done, is often not acknowledged as "leadership." Women are often denied leadership roles because they are seen as less competitive, less hierarchical, less power-driven, and less commanding. Some groups also play into gender stereotypes: women manage collaborative details while men provide the group's public face.

Regardless of gender, effective leaders discourage groups from reaching decisions too quickly and often encourage a period of uncertainty and inquiry. Task requirements and group membership should influence decision making, rather than a one-size-fits-all process.

TECHNOLOGY: COMPUTER-SUPPORTED COOPERATIVE WORK

Technology can make collaboration easy and productive, even at a distance. The disciplinary area called computer-supported cooperative work (CSCW) studies the ways computer systems support collaborative activities. Current trends in CSCW examine a range of strategies that help collaborators in the classroom and the workplace. For example, collaborative creativity benefits from CSCW's development of heuristics to use with high-resolution interactive walls that encourage playfulness and openendedness in brainstorming and decision making (Herrmann 2010). Some CSCW work develops analytical and visualization tools used in collaborative mapping (Haklay 2006) and describes productive ways to use blogs (Broß, Noweski, and Meinel 2011) and mobile technologies (Pering et al. 2010) in collaboration. Other CSCW research examines the value of various representational tools (e.g., argumentative diagrams and textual outlines) in developing strong, collaborative arguments (Munneke et al. 2007).

Regardless of the strategies and tools, CSCW researchers must still address fundamental considerations such as time and place. Table 18.2 shows relationships that generate a need for various kinds of technology

Table 18.2. *Time-space groupware matrix identifying representative examples of computer-supported cooperative work*

	Same time (synchronous)	Different time (asynchronous)
Same place (colocated)	touch screens and multitouch screens, electronic meeting rooms, decision rooms, single-display groupware, public computer displays, wall displays	e-message boards, photo management sites, team rooms, large public computer displays, shift-work groupware, project management tools
Different place (remote)	conference calls, videoconferencing, desktop conferencing, instant messaging, chats, virtual worlds, shared screens, multiuser editors	e-mail, blogs, DVDs, asynchronous conferencing, meeting schedulers, group calendars, version control, forum discussions, wikis

Source: Modified from Johansen 1988.
Note: The time-space matrix can be revised by adding new technologies as they are developed.

to support collaboration: (1) same time/same place, (2) same time/different place, (3) different time/same place, and (4) different time/different place.

Some technologies in table 18.2 can be described as "Web 2.0," referring to changes in user attitudes and behaviors and a noticeable increase in user-generated content. Web 2.0, which emphasizes technologies of sharing and collaboration, has resulted in a range of web-based communities, community practices, and resources: blogs, social-networking sites, wikis, video-sharing sites, and folksonomies (which include social classification, social indexing, and collaborative or social tagging). However, simply using social-networking software, social bookmarking, or a real-time, multiuser editor does not make someone a collaborator. Being a collaborator requires active contribution to shared goals; how you use a technology determines whether the work is collaborative.

At the same time, technology does not always enhance collaboration. While using digital tools may make exchanging and sharing information easier and more efficient, these tools do not necessarily facilitate better communication among collaborators. When working on a complex project that involves multiple components and phases, face-to-face interaction is usually superior to interaction through technological interfaces. Using technology in place of face-to-face interactions on projects that require high levels of collaborator interactions to negotiate tasks and workload can, in fact, present a major obstacle to productivity (Galegher and Kraut 1994).

HEURISTIC: HOW YOU CAN BECOME A BETTER COLLABORATOR

The heuristic in this section (table 18.3) incorporates major points from the preceding literature review, framing questions to help you and your collaborators form strategies for making your collaboration as productive as possible. The discussion of interactions, people, goals, settings, tools, complexity, and perspectives provides terms for the vertical axis of the heuristic's matrix. The discussion of cognition and learning, small-group processes, and technology provides terms for the horizontal axis.

By combining key concepts, the heuristic suggests questions that can help technical communicators plan and implement collaborative work. As a technical communicator, you can use the heuristic to analyze your group dynamics systematically and avoid the many potential pitfalls of collaboration. First, consider your group's cognition and learning as it is reflected in interactions, people, goals, settings, tools, complexity, and perspectives. Then do the same thing for small-group processes and technology. When you complete the heuristic, you will have a thorough understanding of factors that affect your collaboration and can identify places where your group might encounter problems. Familiarity with the heuristic will help you internalize the questions, making them part of the routine assessment you perform when you join a group.

EXTENDED EXAMPLE: APPLYING THE HEURISTIC TO CASSANDRA'S SITUATION

This discussion applies the heuristic in table 18.3 to Cassandra's experience. Rather than answering every question, it selectively demonstrates the process of applying the matrix to a workplace situation.

COGNITION AND LEARNING

During the initial group meeting, Cassandra and her group considered their *interactions*. They outlined each person's roles and responsibilities in order to determine the kinds of interactions they needed to work together productively. The grant that funded the project specified many details related to *people*. While the job title for each person was predetermined by the grant, the group itself needed to articulate and negotiate the major tasks and their steps. Each group member brought different attitudes, behaviors, patterns of thought, and styles of learning that affected group interaction. For example, Patricia tended to be assertive in

Table 18.3. *Matrix posing heuristic questions*

	Cognition and learning	Small-group processes	Technology
Interactions			
Interactions include all forms of communication between and among group members.	What sorts of interactions will you and your collaborators need to be involved in to complete the task? How can these interactions improve thinking and learning?	What guidelines, assumptions, and boundaries govern interactions within your group's processes? How can interactions between you and your collaborators productively challenge these guidelines, assumptions, and boundaries?	What technologies are needed to facilitate interaction? How do technologies influence, shape, or direct the interactions in your group?
People			
You and your collaborators are the primary people involved, but you might also consider people outside the group who can affect the group's activities.	What skills and expertise do you and your collaborators bring to the project? What do you and your collaborators need to know about one another's attitudes, behaviors, thinking, and learning? How will you learn about these aspects of each person? How do different attitudes, behaviors, thinking, and learning affect the group?	What people are necessary for the success of each of the group's processes? What hierarchies are in place? How do you and your collaborators differ in the ways you engage with various processes?	What are your and your collaborators' technological competencies? What preferences or aversions does each group member have toward various technologies? What technologies seem best suited for the project? How do you and your collaborators use various technologies?
Goals			
The primary goal is the completion of the task that brings collaborators together, but other goals might factor into the steps in or results from the task's completion.	What learning is necessary to accomplish the project's goals? How is learning itself a goal of the collaboration?	What processes are necessary to accomplish the project's goals? How might varying—or even flouting—established processes move collaborators closer to achieving these goals?	What technologies are necessary to accomplish the project's goals? How might innovative uses of technology move collaborators closer to achieving these goals?

able 18.3. *(continued)*

	Cognition and learning	Small-group processes	Technology
ettings he settings are he places, both hysical and virtual, vhere collaborative nteractions occur.	What settings are available for the thinking and learning that the collaboration requires? How do different settings affect thinking and learning?	What settings are available for the small-group processes the collaboration requires? How do settings affect small-group processes?	What technologies are available to you and your collaborators in these settings? What alternative settings are available because of the technology? How do different settings affect your group's use of technology?
ools ools are items you se to complete asks during ollaboration—a pecific whiteboard r specific word-rocessing software. echnologies are roader categories f tools, such as vhiteboards or word-rocessing software in eneral.	What tools could facilitate thinking and learning? How can learning to use new tools, or to use familiar tools in new ways, help the group?	What tools could facilitate small-group processes? How can using new tools, or using familiar tools in new ways, improve small-group processes?	What technologies are the most useful tools for you and your collaborators? How can your group use technologies as more than tools?
omplexity task's complexity onsists of the multiple nd sometimes onflicting elements nvolved in its causes, nalysis, resolution, and ffects.	What can you and other collaborators learn to gain a better understanding of the task's complexity? How could learning reduce the task's complexity?	What small-group processes could help you and your collaborators manage the task's complexity? How can the group assess the task's complexity and the steps involved in completing it? How could small-group processes reduce the task's complexity?	What technologies could help you and your collaborators manage the task's complexity? How could technologies reduce the task's complexity?

Table 18.3. *(continued)*

	Cognition and learning	Small-group processes	Technology
Perspectives			
A perspective is a way of seeing, approaching, or articulating an aspect of a task. Each collaborator can offer one or more perspectives on any aspect of a task.	What do you and your collaborators need to learn about one another's perspectives? How can reflecting on these perspectives facilitate the task?	What small-group processes can take advantage of your group's multiple perspectives? What differences in perspective should you acknowledge? How can you use these differences? How can your group use small-group processes to generate new perspectives?	What technologies can help your group take advantage of multiple perspectives? How can your group use technologies to change existing perspectives and generate new ones?

Note: The categories of cognition and learning, small-group processes, and technology along the horizontal axis intersect with the categories of interactions, people, goals, setting, tools, complexity, and perspectives along the vertical axis to provide cells with examples of critical questions for you and your collaborators to ask and answer.

some group meetings when other members were not. The group adapted to her style, but each member responded differently. The overall *goal* of the project was also predetermined in the grant: provide peer support to Korean women recently diagnosed with breast cancer. Ultimately, learning itself was a goal of this collaboration, not only because many group members learned new skills, but because group members also learned strategies for resolving differences and working together to produce a better final product.

The typical *setting* for the group's meetings, the organization's conference room, kept them focused on their project by removing distractions. It also provided a variety of *tools*, including a conference table, whiteboards, a desktop computer, and a projector. Group members often wrote on the board or used the desktop computer to project documents, enhancing their discussions.

Both Cassandra and Diana realized that they needed to learn new skills in order to fulfill their roles on the project. Diana needed to learn how to use the project's database, and Cassandra needed to learn more about layout and design. Once they learned to use these tools, each could better manage the *complexity* of the tasks assigned to her. Patricia, however, disagreed with Cassandra about the design of the brochures. Patricia and Cassandra had to discuss their differing *perspectives*, so that the group could benefit from their disagreement and they could reach a decision about the final design.

SMALL-GROUP PROCESSES

In Cassandra's situation, many guidelines for *interactions* were established in the initial group meeting. Samantha took the lead on the project, and the remaining group members worked in a flat hierarchical structure beneath her. Because *people* worked as peers, they often experienced tension and disagreement during small-group interactions when Samantha was not present. When Samantha returned for a meeting, each person would update the group, and Samantha would lead them to consensus so they could focus on their overall *goals*. In order to resolve disagreements about the brochures, however, Cassandra and Patricia flouted the convention of relying on consensus, instead working with Samantha separately to incorporate multiple viewpoints in the brochures.

When creating the brochures, Cassandra and Patricia often collaborated by phone. This *setting* proved more expedient than relying solely on e-mail. However, e-mail was still a useful *tool* because they e-mailed brochure drafts to each other to refer to during their phone conversation. The process of creating brochures with the software was more *complex* than many of the other group members could understand, so Cassandra relied on Patricia to represent their progress to the group in a way that omitted technical details. This process allowed Patricia to bring her *perspective* to the creation of the brochures as well as to the rest of the group during meetings. Combining perspectives not only ameliorated conflicts, but it also led to new ideas that improved the services the group eventually provided to the project's patients.

TECHNOLOGY

You may have noticed that in addressing questions in the matrix's columns for cognition and learning and small-group processes, technology came up several times, providing information that anticipates some of the questions in the technology column. This repetition reflects the synergy of the matrix's heuristic elements; since many elements related to technology have already been addressed, this application of the heuristic is relatively brief.

The *interactions* of Cassandra and her group relied on digital technologies during collaboration via different software programs—word processing, publishing design, database management—as well as via electronic communication *tools*—e-mail in particular. Each group member also had varying levels of proficiency with these tools. Most *people* in the group knew word processing, whereas only Cassandra knew publishing design software, and only John knew how to program the database management system.

These technologies facilitated sharing *perspectives* while decreasing face-to-face interaction. Cassandra and Patricia, for example, worked on several drafts of an in-take form via e-mail before they discussed the form in person.

As the study progressed, the group began to experiment with other digital technologies using social-networking sites as *settings*. Patricia suggested that they set up a social-networking page that patients and the project volunteers could join. Cassandra proposed that the page might be used to recruit new patients and provide links to online resources about breast cancer. The group could also help reach their *goals* for event attendance by using the page to advertise. At first Samantha was concerned about the *complexity* of maintaining patient privacy and confidentiality, but Cassandra explained that added security could ensure privacy, and Patricia suggested that she could communicate directly with patients and volunteers who opted out of social networking by sending e-mails or posting announcements on a website.

CONCLUSION: THE POTENTIALS AND PERILS OF COLLABORATION

Cassandra's workplace experiences provide details about many challenges she and her group faced during their collaboration. Addressing the heuristic's questions helped them think through many of the situations they encountered, but as often happens in the workplace, not all challenges were completely resolved, and group members did not always get what they wanted in every situation. For example, Cassandra preferred a different design for the group's brochure, but she saw the value of Patricia's perspective and accepted the majority decision for the welfare of the group. Collaborators should be receptive to compromise, putting the collective goal ahead of individual interests. Ultimately, the brochure was successful in recruiting a sufficient number of patients even though it did not use the design that Cassandra preferred.

Group members bring different approaches to collaboration. What works well for one group might not work as well for another. As a professional technical communicator, most of your work will involve collaboration at some point. You must be prepared to change your collaborative strategies to adjust to different workplace contexts. For each project, accounting for the specific rhetorical situation, planning your group's processes, and considering the best possible uses for modes and media will help you increase chances for success.

Despite collaboration's potential for greater successes, the variables involved in collaboration—diverse interactions, people, goals, settings, tools, complexity, and perspectives—involve as many perils as promises. You will share responsibilities with other members of your group, but for that sharing to be successful, the group must reach a common understanding of what the responsibilities are, and each person must carry out responsibilities in keeping with the group's expectations. Technology might seem like a shortcut for building understanding, coordinating responsibilities, and sharing tasks, but each new tool introduces new risks that could have heavy costs in terms of the group's time investments and buy-in.

Reflecting on the example and using the heuristic in this chapter should help you minimize problems and maximize rewards, but expect each collaboration to involve unique challenges, disagreements, and setbacks. Ultimately, your ability to turn moments of difficulty into sites of productivity could make the difference between success and failure.

DISCUSSION QUESTIONS

1. Think about a recent collaborative project. What questions in this chapter's heuristic did your group address explicitly? What questions might you have benefited from addressing?
2. Many chat rooms, online games, and social-networking sites allow people to misrepresent themselves or their situations. Although such digital environments were not necessarily designed for the workplace, technical communicators sometimes adapt them for workplace collaboration. What ethical obligations should govern self-representation when technical communicators use such digital environments for collaborative work?
3. Think about a collaborative project you are planning or currently undertaking. What *nondigital* technologies might be useful for group meetings, and how might you use them? What nondigital technologies might be useful for interactions that aren't face to face?
4. Even in relatively homogeneous groups (consisting, for example, of people who have similar cultural backgrounds), people's differences can affect the ways they work. How can diversity—in terms of culture, ethnicity, race, nationality, regional background, sex, gender, sexual orientation, age, class, ability, and so forth—influence collaboration? Which of these categories are most important for technical commu-

nicators to consider as they plan collaborations? How might different circumstances—different sorts of projects—change the priorities that technical communicators should give to different categories?

5. Think about a major local, regional, national, or international problem with which you are currently engaged or that you would like to address in the future. How will collaboration help you identify and implement solutions?
6. Imagine the type of workplace or organization where you see yourself practicing technical communication in the future, and envision situations to which you might apply this chapter's heuristic. How would you modify questions in the heuristic to create a better, more specific match with your working conditions? Rewrite and redesign the heuristic so that it matches the language and priorities of your future workplace.
7. In this chapter's opening scenario, Cassandra needed to learn publishing design software to complete her portion of the group's task. Other than software, what might collaborators need to learn for a group project, and how would they learn it? Think of three tools, practices, or behaviors that collaborators might need to learn. For the one you expect to be needed most often, describe the process involved in learning. First, list resources that users will need. Second, help users specify learning goals that they can reach with those resources. Third, characterize activities that users should undertake to advance learning. Finally, provide methods or means for self-assessment. Decide how best to design this information for technical communicators.
8. Second Life (http://secondlife.com) is a virtual environment that involves user-designed avatars, verbal and nonverbal communication, and a whimsical take on the laws of physics. How might such an environment affect a group collaborating on a 3-D design project? Meet with a group of classmates in Second Life or a similar environment to discuss such a project. Take notes on your interactions, paying particular attention to the environment's impact on your attitudes, perspectives, and goals.
9. Corporate authors often face the challenge of "speaking with one voice" when they collaborate on documents and presentations. However, a corporate group's multiple voices and perspectives are often valuable. Can organizations achieve a unified voice without sacrificing the value of having diverse perspectives? If so, how? Analyze a variety of pages and documents available on a corporate website. Annotate rhetorical moves that establish a voice and identity for the corporation as well as any moments of heteroglossia. Evaluate the effectiveness

of the corporate voices, making recommendations for improvement. Write your recommendations in a memo as if you were a company employee contacting your supervisor about a proposed revision of the website.

10. Imagine that you have a job at a corporation where most employees work in cubicles while managers work in centrally located offices. Because most of the corporation's work is accomplished through short- and long-term group projects, meetings frequently take place either in conference rooms or at small tables in managers' offices. In addition, less formal interactions occur in a break room where coffee and bottled water are readily available. A task force has recently circulated a proposal that would allow the majority of employees a choice to telecommute up to 60% of the time—that is, to work on computers in remote locations up to three days a week. Employees who favor the proposal believe that the time saved from commuting will increase productivity, generate goodwill, and decrease the carbon footprint from driving to work. Employees who oppose the proposal believe that the decrease in face-to-face contact will decrease the productivity of collaborative work. Your supervisor has asked you to assess the claims of the proposal and its opposition. What aspects of collaboration might be affected by the change? Write a short report that systematically explains the impact of telecommuting on cognition and learning, small-group processes, and technology. Refer explicitly either to this chapter's heuristic or to the heuristic you created in response to question 6. Frame your assessment as an argument that addresses each perspective and then recommends that the corporation either adopt or reject the proposal.

WORKS CITED

Bakhtin, Mikhail M. 1981. *The Dialogic Imagination: Four Essays*. Edited by Michael Holquist. Translated by Caryl Emerson and Michael Holquist. Slavic Series No. 1. Austin: University of Texas Press.

Barthelmess, Paulo, Edward Kaiser, Rebecca Lunsford, David McGee, Philip Cohen, and Sharon Oviatt. 2006. "Human-Centered Collaborative Interaction." In *HCM '06 Proceedings of the 1st ACM International Workshop on Human-Centered Multimedia*. http://citeseerx.ist.psu.edu. Accessed July 27, 2011.

Bell, Chip R., and John R. Patterson. 2006. "Leaders as Bridge Builders." *Leadership Excellence* 23:14–15.

Bemer, Amanda Metz, Ryan M. Moeller, and Cheryl E. Ball. 2009. "Designing Collaborative Learning Spaces Where Material Culture Meets Mobile Writing Processes." *Programmatic Perspectives* 1:139–166.

Broß, Justus, Christine Noweski, and Christoph Meinel. 2011. "Reviving the Innovative Process of Design Thinking." In *ICIW 2011: The Sixth International Conference on*

Internet and Web Applications and Services, 142–149. St. Maarten, Netherlands Antilles: International Academy, Research, and Industry Association (IARIA).

Burnett, Rebecca E. 1991. "Cooperative, Substantive Conflict in Collaboration: A Way to Improve the Planning of Workplace Documents." *Technical Communication* 38:532–539.

———. 1993. "Decision-Making during the Collaborative Planning of Coauthors." In *Hearing Ourselves Think: Cognitive Research in the College Writing Classroom*, edited by Ann Penrose and Barbara Sitko, 125–146. New York: Oxford University Press.

———. 1994a. "Interactions of Engaged Supporters." In *Making Thinking Visible: Writing, Collaborative Planning, and Classroom Inquiry*, edited by Linda Flower, David L. Wallace, Linda Norris, and Rebecca E. Burnett, 67–82. Urbana, IL: National Council of Teachers of English.

———. 1994b. "Productive and Unproductive Conflict in Collaboration." In *Making Thinking Visible: Writing, Collaborative Planning, and Classroom Inquiry*, edited by Linda Flower, David L. Wallace, Linda Norris, and Rebecca E. Burnett, 239–244. Urbana, IL: National Council of Teachers of English.

Conklin, Jeff. 2005. "Wicked Problems and Social Complexity." In *Dialogue Mapping: Building Shared Understanding of Wicked Problems*. New York: Wiley. http://cognexus .org/wpf/wickedproblems.pdf . Accessed November 28, 2011.

Cross, Geoffrey A. 1994. *Collaboration and Conflict: A Contextual Exploration of Group Writing and Positive Emphasis*. Cresskill, NJ: Hampton Press.

———. 2001. *Forming the Collective Mind: A Contextual Exploration of Large-Scale Collaborative Writing in Industry*. Cresskill, NJ: Hampton Press.

Denmark, Florence L. 1993. "Women, Leadership, and Empowerment." *Psychology of Women Quarterly* 17:343–356.

Flower, Linda. 1987. *The Role of Task Representation in Reading-to-Write*. National Center for the Study of Writing and Literacy Technical Report. http://www.nwp.org/cs/public/ print/resource/592. Accessed July 27, 2011.

Galegher, Jolene, and Robert E. Kraut. 1994. "Computer-Mediated Communication for Intellectual Teamwork: An Experiment in Group Writing." *Information Systems Research* 5:110–138.

Haklay, Mordechai. 2006. "Usability Dimensions in Collaborative GIS." In *Collaborative Geographic Information Systems*, edited by Shivanand Balram and Suzana Dragićević, 24–42. Hershey, PA: Idea Group Publishing.

Hardy, Barry, and Douglas Connect. 2008. "Collaboration, Culture and Technology." *Knowledge Management Review* 10:18–23.

Hartwig, Ryan T. 2008. "Facilitating Collaboration: Toward a Facilitative Leadership Style of Top Management Teams." Paper presented at the annual meeting of the National Communication Association, San Diego, CA, November 21–24. http://www.allacademic .com/meta/p257974_index.html. Accessed November 21, 2009.

Healey, Patrick G., Graham White, Arash Eshghi, Ahmad J. Reeves, and Ann Light. 2008. "Communication Spaces." *Computer Supported Cooperative Work* 17:169–193.

Herrmann, Thomas. 2010. "Support of Collaborative Creativity for Co-located Meetings." In *From CSCW to Web 2.0: European Development in Collaborative Design*, edited by D. Randall and P. Salembier, 65–95. London: Springer-Verlag. http://www.springerlink .com/content/u3707727jl374006/. Accessed November 28, 2011.

Hewett, Beth L., and Charlotte Robidoux. 2010. *Virtual Collaborative Writing in the Workplace: Computer-Mediated Communication Technologies and Processes*. Hershey, PA: Global.

Janis, Irving L. 1972. *Victims of Groupthink*. Boston: Houghton Mifflin.
Johansen, Robert. 1988. *Groupware: Computer Support for Business Teams*. New York: Free Press.
Johnson, David W., and Roger T. Johnson. 1998. *Learning Together and Alone: Cooperative, Competitive, and Individualistic Learning*, 5th edition. Boston: Allyn and Bacon.
Kleid, Naomi A. 2004. "Teaching and Practicing Teamwork in Industry and Academia." In *Society for Technical Communication Proceedings*. http://www.stc.org/ConfProceed/2004/PDFs/0038.pdf. Accessed on November 21, 2009.
Lowry, Paul Benjamin, Tom L. Roberts, Nicholas C. Romano Jr., Paul D. Cheney, and Ross T. Hightower. 2006. "The Impact of Group Size and Social Presence on Small-Group Communication: Does Computer-Mediated Communication Make a Difference?" *Small Group Research* 37:631–661.
Lunsford, Andrea, and Lisa Ede. 1990. *Singular Texts/Plural Authors: Perspectives on Collaborative Writing*. Carbondale: Southern Illinois University Press.
Matusak, Larraine R. 2001. "Leadership: Gender Related, Not Gender Specific." In *Concepts, Challenges, and Realities of Leadership: An International Perspective*, edited by James MacGregor Burns, Georgia Sorenson, and Larraine R. Matusak. Academy of Leadership. http://www.academy.umd.edu/Publications/global_leadership/salzburg/chapter3.htm. Accessed June 30, 2006.
Munneke, Lisette, Jerry Andriessen, Gellof Kanselaar, and Paul Kirschner. 2007. "Supporting Interactive Argumentation: Influence of Representational Tools on Discussing a Wicked Problem." *Computers in Human Behavior* 23:1072–1088. http://edu.fss.uu.nl/medewerkers/gk/files/Munneke-etal-CiHB2007.pdf. Accessed November 28, 2011.
Nunamaker, Jay F. Jr., Alan R. Dennis, Joseph S. Valacich, Douglas D. Vogel, and Joey F. George. 1991. "Electronic Meeting Systems to Support Group Work." *Communications of the Association for Computer Machinery* 34:40–61.
Pering, Trevor, Kent Lyons, Roy Want, Mary Murphy-Hoye, Mark Baloga, Paul Noll, Joe Branc, and Nicolas De Benoist. 2010. "What Do You Bring to the Table? Investigations of a Collaborative Workspace." In *Proceedings of the 12th ACM international conference on ubiquitous computing*, 183–192. http://techna.org/~kent/professional/pubs/table-ubicomp10.pdf. Accessed November 28, 2011.
Reece, Gloria A. 2002. "Accessibility Meets Usability: A Plea for a Paramount and Concurrent User-Centered Design Approach to Electronic and Information Technology Accessibility for All." In *STC Proceedings 2002*. http://tc.eserver.org/19263.html. Accessed July 27, 2011.
Rosen, Evan. 2007. "Hierarchy Busters Enable Collaboration." The Culture of Collaboration. http://collaborationblog.typepad.com/collaboration/2007/08/hierarchy-buste.html. Accessed July 27, 2011.
Soller, Amy, Alan Lesgold, Frank Linton, and Brad Goodman. 1999. "What Makes Peer Interaction Effective? Modeling Effective Communication in an Intelligent CSCL." *Proceedings of the 1999 AAAI Fall Symposium: Psychological Models of Communication in Collaborative Systems*, North Falmouth, MA, 116–123. http://www.aaai.org/Papers/Symposia/Fall/1999/FS-99–03/FS99-03-017.pdf. Accessed July 27, 2011.
Stahl, Gerry. 2005. "Group Cognition in Computer-Assisted Collaborative Learning." *Journal of Computer Assisted Learning* 21:79–90.
Stahl, Gerry, Timothy Koschmann, and Dan Suthers. 2006. "Computer-Supported Collaborative Learning: An Historical Perspective." In *Cambridge Handbook of the*

Learning Sciences, edited by R. K. Sawyer, 409–426. Cambridge: Cambridge University Press. http://www.cis.drexel.edu/faculty/gerry/cscl/CSCL_English.pdf. Accessed November 21, 2009.

Suleiman, James, and Richard T. Watson. 2008. "Social Loafing in Technology-Supported Teams." *Computer Supported Cooperative Work* 17:291–309.

Van den Bossche, Piet, Wim H. Gijselaers, Mien Segers, and Paul A. Kirschner. 2006. "Social and Cognitive Factors Driving Teamwork in Collaborative Learning Environments: Team Learning Beliefs and Behaviors." *Small Group Research* 37: 490–521.

Vygotsky, Lev S. 1978. *Mind in Society: The Development of Higher Psychological Processes*. Cambridge, MA: Harvard University Press.

Winsor, Dorothy A. 2000. "Ordering Work: Blue-Collar Literacy and the Political Nature of Genre." *Written Communication* 17:155–184.

Yoder, Janice D. 2001. "Making Leadership Work More Effectively for Women." *Journal of Social Issues* 57:815–828.

KIRK ST. AMANT

19 What Do Technical Communicators Need to Know about International Environments?

SUMMARY

Today, many technical communicators work as part of an international business process. In some cases, they design web interfaces or develop online content for global users. In other cases, they draft documents that will be translated for release into international markets. And in yet other instances, technical communicators work with overseas colleagues to produce informational products. As a result, today's technical communicator needs to understand how to work in a range of international environments.

To participate effectively in such contexts, technical communicators must develop new skills and knowledge bases. Working successfully in global environments, however, often involves one central factor—*credibility*. That is, effective international communication involves developing materials members of other cultures will consider credible, or worth using. Creating such credibility is often a matter of rhetoric, or knowing how to present information in a way that different cultural audiences will consider credible. This chapter examines how cultural rhetorical expectations can affect the perceived credibility (and usability) of informational or instructional materials. The chapter also provides technical communicators with a heuristic they can use to identify and address such credibility expectations when working in international or intercultural contexts.

INTRODUCTION: WHAT WENT WRONG WITH THIS INTERNATIONAL PROJECT?

Pat is a senior technical communicator at Alliance X—a company that develops accounting software for small businesses. Recently, the company won a lucrative contract to provide software to the Ministry of Education in the nation of Riendutout, where English is the official language. Pat has been asked to lead the initiative to develop a help website for providing technical support to clients in both the United States and Riendutout. It is a major project, and if successful, it could lead to contracts with other government agencies in Riendutout.

For the project, Pat's initial thought is to develop a help site that consists of two parts. One part would be static, or simply display information. This static part would consist of a FAQ (frequently asked questions) page providing solutions to problems that commonly arise when using the software for the first time (e.g., setting or resetting the software's preferences to work on a PC or a Mac platform). The other portion of the site would be interactive, or dynamic. For this section, Pat wants to apply a current trend in social media and create an online bulletin board that allows product owners to post questions they might have or problems they might encounter. Other clients could then post solutions based on their own experiences with the software. Once or twice a week, company technicians would check the bulletin board and provide their own answers to items found on the site.

Through such an approach, the dynamic part of the site builds a community of product users who would remain interested in the site, the company, and the company's products. The use of this dynamic site would also reduce the time and energy technical support staff spends answering "simple and easy" questions not covered in the FAQ part of the site. These technicians could then focus their time and attention on more complex issues encountered by users.

Pat presents this plan to the management of Alliance X and receives approval to begin work on the two-part site. In just over a month, the site is completed, and it goes live a week later. Both Pat and the management team at Alliance X are confident this project will lead to future successful ventures with agencies in Riendutout.

However, a week after the site goes live in Riendutout, Pat receives a troubling phone call from his supervisor at Alliance X. It turns out users are not responding to the dynamic help site as Pat and Alliance X had expected. Rather, in the last seven days, the Ministry of Education in Riendutout has been bombarded with complaints and notifications from concerned citizens. Many of these individuals have contacted the ministry to notify officials that the help site has been hacked and imposters are dispensing suspect information and suggestions. Other citizens have contacted Riendutout authorities to make these agencies aware of a scam help website trying to present itself as an extension of the Ministry of Education. Upon closer inspection, representatives of Riendutout's Ministry of Education have determined that the dynamic communication part of the help site—where average consumers provide advice, rather than technical experts with Alliance X—constitutes an act of bad faith on the part of Alliance X. For this reason, the ministry has taken the site offline and is considering filing a legal claim against Alliance X. In short, the help site is a disaster from which Alliance X might not recover. Alliance X is now giving Pat twenty-four hours to produce a full report on the site, and the report is to include strategies on how to mitigate the situation as quickly as possible.

After the phone call, Pat sat in stunned silence for several minutes and tried to determine what had happened. The site was not based on any new concepts, approaches, or technologies. In fact, it was an approach that had been used successfully with a number of other U.S.-based projects across a range of industries. And management had reviewed and approved the idea before Pat began the project. What had gone wrong?

In this case, Pat failed to consider one all-important factor—*culture*—and how that factor might affect the expectations, perceptions, and uses of an informational product. This oversight cost Alliance X dearly. Yet the situation could have been avoided if Pat had made culture one of the central considerations for the project.

This chapter examines how an understanding of rhetorical factors can help you create credible materials for international audiences. It begins by overviewing how cultural expectations of rhetoric and credibility can affect perceptions of and reactions to messages and products. The literature review, for example, summarizes how varying communication expectations can arise in different cultures. That section also explains how such differences influence preferred cultural approaches to presenting information (rhetoric) and expectations of what constitutes a credible way to share ideas in a given culture. Next, the chapter discusses how these ideas of rhetoric and credibility can become the foundation for a heuristic framework that technical communicators could use to examine issues related to creating credible messages in international contexts. The chapter concludes with an extended example that illustrates how a rhetoric-based heuristic could be applied to the case of Alliance X to avoid the problems encountered with the online help site. This application of the heuristic reveals how it can provide technical communicators with the foundational knowledge needed to participate more effectively in today's global work environment.

LITERATURE REVIEW: WHAT IS CULTURE AND HOW DOES IT AFFECT COMMUNICATION?

Effective international communication involves understanding how culture influences behaviors and expectations. Within this context, perhaps the best way to think of culture is as an organizational system, or a *world-*

view, that you use to identify and rank the importance of different items in your environment (Neuliep 2000, 14–16; Varner and Beamer 1995, 2–7). Actions that support what is important to your culture's worldview are considered "acceptable" by the members of your culture. Actions that do not respect this worldview constitute "unacceptable" behavior to the members of your culture (Neuliep 2000, 18–21; Berry et al. 2002, 29–33).

In the worldview of some cultures, for example, status—and status differences—are very important, and communication behavior must reflect this importance (e.g., the use of formal titles when addressing someone; Hofstede and Hofstede 2005, 39–40). In the worldview of a different culture, however, status might not be considered as important, and different communication behaviors (e.g., calling one's superior by his or her given name) might be considered acceptable. Thus, a culture's worldview affects the *rhetoric*—or the communication practices and styles—its members use when interacting.

What is essential to these rhetorical differences is that no culture is correct or incorrect when compared to another. Rather, each culture has its own internal rules for determining what is an appropriate and expected—or credible—way to do things according to the shared worldview of the members of that culture (Berry et al. 2002, 29–33; Neuliep 2000, 18–21; Varner and Beamer 1995, 2–7). Behavior that contradicts such cultural expectations of "appropriate" is generally considered not acceptable and not credible by most members of that culture.

From an international business perspective, such cultural differences—and a failure to recognize them—can be problematic. In some cases, these cultural disconnects can result in consumers in a prospective market viewing a particular product negatively or with skepticism. The practice of critiquing a competitor's products (e.g., Coke critiquing Pepsi), for example, is common in U.S. advertising, but is generally considered inappropriate in India (Kamath 2000, 10–11). In other cases, such differences can result in confusion and miscommunication that lead to lost time and money when coworkers from different cultures collaborate on projects (Ang and Inkpen 2008, 339–340). Effective participation in the global marketplace thus requires organizations to understand and address such differences before they can cause problems.

To work effectively in international environments, technical communicators must be aware of such cultural rhetorical differences. In essence, because an informational or instructional item is a product of the culture that created it, that item contains embedded rhetorical factors specific to that culture (Esselink, 2000, 27–28; Yunker 2003, 16–20). For example, as English-speaking cultures read from left to right, the members of these cultures often create websites with menu bars located on the left-hand side of a web page. Israeli websites in Hebrew, however, have menu bars on the right side of a web page, for readers of Hebrew read from right to left. In these cases, differing cultural expectations of reading direction prompt the members of these cultures to design their websites in two different ways.

For technical communicators, successful intercultural communication involves identifying and addressing culture-specific rhetorical factors that could cause problems in international contexts. Once identified, these rhetorical aspects need to be removed or revised to make a product more credible and acceptable to a wider, international audience (Barnum and Li 2006, 150–158; McCool 2006, 178–181). In the case of website menu bars, this change could involve positioning such features to account for the reading expectations of the largest possible group of international users (e.g., using a centered menu bar located at the top of the pages on an international website).

The process of revising materials to meet the expectations of a wide range of cultural groups is known as *internationalization* (Esselink 2000, 2–3; Yunker 2003, 19–20), and it can take a preemptive or a revisionist approach. In a preemptive approach, you identify the rhetorical factors that might cause credibility problems when materials designed for your culture are used with a greater international audience. You then create materials that avoid these rhetorical problem areas. This preemptive approach can be seen in writing according to standard global English—or a controlled English—that avoids cultural rhetorical factors that could cause confusion (e.g., passive voice; Kohl 2008, 41–44). The objective is to produce documents that can be read—and be seen as credible—by a wide range of international English-speaking audiences. Such preemptive approaches are often more time- and cost-intensive initially. However, the resulting materials can quickly and effectively go from a final draft to a global audience, and thus reduce the time and money an organization might need to spend later on addressing problems related to rhetoric and culture.

In the revisionist approach to internationalization, you create materials according to the rhetorical expectations of your native culture. An expert in internationalization then revises these materials to remove problematic items and make the product more rhetorically acceptable to a greater international audience. The editing-for-translation process is one such revisionist approach to internationalization. It involves creating documents according to the rhetorical expectations of your culture and having a trained editor revise or remove culture-specific rhetorical elements to create a more culturally neutral text (Rude 1998, 247–249). Such texts are often easier for translators to convert from one language to another and can save on translation costs for a product. In this process, creating original documents takes less time and costs less initially, but the later editing increases the time and cost to the overall process.

In some situations, organizations might want to revise materials to meet the rhetorical expectations of a specific culture rather than create products for a general international audience. This process of revising materials to meet the expectations of a specific culture is known as *localization*, and it often requires a certain degree of flexibility in product design (Esselink 2000, 3–5; Yunker 2003, 16–19). If, for example, you design a website only for members of your native culture, it might be difficult to revise the placement of features such as menu bars without making the overall site look odd to members of the culture for which that site is being localized. However, if you know what cultural aspects of site design can be rhetorically problematic and avoid using them, you make it easier to reconfigure/localize that site for other cultures. (Placing a menu bar horizontally across the top of a page avoids having to redesign that page to shift a left-side menu bar to the right side of the page.) In other cases, you can create materials in a way that would make them easy to localize (e.g., inserting images in a way that allows them to be removed without causing a major shift in the flow of text on a web page).

In these instances, what is important is that you know what rhetorical aspects can cause cross-cultural communication problems. Based on this knowledge, you can either avoid using such elements or use them in a way that allows for easy revision/localization or internationalization. Engaging effectively in international technical communication practices, therefore, means you must understand a number of central rhetorical factors that can guide your professional activities in global contexts.

From a rhetorical perspective, credibility begins with one all-important factor: *audience*. The group of individuals to which you present information has expectations of how to convey ideas appropriately, or credibly. These expectations are not random, nor are they uniform across all situations. Rather, audience expectations of credibility are often closely linked to the context, setting, or genre in which information is presented (St. Amant 2006, 56–57). In rhetorical terms, this context/genre is called the *forum*, or the place to which individuals come to find specific information related to achieving a particular objective.

The idea works as follows: I am interested in getting information on whether a particular film is worth seeing. My culture has identified a specific forum—the movie review—as a mechanism I can use to find information related to achieving this objective. So when I read a movie review, I'm looking for the author to mention a number of specific informational areas (e.g., plot, run time, cast, good acting, good editing, etc.) that I need to know about in order to achieve my objective for reading this review. If the author of that review meets my forum expectations and my purpose for reading the review, I'll likely consider the review to be credible and worth my consideration. If, however, the author uses his or her movie review to explain why I should purchase a particular car and fails to mention anything about movies, I'll likely discount the credibility of what the author says. In fact, I might be annoyed with that author for wasting my time by having me read something completely unrelated to my objective for consulting this review. So by not meeting an audience's expectation of the purpose for using a given forum, an author can fail to establish the credibility needed for that audience to consider his or her ideas worthwhile.

Within the context of the film review, I was looking for mention of certain topics, or kinds of information essential to achieving my purpose for reading a movie review. Because these topics help me achieve my forum-related purpose, I tend to view them as markers that indicate the credibility of the related review. So, if these topics are present, I'll likely consider a review to be credible. If they are missing, I'll likely consider the review to be noncredible.

From a rhetorical perspective, these forum-specific items are known as *special topics*, and they are some of the most important factors associated with creating credibility in a given forum (Miller and Selzer 1985, 324–325). Campbell (1998, 39), moreover, notes that special topics are important factors affecting credibility in cross-cultural exchanges. So when creating credible messages—be they for your own culture or another culture—you need to know two key things about your audience:

- what purpose/objective your audience associates with a specific forum; and
- what special topics—or kinds/categories of information—your audience considers essential to achieving the purpose/objective associated with that forum.

Because these rhetorical aspects can vary from culture to culture, they are often the central factors contributing to cross-cultural miscommunication (Kaplan 2001, ix–xiv; Panetta 2001, 4–6). They are also the factors to be aware of when working in international environments.

One classic example used to illustrate these cultural and rhetorical ideas is the genre of a business letter. What is the purpose of this particular forum? If you are an Anglo-American, you generally view this forum as a mechanism for presenting facts related to a particular business situation (Campbell 1998, 35–36; St. Amant 2006, 56–57). For this reason, the topics one must address to achieve this objective (and create a credible presentation) would involve facts, dates, and other information related to a particular business situation. If the writer fails to mention such topics, or addresses topics unconnected to the reader's purpose for using this forum (e.g., questions about the reader's region or family), that audience/reader will likely view the letter as inappropriate and not credible. So how is this example related to working in international environments?

Research indicates other cultures often associate a different purpose with the forum of a business letter (Campbell 1998, 39; Tebeaux 1999, 53–56). In fact, for a number of cultures, the purpose of a business letter is not to convey the facts of a given business situation. Rather, the letter is a mechanism individuals use to initiate or to maintain a long-term relationship essential to the future success of both the letter's sender and its receiver (Campbell 1998, 39; Tebeaux 1999, 62). This difference in purpose means readers from such cultures will expect an author to mention different kinds of topics—those associated with establishing or maintaining such relationships—in order to create a credible presentation in that forum.

As Campbell (1998, 36–41) explains, for many Japanese businesspeople, business letters are about creating or maintaining relationships. For this reason, it is not uncommon for an initial letter to a prospective Japanese partner to mention some fact about the recipient's nation or culture. In one study, for example, Campbell reports that a Chinese author writing to a prospective Japanese client mentions certain aspects of Japanese geography in an initial business letter. Why was this information important? In mentioning such geographic details, the letter's author reveals he or she took the time to do initial research on the reader's culture in order to mention it in the letter. Such an action indicates the author's desire to engage in a long-term relationship with the letter's reader, for one would not expend the effort to learn such information if that person were not interested in having more than an initial exchange with the letter's recipient. Thus, a different kind of special topic—information revealing the sender's knowledge of the recipient—is connected to Japanese expectations of creating credibility in this forum.

For the technical communicator, working effectively in international environments involves knowing of such differences in advance and addressing them. A familiarity with key rhetorical concepts can thus allow technical communicators to understand interactions between cultural groups in order to foster more successful and productive exchanges. To achieve this goal, individuals can use a rhetorical framework—or a heuristic—to ask important questions that can guide the international communication process. Also, in the case of technical communication, different genres tend to represent the forums individuals use to covey ideas and information to a particular audience. For this reason, the remaining sections of this chapter will examine the concept of the forum through this "forum/genre" relationship.

HEURISTIC:
HOW CAN YOU USE A KNOWLEDGE OF RHETORIC TO COMMUNICATE EFFECTIVELY IN INTERNATIONAL ENVIRONMENTS?

The key to working effectively in international environments is doing research on the rhetorical expectations of other cultures. Yet such research must be directed by a particular mechanism or approach to be successful. Rhetorical concepts connected to credibility can provide the direction needed to conduct effective research related to cross-cultural communication. The idea is to use rhetorical concepts in a focused way to identify the credibility expectations that cultures associate with different communication contexts. These concepts are quite foundational and involve two interrelated parts. The first part is knowing the forum, or setting/context, in which information is presented and what purpose a particular cultural audience associates with that forum. The second part involves identifying the specific subjects or topics the members of a given culture expect to encounter in a particular forum.

These rhetorical concepts can serve as a valuable guide for conducting effective research on culture, communication, and credibility, for they let you know what items to look for when examining culture and communication. For example, by using the concepts of the forum and the special topics as they relate to credibility, you can effectively analyze materials produced by individuals from other cultures. This analysis can provide you with a model of what the members of that culture associate with credible presentations in a given forum/genre.

To engage in such research, you could locate and review multiple examples of the same kind of document produced by natives of the *target culture* (i.e., the cultural audience for which you will design materials). Such an analysis would focus on identifying what appear to be the culture-specific credibility expectations associated with that forum. You could also interview the members of a particular culture about credibility expectations for a given forum—including the purpose of the forum and the kinds of special topics expected in that forum. You can then use the results of this research to develop guidelines for how to meet the credibility expectations of the related cultural audience.

For this research process to be successful, you need to ask certain questions connected to rhetoric and credibility. These questions relate to four rhetorical tasks involving the forum and the special topics.

TASK 1: IDENTIFY THE FORUM FOR CONVEYING INFORMATION

While identifying the forum seems an obvious task, it is often overlooked in international communication situations. The problem is the incorrect assumption that all cultures use the same genre-based forums to present information. In fact, what might be considered relatively common forums/genres in one culture might not exist in another. Kristin R. Woolever (2001, 56), for example, notes, "In many high-context cultures [e.g., Japan, China, and Saudi Arabia] people rely on more informal relationships [i.e., personal conversations] and discussions to establish the foundations and standards for proposals." As a result, individuals from these cultures might "view as unnecessary and perhaps as offensive the use of written Requests For Proposals (RFPs)." Thus, some high-context cultures might not use a genre-based forum that is often considered foundational to many low-context (e.g., Anglo-American) business practices.

In Woolever's example, the expectation that a forum/genre of RFP exists in other cultures could result in miscommunication or offense if used with individuals from cultures that might not have such a forum. For example, in cultures where one's word is one's bond, the request for something in writing might be seen as indicating a lack of trust (e.g., "Isn't our word alone good enough? Do you have so little trust in us that you need such details in writing?"). From a cultural-rhetorical perspective, it is a case of credibility being lost through failure to understand the forum in which audiences expect information to be presented. For this reason, you should always begin your research on international communication by asking "Does the genre (i.e., forum) I plan to use to convey information exist in the culture with which I will share that information?" If not, you will need to do more research on the related culture to determine what forums/genres might be used to convey such information. If the related genre does exist, however, you need to move on to a second and equally important task—determining the purpose of that forum/genre.

TASK 2: IDENTIFY THE PURPOSE OF THE FORUM

Individuals use specific forums to achieve a particular purpose. This purpose influences reader expectations of what kind of information should be presented in that forum. However, just because two cultures have the same forum/genre does not mean both cultures see that forum as having a similar purpose. As the earlier example of Anglo-American versus

Japanese business letters illustrates, cultures can associate different purposes with the same forum. So once you have determined a culture has a particular forum, you must next answer the question "What purpose does this forum serve in this culture?" To do so, you could review sample documents (representatives of this forum) produced by the members of the related culture. During this review process, you would look for the mention of certain kinds of topics that could help reveal the purpose of a particular genre (i.e., forum). You could also interview members of that culture to ask what objectives or purposes they associate with the specific genre you are researching. At this point, an important consideration is that common expectations of purpose do not necessarily mean similar expectations of the kind of information needed to achieve that purpose. Thus, once you've identified the purpose of a genre, you must next determine what categories of information—or special topics—a culture associates with achieving that purpose.

TASK 3: IDENTIFY THE SPECIAL TOPICS RELATED TO ACHIEVING THE PURPOSE OF THE FORUM

The research literature on cross-cultural communication indicates that cultures can have different expectations of how to achieve the same objective in a given forum. Many members of both the Japanese and the Mexican cultures, for example, view the purpose of a business letter as establishing and maintaining long-term relationships. Yet each group expects to encounter very different kinds of information—or special topics—in relation to achieving this objective.

As Campbell's (1998, 39) research reveals, displays of information about the reader's culture tend to be a special topic that many Japanese audiences associate with credible presentations in business letters. Such special topics reflect a worldview that places a very high value on the cultural group (i.e., the Japanese as a culture) and the various cultural and historical factors related to the identity of this greater cultural group. Elizabeth Tebeaux's (1999, 53–56) research on Mexican business letters, by contrast, reveals that many Mexican readers look for a different kind of special topic—family ties—when assessing the credibility of the same letter. As John C. Condon (1997, 25–29) explains, in Mexican business practices, an individual's interactions and relationships tend to be governed by family connections. Thus, according to the Mexican worldview, an individual's identity is linked to his or her family more than to the common history or aspects of a larger national culture or ethnic group.

This cultural perspective means family members are generally considered trusted partners for long-term business relations. Nonfamily, by contrast, might be viewed with suspicion and might not gain trust unless a family member can attest to the reliability of the outsider. For this reason, creating credibility in Mexican business letters often involves the author noting some family connection with the reader. The mention of special topics that create credibility with many Japanese readers (e.g., general cultural information) might not create credibility in the eyes of many Mexican readers. Thus, similar cultural expectations of a forum's purpose do not necessarily mean the same special topics can be used to create credibility within that forum.

Once your research has identified the purpose a cultural audience associates with a particular forum, you need to ask an important follow-up question: "What (special) topics does my audience associate with the credible presentation of information in this forum/genre?"

You can use the answer to this question to create more effective materials for different cultural audiences. Identifying special topics, however, is only one part of communicating effectively with people from other cultures. The other—and equally important—part is knowing when and how to present such topics within a given forum/genre.

TASK 4: UNDERSTAND THE PROPER PRESENTATION FORMAT OF THE SPECIAL TOPICS WITHIN A FORUM

While identifying culture-specific special topics might seem relatively easy, using them effectively is another matter. This is because the point at which one is expected to mention a particular special topic can be crucial to establishing credibility within a forum. As Linda Driskill (1996, 28) notes, correspondence to Japanese audiences tend to begin with "polite, solicitous comments"—something that many "U.S. writers would consider unnecessary." Next, the writer might raise the mutual goal that both parties (sender and recipient) can work toward. Finally, toward the conclusion, one can subtly raise the "business" function the letter is designed to serve (e.g., an update or an evaluation or a formal record; Campbell 1998, 36). Failure to follow this expected sequence for raising these culture-specific special topics could result in problems. In the context noted here, one must first display the desire to form a long-term relationship with the letter's recipient before discussing the more "nuts-and-bolts" purposes for sending the letter. To do otherwise might be seen as too brazen an approach.

Thus, your research on cultural expectations of special topics should not merely answer the question "What are the special topics members of this culture expect me to address in this forum?" Rather, you must also ask and answer the follow-up question "In what order do I need to present these special topics so I appear credible to the audience?" Once you identify this presentation order, you can follow it when sharing information with other cultural audiences.

Culture is a complex concept, and cultural rhetorical expectations contain degrees of nuance. Working effectively in international environments therefore requires extensive and continuous research. The heuristic framework presented here is but a starting point in this research process. For this reason, you should not consider this heuristic the only tool needed to engage successfully in international projects. This framework, however, can help you make effective initial choices that can contribute to successful communication practices within a variety of global contexts. To better understand the ways in which you can apply the rhetorical ideas discussed here, let's look at how this heuristic might be used to address the situation of Alliance X that we saw at the start of this chapter.

EXTENDED EXAMPLE: HOW CAN TECHNICAL COMMUNICATORS APPLY THIS HEURISTIC IN GLOBAL CONTEXTS?

Let's say the situation is the same as before. The Ministry of Education in Riendutout commissions Alliance X to develop an online help website to provide technical support to clients in both the United States and Riendutout. As before, Pat is tasked with creating this site. Pat decides to develop a two-part site that allows for static displays of information and includes a dynamic component for obtaining answers to technical questions. Once again, Pat presents this plan to the management of Alliance X, and receives approval to begin work on the two-part help site. This time, however, Pat takes one additional and all-important step: researching culture and communication expectations.

Pat decides to begin this research by looking at Riendutout companies that offer similar kinds of online technical support to businesses in that same country. Once a few companies have been identified, Pat tries to locate the specific web pages those organizations use to provide online support to customers. During this review process, Pat notices an interesting trend: while these sites all contain a static FAQ page, none of them has anything like the dynamic bulletin board Pat had in mind (and that was used by a number of U.S.-based companies).

This discovery prompts Pat to review a wider range of similar websites associated with different businesses located in Riendutout. The results of this wider search reveal trends similar to what Pat noted earlier. The FAQ page seems to be a standard part of such sites and is widely used. The dynamic bulletin board approach to addressing technical questions, however, doesn't appear on any of the sites Pat reviewed.

Based on these results, Pat decides it is time to talk with someone from Riendutout to get a better understanding of these findings. To do so, Pat contacts a Riendutout colleague to discuss ideas for this project and to discuss the results of Pat's initial research on online forums for sharing information. The discussion of these ideas begins with Pat asking if online help sites (forums) are commonly used in Riendutout. Pat's colleague replies, "Yes, they are, and they are very popular here." Pat next asks about the common reason (purpose) for which individuals in Riendutout use such sites. Pat's colleague responds that reasons for use are twofold:

- to get standard answers to common questions; and
- to get answers to specific technical questions or problems.

At this point, it appears that the development of a single help site for both cultures could still be a relatively easy process.

Pat then asks an all-important question: "What features (special topics) do users look for to determine if the online help site is good or credible?"

The colleague in Riendutout responds: "Well, presenting a credible answer requires you to note three things in your responses to user questions: (1) the job title of the technician providing the answer; (2) the amount of time the technician has worked for the company; and (3) the university at which the technician obtained his or her degree in computer science."

Pat's colleague then ends this explanation with: "In Riendutout, that's pretty standard stuff, but if you don't include it, the users will think you're disregarding them, for you didn't put in the effort to get a *real* technical expert to answer their questions. Moreover, if such details are not provided, users in Riendutout will likely also be suspicious of the source and the legitimacy of the site providing such information."

At this point, Pat begins to realize that the initial assumption of working on an easy international project is more of a dream than a reality. Pat explains the proposed customer-driven, social-media site to the colleague in Riendutout. During the course of this explanation, Pat also discusses the results of the initial rounds of research conducted on websites in Riendutout. Upon hearing of these items, the colleague replies, "As you've observed, a system like that wouldn't work here. Users need to know they are working with trained company employees—not receiving suggestions from just another consumer. If you used such a customer-driven approach to online help in Riendutout, you'd offend folks and might create the impression of a fraudulent site. You'd certainly lose a major part of your customer base."

The results of this conversation and the previous research on online forums prompts Pat to ask the colleague from Riendutout to collaborate on the development of a help site that will be considered credible by users in both the United States and Riendutout. The system contains the original static FAQ page that both United States and Riendutout audiences consider a credible approach to providing online help. The splash (initial) page for the site's dynamic portion, however, offers users a choice:

- E-mail a Technician
  (A response will be sent within 24–48 hours of receiving this message.)
- Contact the Community
  (Post a question to a community message board and receive quick answers and advice from technicians and from other users.)

Such an approach allows users from the two cultures to select the method of technical support that best appeals to their cultural expectations. In addition, Alliance X technicians have been informed that, when responding to customer questions, they must conclude their response with a signature line that includes

- their name,
- their job title with the company,
- the number of years they have been with the company, and
- the university at which they obtained their degree.

In this way, technicians at Alliance X can provide users from Riendutout with the information they expect to find when assessing the credibility of a response. At the same time, this approach will likely not be distracting to U.S. users, the majority of whom are accustomed to such signature lines appearing at the end of professional e-mails.

While the development of the interactive help system took more time than anticipated, it met with approval from clients in both cultures. Moreover, Alliance X's willingness to meet the rhetorical expectations of the other culture helped the company gain credibility as a vendor in Riendutout. This credibility allowed Alliance X to secure contracts with other organizations in the country. Thus, Pat's initial series of rhetorically based questions provided the forum and the special topics information essential to the successful execution of this international project.

CONCLUSION: WHAT ARE THE NEXT STEPS?

Working in international environments is not a simple process, but it is an essential one in today's global economy. Effective international communication skills are therefore a mechanism for adding value in an increasingly competitive global marketplace. For this reason, technical communicators who can interact successfully with overseas clients and colleagues have a great deal to offer an organization. These technical communicators also possess skills that can increase their job security today and for the future.

An understanding of rhetorical concepts can thus help technical communicators participate more effectively in the global distribution of information and ideas. Success in international projects, however, is a matter of engaging in focused research that examines the connections between rhetorical ideas and cultural expectations of credibility. The rhetorical heuristic presented here can help you participate more effectively in global activities related to technical communication. For such research to be effective, however, it must be a continual process that allows you to refine your approach to cross-cultural communication over time. You should therefore consider the ideas presented in this chapter as a first step toward developing your international communication skills in order to remain a valuable contributor to an organization's activities.

DISCUSSION QUESTIONS

1. What characteristics or ideas do the members of your culture value highly? How do these factors reflect your culture's worldview? How does this worldview affect your behavior?
2. Collect three to five examples of the most commonly used forums/genres in your culture. Next, review these examples to determine the purpose for which members of your culture use these forums. Once you've identified this purpose, review these materials a second time to identify common special topics that seem to appear in these forums. During this second review, try to determine how these special topics help the members of your culture achieve the purpose for which they use that particular forum.
3. How might nationally centralized education (versus state-by-state standards on education) affect forum expectations within a culture? What might this factor mean for the topics you should research when engaging in international communication projects?
4. What rhetorical aspects or factors do you think are unique to your particular culture? What rhetorical aspects or factors do you think are commonly used across multiple cultures?
5. Identify a particular forum that has both print and online versions in your culture (e.g., a newspaper). Next, review both the print and the online versions of this item to determine if you could use the same approach to revise these items in order to make them appear more credible to an international audience. During this process, keep a list

of the similarities and the differences in the approaches you would use to internationalize these items, and compare your findings with those of your classmates.

6. Do you think it is possible to revise a printed document to meet the rhetorical needs of all international reading audiences? Why or why not?
7. Locate a printed software manual for a product you use on a regular basis. Next, apply the heuristic presented in this chapter to try to convert one section of this manual into a website used to convey the same instructional information to users from another culture. As you perform this conversion process, consider the kinds of rhetoric-based questions you will need to ask for this process—and the resulting product—to be successful.
8. In your opinion, do more companies engage in preemptive or revisionist approaches to creating materials for international audiences? Why? What argument might you use to convince a company to shift its international communication practices from a revisionist to a preemptive approach?

WORKS CITED

Ang, Soon, and Andrew C. Inkpen. 2008. "Cultural Intelligence and Offshore Outsourcing Success: A Framework of Firm-Level Intercultural Capability." *Decision Sciences* 39:337–358.

Barnum, Carol M., and Li Huilin. 2006. "Chinese and American Technical Communication: A Cross-Cultural Comparison of Difference." *Technical Communication* 53:143–166.

Berry, John W., Ype H. Poortinga, Marshall H. Segall, and Pierre R. Dasen. 2002. *Cross-Cultural Psychology: Research and Applications*. 2nd ed. New York: Cambridge University Press.

Campbell, Charles C. 1998. "Rhetorical Ethos: A Bridge between High-Context and Low-Context Cultures?" In *The Cultural Context in Business Communication*, edited by Susanne Niemeier, Charles P. Campbell, and Rene Dirven, 31–47. Philadelphia: John Benjamins Publishing Company.

Condon, John C. 1997. *Good Neighbors: Communicating with the Mexicans*. Yarmouth, ME: Intercultural Press.

Driskill, Linda. 1996. "Collaborating across National and Cultural Borders." In *International Dimensions of Technical Communication*, edited by Deborah C. Andrews, 23–44. Alexandria, VA: Society for Technical Communication.

Esselink, Bert. 2000. *A Practical Guide to Localization*. Philadelphia: John Benjamins Publishing Company.

Hofstede, Geert, and Gert Jan Hofstede. 2005. *Cultures and Organizations: Software of the Mind*. New York: McGraw-Hill.

Kamath, Gurudutt R. 2000. "The India Paradox." *Intercom*, May, 10–11.

Kaplan, Robert B. 2001. "What in the World Is Contrastive Rhetoric?" In *Contrastive Rhetoric Revisited and Redefined*, edited by Clayann Gilliam Panetta, vii–xx. Mahwah, NJ: Lawrence Erlbaum Associates.

Kohl, John R. 2008. *The Global English Style Guide: Writing Clear, Translatable Documentation for a Global Market*. Cary, NC: SAS Institute.

McCool, Matthew. 2006. "Information Architecture: Intercultural Human Factors." *Technical Communication* 53:167–183.

Miller, Carolyn R., and Jack Selzer. 1985. "Special Topics of Argument in Engineering Reports." In *Writing in Non-academic Settings*, edited by Lee Odell and Dixie Goswami, 309–341. New York: Guilford Press.

Neuliep, James W. 2000. *Intercultural Communication: A Contextual Approach*. Boston: Houghton Mifflin.

Panetta, Clayann Gilliam. 2001. "Understanding Cultural Differences in the Rhetoric and Composition Classroom: Contrastive Rhetoric as Answer to ESL Dilemmas." In *Contrastive Rhetoric Revisited and Redefined*, edited by Clayann Gilliam Panetta, 3–13. Mahwah, NJ: Lawrence Erlbaum Associates.

Rude, Carolyn. 1998. *Technical Editing*. 2nd ed. Boston: Allyn and Bacon.

St. Amant, Kirk. 2006. "Globalizing Rhetoric: Using Rhetorical Concepts to Identify and Analyze Cultural Expectations Related to Genres." *Hermes: Journal of Language and Communication Studies* 37:47–66.

Tebeaux, Elizabeth. 1999. "Designing Written Business Communication along the Shifting Cultural Continuum: The New Face of Mexico." *Journal of Business and Technical Communication* 13:49–85.

Varner, Iris, and Linda Beamer. 1995. *Intercultural Communication in the Global Workplace*. Boston: Irwin / McGraw-Hill.

Woolever, Kristin R. 2001. "Doing Global Business in the Information Age: Rhetorical Contrasts in the Business and Technical Professions." In *Contrastive Rhetoric Revisited and Redefined*, edited by Clayann Gilliam Panetta, 47–64. Mahwah, NJ: Lawrence Erlbaum Associates.

Yunker, John. 2003. *Beyond Borders: Web Globalization Strategies*. Boston: New Riders.

Contributors

Ann M. Blakeslee is professor of English and director of the University Writing Center and Writing across the Curriculum at Eastern Michigan University. Her publications focus on qualitative research, audience, and writing in disciplinary and workplace contexts. She received the Society for Technical Communication Ken Rainey Award for Excellence in Research in Technical Communication in 2012. She also serves as treasurer for the Association of Teachers of Technical Writing (ATTW).

Rebecca E. Burnett, professor of rhetoric and director of writing and communication at Georgia Institute of Technology, completed her PhD at Carnegie Mellon University. She also works as an industry consultant and trainer and as an expert witness. Her research includes risk communication, collaboration, technical communication, and multimodality.

Antonio Ceraso is assistant professor in the Department of Writing, Rhetoric, and Discourse and director of the Master of Arts in New Media Studies program at DePaul University. His research interests include open-source culture and history, and user-generated content in technical communication. His recent work includes "Open Source Culture and Aesthetics," a co-edited special issue of the journal *Criticism*.

Kelli Cargile Cook is associate professor of technical communication and rhetoric at Texas Tech University. She has taught technical communication in high school, undergraduate, and graduate programs. Her research focuses on online education in technical communication, but she also studies technical communication pedagogy, program development, and program assessment.

Emily Cook is a proposal manager with GC Services Limited Partnership in Houston, Texas. She has also worked as a proposal analyst, technical writer, and training curriculum development specialist in the aerospace and electronic recording industries. She earned master's and bachelor's degrees in English, with an emphasis in technical communication, from Utah State University.

L. Andrew Cooper, assistant professor of film and digital media at University of Louisville, completed his PhD at Princeton University. His most recent book, *Gothic Realities: The Impact of Horror Fiction on Modern Culture*, examines rhetorics surrounding violent fictions. His research includes gothic literature, horror films, technical communication, and multimodality.

R. Stanley Dicks is associate professor and director of the master of science in technical communication at North Carolina State University. His book *Management Principles and Practices for Technical Communicators* (2004) is part of the Allyn and Bacon series in technical communication.

T. Kenny Fountain is assistant professor of English at Case Western Reserve University. His research focuses primarily on the visual representations and practices of biomedical science. He has published work in the *Journal of Technical Writing and Communication* and *Medicine Studies*.

William Hart-Davidson (PhD 1999, Purdue) is associate professor of rhetoric and writing at Michigan State University, where he directs the rhetoric and writing graduate program and is codirector of the Writing in Digital Environments (WIDE) Research Center. He teaches courses in rhetoric, technical communication, and interaction design.

Jim Henry has been coupling technical writing courses with service learning since the mid-nineties and is currently developing an e-book to support technical writers in organizational analysis. He directs the Mānoa Writing Program, which administers more than five hundred writing-intensive courses per semester. His current research includes mentoring writers, writing as performance, and place-based writing across the curriculum.

Brent Henze is associate professor and area coordinator for technical and professional communication at East Carolina University. He directs the internship program and teaches courses in science writing, research methods, technical editing, and grant writing. His current scholarship considers how scientists enlist and regulate the participation of nonspecialists in data collection and interpretation.

Johndan Johnson-Eilola works at Clarkson University as professor of communication and media, where he teaches classes in audio production, typography, and design. He is the author or editor of numerous journal articles, book chapters, edited collections, and books, including *Nostalgic Angels*, *Datacloud*, *Central Works in Technical Communication*, and *Writing New Media*.

Bernadette Longo is associate professor at the University of Minnesota, teaching technical communication and information design. In collaboration with the University of Minnesota Libraries, she recently developed a digital collection, *Technological Emergence*, exploring communication and social issues relating to technology diffusion between the global North and South.

Brad Mehlenbacher is associate professor of distance learning in the Department of Leadership, Policy, and Adult and Higher Education and author of the award-winning *Instruction and Technology: Designs for Everyday Learning* (2010). He has published extensively on human-computer interaction, communication design, and online instruction.

Ben Minson is a senior technical writer for the Church of Jesus Christ of Latter-Day Saints. He creates quick-reference guides, screencasts, help, and other material for audiences of custom software. He graduated from Utah State University with a bachelor's degree in English with an emphasis in professional and technical writing.

Barbara Mirel is a research scientist in the School of Education at the University of Michigan. She collaborates with the Medical School and its Biomedical Informatics Department to develop and evaluate user-centered systems for biomedical research. She is the author of *Interaction Design for Complex Problem Solving* (Elsevier, 2004) and over fifty refereed articles.

James E. Porter is professor in English and in the Armstrong Institute for Interactive Media Studies at Miami University. His research focuses on how digital technologies change rhetoric, writing, and research. His book *The Ethics of Internet Research,* coauthored with Heidi McKee, shows how rhetoric theory can inform ethical decisions about Internet research practices.

Kirk St. Amant is professor of technical and professional communication and of international studies at East Carolina University.

Gerald J. Savage recently retired from Illinois State University, where he was professor of rhetoric and technical communication and directed the technical communication program and English studies internships. His current research focuses on literacy practices and social justice in technical communication. He received the Society for Technical Communication's J. R. Gould Award for Excellence in Teaching Technical Communication in 2008.

Karen Schriver is president of KSA Communication Design and Research, a firm that specializes in improving the quality of communications about technology, science, education, and health. Winner of ten national awards for her research, Karen helps organizations around the globe realize the social and economic benefits of excellence in information design.

J. Blake Scott is associate chair for degree programs in the Department of Writing and Rhetoric at the University of Central Florida, where he teaches courses in rhetoric and professional communication. His most recently published book project, coedited with Rebecca Dingo, is titled *The Megarhetorics of Global Development*.

Stuart A. Selber is associate professor of English at Penn State. He is a past president of the Association of Teachers of Technical Writing and the Council for Programs in Technical and Scientific Communication, and a past chair of the CCCC Committee on Technical Communication. His work has won awards from ATTW, NCTE, and *Computers and Composition*.

Cynthia L. Selfe is Humanities Distinguished Professor in the Department of English at Ohio State University. She is cofounder and executive editor of Computers and Composition Digital Press / Utah State University Press (with Gail Hawisher) and cofounder and codirector of the Digital Archive of Literacy Narratives (with H. Lewis Ulman).

Richard (Dickie) J. Selfe directs the Center for the Study and Teaching of Writing (CSTW) at The Ohio State University. Selfe's academic interests lie at the intersection of communication pedagogies, programmatic curricula, and the social/institutional influences of digital systems.

Clay Spinuzzi is associate professor of rhetoric at the University of Texas at Austin. His interests include research methods and methodology, workplace research, and computer-mediated activity. His books include *Tracing Genres through Organizations* (2003) and *Network* (2008).

Jason Swarts is associate professor of technical communication at North Carolina State University. He researches networked communication, computer-supported cooperative work, and mobility. His book, *Together with Technology*, was published in 2008.

Candice A. Welhausen, assistant professor of technical communication at the University of Delaware, completed her PhD at the University of New Mexico. She has worked as a technical communicator in nonacademic settings, particularly biomedicine, for a number of years. Her research includes visual rhetoric, technical communication, and rhetorical theory.

Stephanie Wilson studies technical communication and rhetoric at Texas Tech University, focusing on visual rhetoric, new media, and user-centered design. She also works as an information-design specialist at the University of Illinois, where her main job responsibilities include usability testing, software prototyping, and usability consulting for campus IT.

Anne Frances Wysocki teaches written, visual, and digital rhetorics at the University of Wisconsin–Milwaukee. Lead author of *Writing New Media: Theory and Applications for Expanding the Teaching of Composition*, she has also designed and produced software to teach 3-D visualization and geology. Her new-media pieces have won the *Kairos* Best Webtext award and the Institute for the Future of the Book's Born Digital Competition.

Index